Lecture Notes in Social Networks

More information about this series at http://www.springer.com/series/8768

Alejandro Peña-Ayala

Editor

Educational Networking

A Novel Discipline for Improved Learning
Based on Social Networks

 Springer

Editor
Alejandro Peña-Ayala 🆔
WOLNM, Artificial Intelligence on Education Lab
Mexico City, Mexico

Instituto Politécnico Nacional, Escuela Superior
de Ingeniería Mecánica y Eléctrica, Zacatenco
Mexico City, Mexico

ISSN 2190-5428   ISSN 2190-5436   (electronic)
Lecture Notes in Social Networks
ISBN 978-3-030-29975-0   ISBN 978-3-030-29973-6   (eBook)
https://doi.org/10.1007/978-3-030-29973-6

This Springer imprint is published by the registered company Springer Nature Switzerland AG
The registered company address is: Gewerbestrasse 11, 6330 Cham, Switzerland

Preface

In the education domain, *21st Century* represents a breaking point to highlight the need for considering disruption and innovation as the main policies to conceive, design, and develop the 21st Century: Education, Teaching, Learning, Skills, and Students. All of this with the aim of enabling people, who pertain to the Millennials, Z, Alpha, and future Generations, to cope, survive, and succeed in a complex and demanding world, where information and communication technologies (ICTs), particularly recent versions of the Web, transform human daily life.

On the other hand, a *social network* (SN) represents the natural surroundings of individuals since they are conceived and born, through their life people use to pertain to diverse kinds of SNs, develop diverse aims, and conclude their affairs. Such a social behavior has been fostered by transport and communication facilities through time, but recently as result of the emergence of ICTs, personal SNs have been transformed to *online social networks* (OSNs), which facilitate (a)synchronous worldwide broadcasting of communication and social media.

In this sense, *educational networking* (EN) burst as a vigorous trend that complements, and even replaces, traditional face-to-face, distance, and virtual education. EN is a sort of crossroad, where twenty-first-century education, ICTs, and OSNs stake their best advances, technologies, resources, and practice to socialize learning for all, where in and out classroom students construct knowledge and increase social capital through the organization of communities of learning practice, where networking, collaboration, and social constructivism are part of daily affairs.

The present edited book outlines a landscape of the evolution and status of EN, as well as a glance at trends to inspire future work to spread and enhance the field. Hence, this book gathers a sample of labor recently achieved by groups of researchers in four continents, who are engaged in examining the progress of the domain, proposing new constructs to ground the EN work, sharing project advances, describing ways of practice, and revealing field studies. With this in mind, the following five topics cluster the contributions accepted for this volume.

1. *Reviews* gather related works to shape a broad or specialized landscape of the EN arena, providing a profile of the reported labor, characterizing features, analyzing achievements, and discussing the state and future trends.

2. *Conceptual* introduces a logistic proposal to ground the development of EN duties, as well as tailor a specific guideline to lead the labor with the aim of contributing to strengthen the baseline of the domain.
3. *Projects* brief the aims, resources, and activities accomplished for investing in EN to reach a specific objective, whose achievement represents a valuable benefit for the institution and users, where the outcome is implemented.
4. *Approaches* focus on a specific problem or goal to propose a solution, whose development is thoroughly described, adding empirical evidence that shows the results of their application and the way they contribute to reach a purpose.
5. *Study* examines how SNs practice, and particularly SN analysis, reveal, and visualize interesting features and relations of academic labor, which are interpreted to understand how universities impact society.

This volume is the outcome of the research, development, and practice fulfilled by authors, who generously share their work with the community of practitioners, educators, computer scientists, pedagogues, psychologists, sociologists, communicators, academics, and students interested in exploring, participating, and collaborating to spread education, advance teaching, and facilitate learning through EN.

As a result of the cycle process composed of the submission of proposals and their respective evaluation, as well as the edition of the complete manuscript with the corresponding revision, tuning, and decision according to the Springer quality criteria, 10 works were accepted, edited as chapters, and organized as follows.

Chapter 1: Outlines a systematic account of the academic labor on K-12, specifically teaching and teacher learning with social media during the period of 2004 to 2019, to shape a view about social networking in education, where one finding reveals how Twitter fosters teachers' professional development.

Chapter 2: Analyzes mixed methods approaches related to social network analysis for learning and education through systematic literature review, objective social network analysis with content analysis, and an approach with follow-up in class discussion, whose conclusion claims ethical concerns when gathering such mixed methods data from diverse data sources.

Chapter 3: Traces a review of the background, evolution, and recent labor carried out in the educational networking arena, in addition to proposing a taxonomy to organize the related work and a pattern to characterize their main features, producing an analysis of the current state and a vision for future trends.

Chapter 4: Designs the Platform Adoption Model to spread the harnessing of educational networking through the identification of constrains towards adoption and increased use, underlying theories of behavior change, and a framework for overcoming barriers and improving the platforms' contribution.

Chapter 5: Proposes the Social Network Site *Academic-match* to boost the development of a virtual scholar community of practice among teachers of the National Autonomous University of Mexico, who are involved in academic projects with the aim of stimulating faculty recruitment, teacher selection, and collaboration through forums and spaces of participation.

Chapter 6: Builds a scholar network by means of a YouTube course channel, where students author didactic social media and post videos to progressively construct knowledge among peers, stimulating their satisfaction and engagement due to they perceive the usefulness of being involved in generating content and the benefits of performing active learning instead of the passive.

Chapter 7: Boosts the use of Wiki and Blogs for e-assessments as a way of enhancing suitable skills for students' academic and professional carrier practice, producing as a consequence the acquisition of literacy, cultural awareness, and independent learning strategies through group and personal experiences.

Chapter 8: Aims at stimulating the participation of online learning communities of users, particularly for those who behave as active or passive before the post and repost of valuable information for the academic group, by means of a method that proposes attractive posts for a given student according to his/her profile and three measures of similarity, topic, content, and learner.

Chapter 9: Tailors a Context-Aware System Architecture to recreate learning spaces based on educational networking that engages students to interact offline and online with peers for collaborative and self-directed learning, all of this in order to enhance pedagogical strategies that further adaptive learning.

Chapter 10: Proposes a methodology for university rankings based on the maximum clique problem in networks, which is applied to Higher Education institutions in Mexico giving as a result complex networks that graphically relate key academic features that reveal the influence and importance of social networks and educational networks in the scholar system of a country.

I express my gratitude to authors, reviewers, Springer editorial team, Series Editors Professor Reda Alhajj, Professor Uwe Glässer, Professor Huan Liu, Professor Rafael Wittek, and Professor Daniel Zeng; Publishing Editors Christopher T. Coughlin and Christoph Baumann; Senior Production Editor Jeffrey Taub; and the Production Editor Chandhini Kuppusamy for their support and valuable collaboration to develop this work.

I also acknowledge the support given by the Consejo Nacional de Ciencia y Tecnología (CONACYT) and the Instituto Politécnico Nacional (IPN), both institutions pertain to the Mexican Federal Government, through the grants: CONACYT-SNI-36453, IPN-SIP 2018-1407, IPN-SIP 2019-5563, IPN-SIP/DI/DOPI/EDI-154/18, and IPN-SIBE-2019-2020.

Last but not least, I acknowledge and provide testimony of the unction and strength given by my Father, Brother Jesus, and Helper, as part of the research projects performed by World Outreach Light to the Nations Ministries (WOLNM).

Mexico City, Mexico   Alejandro Peña-Ayala
July 2019

Contents

Part IV   Approaches

Part V   Study

Contributors (Reviewers)

Ghazi Alkhatib Faculty of Economics and Administrative Sciences, Hashemite University, Zarqa, Jordan

Jackeline Bucio UNAM-CUAED-B@UNAM, Mexico City, Mexico

D. Yuh-Shy Chuang International Business Department, Chien Hsin University, Taoyuan City, Taiwan

Eleni Dermentzi Newcastle Business School, Northumbria University, Newcastle upon Tyne, UK

Shazia K. Jan Faculty of Business & Economic, Macquarie University, North Ryde, NSW, Australia

Sun Joo Yoo Principal Consultant Samsung SDS, Seowon University, Tucson, AZ, USA

Esra Keles Department of Computer Education and Instructional Technology, Trabzon University, Trabzon, Turkey

Yasemin Koçak Usluel Computer Education And Instructional Technologies Department, Yasemin Koçak Usluel of Hacettepe University, Ankara, Turkey

Utku Kose Faculty of Engineering, Suleyman Demirel University, Isparta, Turkey

Ines Kožuh Faculty of Electrical Engineering and Computer Science, University of Maribor, Maribor, Slovenia

Lee Kwang-Soon Department of English Language and Literature, Mokpo National University, Muan-gun, South Korea

Stefania Manca Institute of Educational Technology, National Research Council of Italy, Rome, Italy

Paula Miranda Escola Superior de Tecnología de Setúbal, Instituto Politécnico de Setúbal, Setúbal, Portugal

Agostino Poggi Dipartimento di Ingegneria dell'Informazione, Università degli Studi di Parma, Parma, Italy

Idrissa Sarr Sciences & Technics Faculty, Université Cheikh Anta Diop Dakar, Dakar, Senegal

Kai-Yu Tang Ming Chun University, Taipei, Taiwan

Lau Wing-fat. Wilfred Department of Curriculum and Instruction, The Chinese University of Hong Kong, Hong Kong, China

Cheng-Huang Yen Department of Management & Information, National Open University, Taipei, Taiwan

Peter Youngs Dept. of Curriculum, Instruction and Special Education, University of Virginia, Charlottesville, VA, USA

Contributors (Authors)

Daniel Belanche Facultad de Economía y Empresa, Universidad de Zaragoza, Zaragoza, Spain

Jacqueline Bourdeau TÉLUQ University, Montréal (Québec), Quebec, Canada

Dawn B. Branley-Bell Northumbria University, Newcastle, UK

Jackeline Bucio UNAM-CUAED-B@UNAM, Mexico City, Mexico

Luis V. Casaló Facultad de Empresa y Gestión Pública, Universidad de Zaragoza, Huesca, Spain

Francisco Cervantes-Pérez Universidad Nacional Autónoma de México (UNAM), CUAED and ENES-Morelia, Mexico City, Mexico

Sergio G. de-los-Cobos-Silva Departamento de Ingeniería Eléctrica, UAM Iztapalapa, Mexico City, Mexico

Ben K. Daniel Otago University, Dunedin, New Zealand

Dominik Froehlich University of Vienna, Vienna, Austria

Sarah Galvin Michigan State University, East Lansing, MI, USA

Ardalan Ghasemzadeh Department of Information Technology and Computer Engineering, Urmia University of Technology, Urmia, Iran

Christine Greenhow Michigan State University, East Lansing, MI, USA

Miguel Ángel Gutiérrez-Andrade Departamento de Ingeniería Eléctrica, UAM Iztapalapa, Mexico City, Mexico

Tomayess Issa School of Management, Curtin University, Perth, WA, USA

Pedro Lara-Velázquez Departamento de Ingeniería Eléctrica, UAM Iztapalapa, Mexico City, Mexico

Edwin Montes-Orozco Posgrado en Ciencias y Tecnologías de la Información, UAM Iztapalapa, Mexico City, Mexico

Roman Anselmo Mora-Gutiérrez Departamento de Sistemas, UAM Azcapotzalco, Mexico City, Mexico

Bibiana Obregón-Quintana Departamento de Matemáticas, Facultad de Ciencias, UNAM, Mexico City, Mexico

Carlos Orús Facultad de Economía y Empresa, Universidad de Zaragoza, Zaragoza, Spain

Alejandro Peña-Ayala WOLNM, Artificial Intelligence on Education Lab, Mexico City, Mexico

Instituto Politécnico Nacional, Escuela Superior de Ingeniería Mecánica y Eléctrica, Zacatenco, Mexico City, Mexico

Alfredo Pérez-Rueda Facultad de Ciencias Sociales y Humanas, Universidad de Zaragoza, Teruel, Spain

Valéry Psyché TÉLUQ University, Montréal (Québec), Quebec, Canada

Martin Rehm Pädagogische Hochschule Weingarten, Weingarten, Germany

Alireza Rezvanian Department of Computer Engineering, University of Science and Culture, Tehran, Iran

School of Computer Science, Institute for Research in Fundamental Sciences (IPM), Tehran, Iran

Bart Rienties Open University, Milton Keynes, UK

Eric Alfredo Rincón-García Departamento de Ingeniería Eléctrica, UAM Iztapalapa, Mexico City, Mexico

Ali Mohammad Saghiri School of Computer Science, Institute for Research in Fundamental Sciences (IPM), Tehran, Iran

Omid Reza Bolouki Speily Department of Information Technology and Computer Engineering, Urmia University of Technology, Urmia, Iran

S. Mehdi Vahidipour Faculty of Electrical and Computer Engineering, Department of Computer, University of Kashan, Kashan, Iran

Guadalupe Vadillo UNAM-CUAED-B@UNAM, Mexico City, Mexico

Part I
Reviews

Chapter 1
Educational Networking: A Novel Discipline for Improved K-12 Learning Based on Social Networks

Sarah Galvin and Christine Greenhow

Abstract Based on a systematic review of over a decade (2004–2019) of educational research on K-12 teaching and teacher learning with social media, this chapter reports on the state-of-the-art of social networking in education. Through analyzing 56 research articles, we address the following questions: (1) What social networking platforms are used by K-12 educators? (2) What specific features of those platforms are used by teachers and/or their students? (3) For what educational purposes do teachers adopt social networking? (4) What does learning look like in these spaces? This work contributes insights towards defining how educational networking benefits teachers both in the classroom and as a part of their professional development and challenges our perceptions of the role of educational networking in teachers' informal learning. Few platforms are covered in depth in the literature (i.e., Twitter, Facebook, Edmodo) with inconsistent exploration of specific features (e.g., liking, retweeting). We describe trends in depictions of teacher and student learning, noting the high reliance on self-reported changes in practice, motivation, and attitude. Further research is needed to explore the application of specific social networking features in educational contexts and to expand our understanding of meaningful learning and participation in educational networking spaces.

Keywords Educational networking · Facebook · K-12 education · Social media · Social networking · Twitter

Abbreviations

#	Hashtag symbol
@	Mention symbol
EFL	English as a Foreign Language
ERIC	Education Resources Information Center

S. Galvin (✉) · C. Greenhow
Michigan State University, East Lansing, MI, USA
e-mail: galvins1@msu.edu; greenhow@msu.edu

© Springer Nature Switzerland AG 2020
A. Peña-Ayala (ed.), *Educational Networking*, Lecture Notes in Social Networks, https://doi.org/10.1007/978-3-030-29973-6_1

GPA   Grade point average
GRA   Global Read Aloud
ICT   Information and communication technologies
PD   Professional development
PLN   Professional learning network
PRISMA   Preferred Reporting Items for Systematic Reviews and Meta-Analyses
RT   Retweet
SN   Social networking
SNS   Social networking sites

1.1 Introduction

Social media and cloud computing are ubiquitous in informal learning settings around the world. Many learners regularly communicate in unregulated informal spaces to share resources, ideas, and interests. Globally, people are increasingly turning to social media for real-time information and connection in their everyday lives. Approximately two and a half billion people, or one-third of the world's population, are using social networking sites (SNS) to find other people and resources across geographical, cultural, and economic borders (Statista, 2017). In fact, social media users on platforms including Facebook, Twitter, Pinterest, and Instagram find broad engagement in personal as well as professional and political domains.

Policymakers, administrators, teachers, students, and parents must better understand the impact of these growing, global social media spaces have on education in order to acknowledge and leverage the social capital (i.e., resources, information, people) they make available to all users and, in particular, to teachers and students. Through increased understanding, we can work to effectively facilitate the integration of educational networking on social media into teachers' collaboration, professional development (PD), and classroom teaching. To advance the field of educational research, we must not only be up to date on the development of the evolving social phenomena on social media but also critically evaluate the new perspectives, findings, and methodologies that arise as these platforms inform and shape teaching and learning.

This chapter will report on themes derived from a systematic review of 15 years (2004–2019) of educational research from around the world to present the state of the art of social networking in education. Following established standards for quality in systematic literature reviews, this chapter discusses how social media are perceived and used by K-12 teachers and how social media have impacted pedagogy and learning. Drawing from these insights, we set a course for future research on educational and social networking.

Educational networking combines social networking (SN) functionalities with educational content (both online and off-line) and cognitive tools that drive social learning. Educational networking occurs where social networking is applied towards educational aims to produce individual and collective domain knowledge among peers (e.g., students, teachers, others) as a result of their mutual efforts.

Having reviewed the educational research on K-12 teaching and teacher learning with various types of social media, this chapter provides a synthesis of trends in social networking for education. Specifically, after analyzing a final sample of 56 research articles, we present findings on the following questions: (1) What social networking platforms are used by K-12 educators? (2) What specific features of those platforms are used by teachers and/or their students? (3) For what educational purposes do teachers adopt social networking? (4) What does learning look like in these spaces?

The knowledge base pertaining to educational social networking is varied and interdisciplinary (Greenhow, Cho, Dennen, & Fishman, 2019; Rehm, Manca, Brandon, & Greenhow, 2019), with a growing number of studies having been published within the last 5 years (Greenhow & Askari, 2017; Greenhow, Galvin, Brandon, & Askari, in press; Rehm et al., 2019). Existing literature reviews related to social networking in education have focused mainly on learning or teaching with a particular social media type (e.g., social networking sites) (Greenhow & Askari, 2017; Rodríguez-Hoyos, Salmón, & Fernández-Díaz, 2015) or platform (Alias et al., 2013; Manca & Ranieri, 2013, 2016; Tang & Hew, 2017; Wilson, Gosling, & Graham, 2012).

Additionally, the majority of studies in these reviews focus on the perceptions and experiences of college students (Aydin, 2012; Manca & Ranieri, 2013) and higher education faculty (Forkosh-Baruch & Hershovitz, 2012). For instance, Manca and Ranieri (2013, 2016) reviewed research on teaching and learning with Facebook (2007–2015) and found only 17 studies that addressed teacher education or secondary education. Table 1.1 offers an overview of past, related literature reviews informing this chapter.

Furthermore, the benefits of appropriating social networking into teaching and learning contexts are contested in the research literature. Some studies on the integration of social networking sites like Facebook in higher education suggest their affordances for interaction, collaboration, information, and resource sharing (Mazman & Usluel, 2010), encouraging participation and critical thinking (Ajjan & Hartshorne, 2008), and increasing peer support and communication about course content and assessment (DiVall & Kirwin, 2012). Others warn against exploiting social media for learning because of its potentially negative impact on student outcomes, such as college grade point average (GPA) (Junco & Cotton, 2013; Kirschner & Karpinski, 2010).

Determining whether and how social networking can be applied for K-12 educational aims remains problematic in the field. To resolve this, it is necessary to expand on the work of prior literature reviews by synthesizing the existing educational research literature pertaining to social networking in secondary and elementary education, rather than continuing to focus on singular social media platforms (e.g., Facebook and Twitter) or on higher education contexts. We undertook such a project, investigating social media use in elementary and secondary educational settings. The research questions guiding our literature review were as follows: (1) How are social media used by K-12 teachers for their teaching or professional learning? (2) How does such use impact teachers' or students' learning?

Table 1.1 Overview of related literature reviews

Citation	Platform(s)	Educational level included in sample	Research focus
Alias et al. (2013)	Twitter	All levels (Mostly higher education)	What are the main topics, samples, and settings for research on Twitter? What methods and types of analyses are being used?
Aydin (2012)	Facebook	All levels (Mostly higher education)	Who uses Facebook and for what purposes? Are there harmful effects of Facebook, how it is used in an educational environment, and how does it affect culture and language? How does Facebook impact socialization and affective states (e.g., self-esteem)?
Buettner (2013)	Twitter	All levels (Mostly higher education)	What are current trends in research covering Twitter in educational contexts?
Gao, Juo, and Zhang (2012)	Micro-blogging	All levels (Mostly higher education)	What types of research on microblogging are being done? How is microblogging being used for teaching and learning and with what benefits? What implications for research and practice have been found?
Greenhow and Askari (2017)	Social networking sites (SNS)	K-12 education	How do K-12 teachers and students use SNSs? What impact(s) does using SNSs have on teachers' pedagogy or students' learning?
Hew (2011)	Facebook	All levels (Mostly higher education)	How are students using Facebook? How is Facebook impacting their education? What are students' attitudes towards Facebook?
Manca and Ranieri (2013)	Facebook	All levels (Mostly higher education)	How effectively are the affordances of Facebook being transferred into pedagogical practice?
Manca and Ranieri (2016)	Facebook	All levels (Mostly higher education)	An update on their 2013 study: How effectively are the affordances of Facebook being transferred into pedagogical practice?
Rodríguez-Hoyos et al. (2015)	Social network sites (SNS)	All levels (Mostly higher education)	What types of educational research are being done on SNS, what methods are being used, and what educational levels are being studied?
Tang and Hew (2017)	Twitter	All levels (Mostly higher education)	What types of educational research are being done on Twitter? How is Twitter being used in education? How is Twitter impacting interactions and learning outcomes?
Wilson et al. (2012)	Facebook	All levels	What research is being done on Facebook? What types of participants are being studied, what are their motivations, and how are they presenting their identities on Facebook? What is Facebook's role in social interactions? What research exists on privacy issues?

The three main objectives of this chapter are (1) to present the state of the art in social networking in K-12 educational research from 2004 up to 2019, (2) to advance understanding of the knowledge base on educational networking beyond higher education contexts, and (3) to evaluate and synthesize trends in the research arising from these evolving social phenomena.

This literature review is the first of its kind to present the state of the art of social networking in K-12 education up to the present day in 2019. It examines teaching and learning with this technology both in classrooms and outside of them in people's everyday lives. As previously described, existing published literature reviews related to social networking in education are limited in that they have focused mainly on learning or teaching with a particular social media type (Greenhow & Askari, 2017), such as wikis (e.g., Reich, Willet, & Murnane, 2012), social networking sites (e.g., Manca & Ranieri, 2013, 2016), or microblogs (e.g., Tang & Hew, 2017), or a particular platform such as Facebook (e.g., Aydin, 2012), or have focused mainly on trends in higher education and not in K-12 school-related contexts (Manca & Ranieri, 2013, 2016). See Table 1.1 below for further summary of literature reviews related to our work.

This review was undertaken to address limitations in prior related reviews; it reports on the *various* social media discussed in educational and educational technology research, rather than focusing on one type of platform or one single platform, and it targets research on social networking within *K-12* education contexts, rather than the commonly studied higher education settings. The main contribution of this work is that it not only highlights how educational networking benefits K-12 teachers or students when used in the classroom or as a part of teachers' professional development but also points to how educational networking stretches and challenges our perceptions of social media and informal learning in teachers' lives.

Over the course of this chapter, we describe the process, findings, and insights gathered from our review. In Sect. 1.2 we summarize key terminology and features of social media. Section 1.3 introduces the materials and methods used, and in Sect. 1.4 we elaborate on the systematic literature review methodology that led to the final corpus of 56 articles. Our findings, presented in Sect. 1.5, highlight themes related to the four focal topics of the chapter: (1) What social networking platforms are used by K-12 educators? (2) What specific features of those platforms are used by teachers and/or their students? (3) For what educational purposes do teachers adopt social networking? (4) What does learning look like in these spaces? Finally, in Sect. 1.6, implications for research and practice draw attention to how understanding specific features of social media, broader definitions of learning, and the complexities of participation in online networked spaces are critical new directions.

1.2   Body of the Approach

Social media or social networking sites have attracted considerable scholarly interest both within the education sector and beyond it, especially within the last 5 years
(Greenhow & Askari, 2017; Rehm et al., 2019). In Sect. 1.1, we briefly reviewed
prior literature and existing literature reviews of social media and educational
research and identified gaps in the knowledge base that this chapter seeks to address.
     Now, in this section, we define key terminology in the literature and provide an
overview of several unique features of social networking and prominent social
media in order to orient the reader. First, we situate our work in the current literature
that defines social media (Sect. 1.2.1). Then we describe the two platforms most
prevalent in the reviewed studies, Facebook and Twitter, highlighting main features
and terms (Sect. 1.2.2).

1.2.1   Prior Literature and Key Terminology

Although defining social media is a challenge as technology continues to evolve,
scholars have identified key aspects of social media as Web 2.0 Internet-based applications that feature user-generated content, user profiles, and the development of
online social networks by connecting a user's profile with those of other individuals
or groups within the system (Obar & Wildman, 2015). We use the terms *social
media* and *social networking* interchangeably and define these technologies as web-
based services through which individuals can create user profiles, contribute user-
generated content, and develop social connections, either maintaining existing ties
or developing new social ties (Greenhow, Robelia, & Hughes, 2009).
     This broad category of social media encompasses various types, including social
networking sites (SNS), microblogging services, and others that have the above
characteristics. Ellison and Boyd (2013), for instance, define SNSs as technologies
with the following aspects: (1) uniquely identifiable profiles that consist of user-
supplied content and/or system-provided data; (2) (semi-) public display of connections that can be traversed by others; and (3) features that allow users to consume,
produce, and/or interact with user-generated content provided by their connections
on the site (p. 7). Examples include Facebook, Edmodo, and Academia.edu, to name
just a few.
     In Sect. 1.1 we presented an overview of the current knowledge base on educa-
tional networking. Specifically, we summarized findings from existing literature
reviews of social networking in education to emphasize that these reviews have
largely focused on individual platforms (i.e., Facebook and Twitter) and covered
research on higher education rather than elementary and secondary education.
Moreover, as mentioned in Sect. 1.1, the results of integrating educational networking into college and university education show mixed results, with some studies
extolling its benefits while others reveal harmful impacts on grades (see Sect. 1.1).

1.2.2   Key Features of Prominent Social Media

As a type of social media, Facebook is the most popular social networking site worldwide. Its features include a personal profile, with which users can signal their interests and identity, such as their job title and professional affiliation, educational background, relationship status, and more. Facebook users can create posts in text, photo, or video, tag friends in posts, and interact with others' posts through the comment, like, and share buttons. Facebook displays the activities of one's Friends List (i.e., social connections on the site) or one's friends of friends in the News Feed feature. The site also offers a private messaging system and inbox, like email but internal to the Facebook platform. Additionally, users can form Facebook groups, which can be public, semi-public, or private groups of Facebook users.

Another type of social networking, microblogging, consists of "data entries," in which users can share text, images, video, links, hashtags, and mentions (Ebner, Lienhardt, Rohs, & Meyer, 2010, p. 93). The most popular microblog globally is Twitter, which originally constrained its posts (tweets) to 140 characters but expanded the limit to 280 characters in 2017. Design features that distinguish Twitter from other social media are its follower structure, shortened link sharing, and use of hashtags (Boyd, Golder, & Lotan, 2010). Similar to Facebook friends, followers on Twitter are other users that subscribe to one's tweets. A list of who follows a particular user (i.e., followers) and who that user follows are both linked on individual profiles.

Twitter users can read, favorite, reply to (i.e. leave a comment on), and retweet (i.e., re-share from someone else's shared content) from any public account, and since most Twitter users make tweets public (Madden, 2012), it is easy to see links to other web content. Common practices on Twitter to navigate and create content include the use of @ plus the username (e.g., @SpringerPub) to indicate a mention or reply to that user, Twitter's link-shortening feature to produce an abbreviated version of the hyperlink, and the favoriting and bookmarking features, which allow users to save others' tweets for later reference. Authors are notified when their tweets are favorited, and users can see others' favorited tweets. In 2018 Twitter added the bookmarking feature, which allows users to privately collect and save tweets. Lastly, the hashtag (#) is used to organize tweets, designating them as part of a particular category or conversation. For instance, the hashtag #Edtech designates tweets as part of the Twitter stream for those interested in educational technology. Through the use of these features, Twitter facilitates a dynamic space where users can connect and interact in a variety of both highly active (e.g., tweeting, replying, mentioning) and more passive (e.g., observing, favoriting, bookmarking) ways.

1.3   Materials and Method

In this section we explain our methodological approach to conducting a systematic review of the literature on social networking in education. First, we present the materials used in our literature review (Sect. 1.3.1), followed by an overview of the literature review method (Sect. 1.3.2). An account of our methodology process is presented in Sect. 1.4, which includes the procedures used to identify, screen, assess, and determine whether articles were eligible for inclusion in this review.

1.3.1   Materials

The materials used in this review were a final sample of 56 empirical articles published in peer-reviewed journals. As described below, these materials were identified from a search of four educational databases and manual table of contents searches of prominent, selected journals (e.g., educational research journals, educational technology journals, technology-enhanced learning journals).

With the assistance of our university's social sciences research librarian, who specializes in education, we identified four prominent electronic databases for educational research: Education Full Text, Education Resources Information Center (ERIC), Scopus, and Web of Science. These were chosen as the databases for our search for their focus on the field of education (as opposed to PsycInfo, which specializes in psychology, or other non-education specialized databases) and the comprehensiveness of their holdings.

1.3.2   Method

The methodology for this literature review followed the Preferred Reporting Items for Systematic Reviews and Meta-Analyses (PRISMA) standards, which emphasize transparent and detailed reporting of search procedures (Moher, Liberati, Tetzlaff, Altman, & The PRISMA Group, 2009). These procedures include (1) *identification* and (2) *screening* of records, (3) *assessing* articles for eligibility, and (4) determining whether a given study is to be included or excluded from the sample based on *inclusion* and *exclusion* criteria.

Informed by our overview of the existing literature, we sought to identify studies that addressed K-12 teaching or teacher learning with social media in countries across the globe. Figure 1.1 depicts a flow chart summary of the search and selection process we describe below. That is, the flow chart, as indicated by the labels on the left-most side of the diagram, depicts our search procedures in accordance with PRISMA guidelines (Moher et al., 2009).

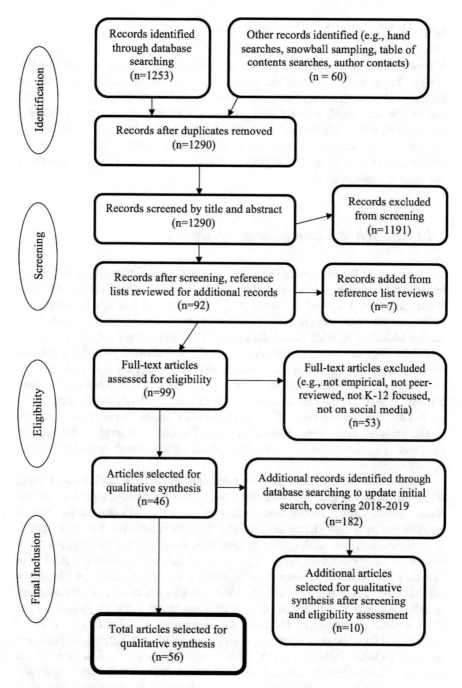

Fig. 1.1 Four phases of search procedures. (Figure adapted from PRISMA flow chart (Moher et al., 2009))

1.4   Development of the Approach

In accordance with the PRISMA standards introduced above, we completed the following phases during our literature search: identification, screening, assessment of articles, and determination of article inclusion or exclusion. In Sect. 1.4.1 we detail and justify our search strategy and process for identifying potentially relevant records (i.e., articles) from four electronic databases. In Sect. 1.4.2 we outline the steps we took to screen, evaluate, and determine which articles met our inclusion criteria and how our 56-article corpus was finalized (see Fig. 1.1 for an overview of our PRISMA procedures).

1.4.1   Search and Identification of Articles

Our search and identification of records occurred in three phases. In the first phase, begun at the end of 2017, we concentrated on identifying studies from the four electronic databases described in Sect. 1.3.1. To address our research questions, we performed searches using the following initial search terms: "(K-12 teaching OR teacher training OR professional development OR elementary school OR middle school OR high school) AND (social media OR social network sites OR social networking sites OR Facebook OR Twitter OR microblogging OR Pinterest OR Instagram) NOT (higher education)." These were the initial search terms for each database. Based on the relevancy of results in each database, search terms were revised to target various types of social media individually. During these searches, results included only peer-reviewed works in English, published within the last 13 years (i.e., 2004–2017). This initial phase of the search resulted in 1253 documents.

In the second phase of the search, we conducted a variety of hand searches. First, we reviewed lists of journals dedicated mostly or entirely to our topic (i.e., educational research journals, educational technology journals, and technology-enhanced learning journals) (see Manca and Ranieri (2013) for a complete list of journals in these topical areas) or high-quality general education journals, as indicated by an impact factor greater than one (e.g., *Review of Education Research, Review of Research in Education, Teachers College Record, American Educational Research Journal*). To surface any relevant articles in journals not covered in the database search, we conducted a manual table of contents search for each journal for the years available: *Journal of Literacy and Technology* (2007–2017); *Journal of Educational Technology Development and Exchange* (2011–2017); and *International Journal of Online Pedagogy and Course Design* (2011–2017). Two journals not covered in the database search, but which appeared on our list of relevant journals, were unavailable through our university library and, therefore, excluded: *International Journal of Continuing Education* and *Lifelong Learning* and *International Journal of Technology in Teaching and Learning*.

Second, we created a snowball sample from the cited references in all of the literature reviews that appeared in our search, as well as from the cited references of any literature reviews, empirical articles, conference proceedings, books, or other publications from our own personal collections. Citation entries were reviewed by title and abstract to determine initial relevance.

Third, we analyzed the database search results and identified all journals with more than five relevant entries. For each of these journals, we then did a manual table of contents search on our focal topic for the years 2004–2017 to ensure that we did not miss any potentially relevant articles.

Fourth, we contacted authors in the field whose articles appeared relevant on our initial screening of records or who we knew to be working in our focal area; we requested articles from them, or from others they knew, that might be relevant. Together, these four hand searches in the second phase of the overall search surfaced 60 additional documents.

1.4.2   Screening, Assessing, and Determining Article Inclusion

From this combination of searches, we removed 30 duplicate records, leaving a total of 1290 records for screening. (Duplicate records that appeared in two or more databases were scanned and then either removed if they were irrelevant or retained until full-text screening.) Screening occurred based on title and abstract review. To remain in our pool, articles had to include social media use by K-12 teachers. They had to be empirical and published in English in peer-reviewed journals. Non-peer-reviewed reports, book chapters, conceptual pieces, unpublished dissertations, conference papers, or masters' theses were excluded. We also excluded literature reviews from our final pool of articles because our focus was on original research articles; however, as mentioned above, we scanned the reference lists of existing literature reviews to surface research articles that met our inclusion criteria, and these were included in the final pool of articles, if applicable. During this screening, 1191 articles were excluded, leaving a total of 92 studies. The references of these articles were also checked for any potentially relevant articles not identified in previous search procedures. This step resulted in an additional seven articles, bringing the total number of identified articles to 99 articles, which then underwent a full review.

The final screening of the 99 studies included a full-text review of each article by two authors who assessed each for eligibility. These articles were screened for inclusion by the two author-reviewers and marked as relevant or irrelevant with a written rationale. Any discrepancies were resolved by a third rater. Inter-rater reliability was calculated at 98%. From this rating process, 53 articles were excluded. These articles were excluded for one of three reasons: the social media in the article was outside the parameters of our review; the quality of the research was insufficient or not sufficiently described to judge its quality (e.g., the article lacked a methods section, data sources were not identified, data analysis procedures were not described, etc.), or the focus was not on teachers (e.g., the study focused on students or school administrators). Thus, a sample of 46 studies was left for qualitative analysis.

In the third phase, which took place in the winter of 2018–2019, we sought to uncover any research articles that may have appeared in peer-reviewed educational research journals or educational technology journals within the 18-month period since our original search, identification, screening, and selection of articles first occurred. Therefore, we again consulted the educational databases (i.e., ProQuest, Education Full Text, and Scopus), using the same search terms, but set the year limits for only the period 2018–2019. This second phase of the search resulted in an additional 182 documents. Title and abstracts of these articles were screened for relevance or irrelevance, and articles deemed relevant then underwent a full-text review where again criteria for inclusion and exclusion were applied. In addition, the table of contents of a select group of prominent educational and educational technology research journals were consulted for the years 2018–2019 (e.g., *Computers & Education, Computers in Human Behavior, Review of Educational Research*, etc.). Through this process, an additional 10 articles were deemed relevant and incorporated into the final resulting corpus of 56 articles.

1.5 Results

Next, we will report findings related to our focal topics: (1) What social networking platforms are used by K-12 educators? (2) What specific features of those platforms are used by teachers and/or their students? (3) For what educational purposes do teachers adopt social networking? (4) What does learning look like in these spaces? We begin by identifying the most prominent social networking platforms used in K-12 education and then expound on particular features studied (or not studied) in the literature (Sect. 1.5.1). In Sect. 1.5.2, we characterize the purposes of social networking in K-12 education using six themes that reveal its function both outside of the classroom as a form of professional development and inside the classroom as a tool for teachers and students. Finally, we turn to the depictions and analysis of learning present in the reviewed articles, summarizing trends in Sect. 1.5.3.

1.5.1 Materials

Our review of the educational research literature concerning social networking in elementary and secondary education revealed that the most prevalent social networking platforms studied in K-12 settings (see Table 1.2) were the widely popular, global social networking sites: Twitter and Facebook, as well as Edmodo, a social networking site designed for education. Of the 56 reviewed studies, 20 concerned

Table 1.2 Platforms studied

Citation	Twitter	Facebook	Edmodo	Other	Various
Asterhan and Rosenberg (2015)		X			
Bartow (2014)					X
Bicen and Uzunboylu (2013)		X			
Blonder et al. (2013)				YouTube	
Blonder and Rap (2017)		X			
Britt and Paulus (2016)	X				
Carpenter and Green (2017)				Voxer	
Carpenter and Justice (2017)					X
Carpenter and Krutka (2015)	X				
Carpenter and Krutka (2014)	X				
Cinkara and Arslan (2017)		X			
Davis (2015)	X				
Forkosh-Baruch, Hershkovitz, and Ang (2015)		X			
Forkosh-Baruch and Hershkovitz (2018)		X			
Gao and Li (2017)	X				
Goodyear, Casey, and Kirk (2014)	X	X			
Greenhalgh and Koehler (2017)	X				
Holmes, Preston, Shaw, and Buchanan (2013)	X				
Hunter and Hall (2018)					X
Hur and Brush (2009)				WeTheTeachers, Teacher Focus, LiveJournal	
Jiang, Tang, Peng, and Liu (2018)				Created for study	
Kelly and Antonio (2016)		X			
Kerr and Schmeichel (2018)	X				
Krutka and Carpenter (2016)	X				
Kuo, Cheng, and Yang (2017)		X			
Lantz-Andersson (2018)		X			
Macià and Garcia (2017)	X				

(continued)

Table 1.2 (continued)

Citation	Twitter	Facebook	Edmodo	Other	Various
Noble, McQuillan, and Littenberg-Tobias (2016)	X				
Prestridge (2019)					X
Rashid (2017)		X			
Risser and Waddell (2018)	X				
Robson (2017)		X			
Rodesiler (2014)					X
Rodesiler (2015)					X
Rodesiler and Pace (2015)					X
Rosell-Aguilar (2018)	X				
Ross, Maninger, LaPrairie, and Sullivan (2015)	X				
Rutherford (2010)		X			
Schwarz and Caduri (2016)		X		Google+	
Shin (2018)			X	Glogster	
Sumuer, Esfer, and Yildirim (2014)		X			
Thibaut (2015)		X			
Trust (2015)		X			
Trust (2016)		X			
Trust (2017a)		X			
Trust (2017b)		X			
Trust, Krutka, and Carpenter (2016)					X
Van Vooren and Bess (2013)	X				
Vazquez-Cano (2012)	X				
Veira, Leacock, and Warrican (2014)		X			
Visser, Evering, and Barrett (2014)	X				
Vivitsou, Tirri, and Kynäslahti (2014)		X			
Wang, Hsu, Reeves, and Coster (2014)			X		
Wesely (2013)	X				
Xing and Gao (2018)	X				
Zhang, Liu, Chen, Wang, and Huang (2017)				Created for study	
	Total 20	*Total* 17	*Total* 7	*Total* 7	*Total* 8

Twitter, 17 concerned Facebook, and 7 concerned Edmodo. The remaining 15 articles covered either various social media (i.e., nonspecific, multiple platforms) ($n = 8$) or other specific platforms not investigated in more than one study ($n = 7$). Table 1.2 summarizes the platforms used across the studies, indicating which platform(s) were used in each.

1.5.1.1 Twitter Features Used by Educators

Of our reviewed studies that focused on Twitter, all but two (Van Vooren & Bess, 2013; Vazquez-Cano, 2012) studied teachers' use of Twitter for professional development (PD). Neither of the classroom-based studies offered significant discussion of specific features used on Twitter for teaching and learning, although students did actively engage in class Twitter assignments in Vazquez-Cano's (2012) work. Further description of these studies can be found in Sects. 1.5.3.1. Here we summarize the features of Twitter as they were studied in the 13 articles related to PD.

The use of hashtags and synchronous chats, two common practices tied to tweeting, were repeatedly found to be important to teachers. Prestridge (2019) noted that Twitter was the most commonly mentioned platform for PD by the instructors interviewed and that the teachers spent anywhere from 4 to 20 hours per week engaging in chats, following hashtags, or following experts in their fields on Twitter. In Carpenter and Krutka's survey of educators (2014, 2015), 19% specifically mentioned synchronous chats as a valued affordance of Twitter.

Teachers mostly used the original post, or tweeting feature, of Twitter, but retweeting, replying, and using mentions were also utilized for professional learning purposes in some studies. The majority of the reviewed studies focused on the platform's main content creation activity: tweeting. Three studies also integrated social network analysis (Greenhalgh & Koehler, 2017; Macià & Garcia, 2017; Zhang et al., 2017) to show connections between users based on their followers. Activities related to tweeting, namely, retweeting, replying, mentioning, or favoriting, were often left out of data collection. Four studies tracked participants' retweeting habits (Gao & Li, 2017; Greenhalgh & Koehler, 2017; Ross et al., 2015; Xing & Gao, 2018), and one of the reviewed studies tracked participants' favoriting habits (Greenhalgh & Koehler, 2017). Replies or mentions were included in only seven studies, and such data often enriched the insights researchers were able to conclude (e.g., Kerr & Schmeichel, 2018; Risser & Waddell, 2018). In Risser and Waddell (2018), for example, tweets from two conference hashtags were analyzed for content as well as for the usage of the reply and mention features. The authors found that higher numbers of replies and mentions reflected a higher sense of community among teacher-participants in the conference hashtag.

Although Twitter was lauded across the studies as a resource well-suited to meet teachers' PD needs because it provided personalization and just-in-time resources, some studies found that educators' active participation on the platform was limited. Gao and Li (2017), for instance, found that half of the participants in the #Edchat Twitter stream tweeted only once, while only a few participants were highly active.

Similarly, Greenhalgh and Koehler's (2017) analysis of the #educattentats Twitter stream after a terrorist attack in Paris showed that 25% of participants did not contribute original tweets and instead only liked others' tweets. Furthermore, the most active 1% of participants was responsible for 15% of the total content, accounting for more than 50% of all original tweets.

Low active participation, yet high professional value seems to define the typical Twitter PD experience; however, the literature has yet to consider why this is the case. Britt and Paulus (2016) mention that many #Edchat participants were likely lurking, or observing online during the conversation, but data on the observers was not collected. Xing and Gao (2018) attempted to understand what factors indicated teachers' continued participation in #Edchat, analyzing tweets from the #Edchat Twitter stream over the course of 6 years. They determined that participating in chat sessions characterized as having more cognitive tweets (i.e., personal ideas, experiences, or opinions; questions or conversation starters) and interactive tweets (i.e., responses to others' tweets) was positively correlated with continued participation, while engaging in chats characterized as having more social tweets (i.e., greetings or shows of appreciation or courtesy) was negatively correlated with continued participation. The authors suggest that perhaps teachers perceived the content from cognitive and interactive tweets to be more interesting and valuable, while social tweets may have been seen as distractions.

Tweeting and the use of hashtags to organize and spark PD conversations among educators have been well-studied, yet the field continues to lack clarity in understanding the value and impact of highly active compared to more passive or observation-based participation habits. More in-depth attention to teachers' practices with *all* Twitter features—especially those associated with passive participation (e.g., observing, favoriting)—is warranted to gain a comprehensive picture of how teachers are adopting the functionalities of Twitter for educational purposes.

1.5.1.2 Facebook Features Used by Educators

Of the 17 studies that investigated teachers' Facebook use, eight studies covered teachers' use of Facebook for PD and eight described how Facebook was used in the classroom. One study fell in both the PD and classroom categories (see Table 1.3). Among the eight studies depicting teachers' PD practices on Facebook, only three discussed features other than the traditional Facebook post. Most studies were centered around particular Facebook groups and, therefore, focused on posts within the group. Rashid (2017), however, looked instead at teachers' Facebook status posts, finding that teachers were benefiting from the dialogic reflection afforded through their posted updates. Two studies also tracked responses to posts in Facebook groups (Cinkara & Arslan, 2017; Kelly & Antonio, 2016). Cinkara and Arslan (2017) analyzed which categories of topics were most discussed in a Facebook group for English Language teachers, finding that although the most posts were made in the resource-sharing category, the most replies and the highest word counts were found in the career development category. No studies addressed teachers' liking habits on

Table 1.3 Purpose for social media use

Citation	Informal PD	Formal PD	Classroom
Asterhan and Rosenberg (2015)			X
Bartow (2014)			X
Bicen and Uzunboylu (2013)		X	
Blonder et al. (2013)		X	
Blonder and Rap (2017)		X	
Britt and Paulus (2016)	X		
Carpenter and Green (2017)	X		
Carpenter and Justice (2017)			X
Carpenter and Krutka (2015)	X		
Carpenter and Krutka (2014)	X		X
Cinkara and Arslan (2017)	X		
Davis (2015)	X		
Forkosh-Baruch et al. (2015)			X
Forkosh-Baruch and Hershkovitz (2018)			X
Gao and Li (2017)	X		
Goodyear et al. (2014)		X	
Greenhalgh and Koehler (2017)	X		
Holmes et al. (2013)	X		
Hunter and Hall (2018)	X		
Hur and Brush (2009)	X		
Jiang et al. (2018)			X
Kelly and Antonio (2016)	X		
Kerr and Schmeichel (2018)	X		
Krutka and Carpenter (2016)	X		X
Kuo et al. (2017)			X
Lantz-Andersson (2018)			X
Macià and Garcia (2017)	X		
Noble et al. (2016)	X		
Prestridge (2019)	X		
Rashid (2017)	X		
Risser and Waddell (2018)	X		
Robson (2017)	X		
Rodesiler (2014)	X		
Rodesiler (2015)	X		
Rodesiler and Pace (2015)	X		
Rosell-Aguilar (2018)	X		
Ross et al. (2015)	X		
Rutherford (2010)	X		
Schwarz and Caduri (2016)			X
Shin (2018)			X
Sumuer et al. (2014)	X		X
Thibaut (2015)			X

(continued)

Table 1.3 (continued)

Citation	Informal PD	Formal PD	Classroom
Trust (2015)	X		
Trust (2016)	X		
Trust (2017a)	X		
Trust (2017b)	X		
Trust et al. (2016)	X		
Van Vooren and Bess (2013)			X
Vazquez-Cano (2012)			X
Veira et al. (2014)			X
Visser et al. (2014)	X		X
Vivitsou et al. (2014)			X
Wang et al. (2014)		X	
Wesely (2013)	X		
Xing and Gao (2018)	X		
Zhang et al. (2017)		X	
	Total 35	*Total* 6	*Total* 19

Facebook, and only one study (Goodyear et al., 2014) integrated analysis of teachers' direct messages. In Goodyear et al. (2014), the participants engaged with one researcher through social media for support when making changes in their practices. The messaging conversations between the researcher and participants became part of the corpus of data.

Articles that focused on Facebook's applications in classrooms, rather than on teachers' professional learning, mostly considered how teachers used Facebook to communicate with students outside of class. (See Sects. 1.5.2.4, 1.5.2.5 and 1.5.2.6 for themes in how social networking functioned in the classroom and to read more about how the features of Facebook facilitated teacher-student connections.) Only two classroom-focused articles described specific features of Facebook used by students. For instance, in Veira et al. (2014), students used Facebook group posts to interact, ask questions, and share resources. Although the students expressed concern over appearing less informed than their peers, students valued the Facebook group as a space to socialize. Lantz-Andersson (2018) similarly explored how students used a Facebook group to interact with each other, but placed more emphasis on how social media promoted informal communication styles, ideal for language development skills.

Thus, Facebook's group feature, which can facilitate either a private group of participants or an open community's ability to collaborate, is the most studied and utilized according to our review of the educational research literature. Consideration of how other Facebook features, such as timeline posts, likes, and direct messaging tools, can be used for educational purposes is less understood.

1.5.1.3 Edmodo Features Used by Educators

Seven articles studied Edmodo, and all but two (Shin, 2018; Thibaut, 2015) looked at Edmodo as a tool for professional development. The two classroom-based studies, despite clearly involving students' active engagement on the platform, do not focus on specific Edmodo features used and are further described in Sects. 1.5.2.4 (Shin, 2018) and 1.5.3.1 (Thibaut, 2015). Of these five studies concerning teachers' use of Edmodo in professional development, only two studies discussed the specific features teachers adopted. These included Edmodo discussion threads that went beyond posting and browsing posts. For instance, Trust (2015) tracked teachers' responses to community posts and found that 24% of teachers made new professional connections through the Edmodo online community; 13% furthered the connection by reaching out through direct messages. Wang et al. (2014) described how, over the course of 4 years, teachers engaged in professional training via Edmodo on how to integrate information and communication technologies (ICTs) in classrooms in ways that translated to changes in their practice. After the PD on Edmodo, teachers altered their lessons: students actively used more technology to complete assignments, demonstrating a range of new literacies and promoting student-centered learning.

From the 23 out of 56 reviewed studies that gave significant attention to which social networking features educators are adopting in K-12 education, we see that regardless of platform (i.e., Twitter, Facebook, Edmodo), most of this literature examined only one particular feature of the social media (e.g., tweets, Facebook group posts, Edmodo discussion thread posts). Rarely did the extant literature provide a comprehensive view of whether teachers utilized the range of various social networking features at their disposal on a given platform and *which* features they adopted for particular purposes. We find that the features and functionalities of social networking platforms in use within education, as part of teachers' learning or teaching experiences, remain largely under explored.

1.5.2 Educational Networking: Social Networking for Educational Purposes

Teachers' use of social media for educational purposes fell into two categories: professional development uses (informal and formal PD combined total; $n = 43$) and classroom uses ($n = 19$). Of our 56 reviewed articles, 4 covered both PD and classroom uses and were therefore considered in both categories. Table 1.3 provides an overview of the articles as they fit into these categories, also indicating which studies investigated teachers' formal versus informal PD practices.

Professional development was the most common educational purpose for social networking identified in the reviewed literature. Forty articles ($n = 40$), or 71% of the total corpus, described how teachers utilized social media for PD, and several studies surveying teachers reported that PD was their most frequent reason for accessing social networking platforms (e.g., Carpenter & Krutka, 2014). Teachers' professional development needs are complex: teachers look for new, timely, and individualized resources, but also seek emotional reassurance and support in their efforts to improve practice. We identified three main themes in teachers' uses of social media for professional development, each of which is further elaborated in the first three sections that follow: (1) Section 1.5.2.1, to find and share information or resources; (2) Section 1.5.2.2, to build community and combat isolation; and (3) Section 1.5.2.3, to meet diverse professional development needs not met by traditional professional development opportunities.

After sections describing teachers' social media use for professional development, we detail themes in the 20 studies ($n = 20$) covering teachers' classroom uses of social networking: (1) Section 1.5.2.4, social media as platforms for student learning; (2) Section 1.5.2.5, social media as platforms for posting reminders to students; and (3) Section 1.5.2.6, social media as platforms for student-teacher connections.

1.5.2.1 Professional Development: Networking to Find and Share Resources

Finding and sharing information or resources was a nearly universal reason for teachers' pursuing PD on social media. Twenty-seven ($n = 27$) of the PD-focused articles explicitly covered this purpose. Hunter and Hall (2018) confirmed in their survey that teachers most commonly accessed social networks seeking new practice-related information.

Although sharing resources was a common purpose for teachers' social networking, the literature showcases diverse approaches to understanding teachers' resource-cultivating habits. For example, Ross et al. (2015) surveyed teachers about their use of Twitter using a conceptual framework that emphasized Twitter as professional learning networks (PLNs). They found that the depth of resources and information available on Twitter made it an invaluable PLN for educators. Not only did 90% of participants indicate that they were extremely likely to use Twitter for PD, but 69% indicated that they intended to increase their use of the platform in the coming year. Offering another unique perspective on teachers' use of Twitter for PD, Kerr and Schmeichel (2018) considered gender differences in how educators participated in #sschat, noting that men were more likely than women to send tweets sharing resources. Although both genders were equally active in the Twitter chats, men were also more likely to critique, boast, advise, and promote through tweeting than were women. The reviewed literature strongly supports social networking as an effective practice for—and one already commonly used by—teachers to both find and share resources.

1.5.2.2   Professional Development: Networking for Emotional Support

Teaching can be draining and isolating; in addition to seeking new knowledge and information to improve their practice, teachers look for community support from their training experiences. Twenty-two ($n = 22$) of the reviewed studies reflected this theme. Through social media PD, teachers positioned themselves within a professional network, but also sought emotional reassurance and a sense of belonging. The community that can be created through social media PD is one that teachers valued for its richness both in educational content and supportive camaraderie.

Teachers reported that collaborating and networking with other educators was an affordance of PD through social media (e.g., Rosell-Aguilar, 2018; Ross et al., 2015). In Cinkara and Arslan's (2017) analysis of interactions in a Facebook group for English as a Foreign Language (EFL) teachers, the relationships between participants were described as forms of mentorship. Social media facilitated teachers' connections not just with other educators, but with knowledgeable, experienced peers (Trust et al., 2016; Wesely, 2013).

Beyond the professional networking aspects, teachers built genuine friendships through social media (Noble et al., 2016). Rodesiler's (2014) investigation of secondary English teachers' use of social media found that the teachers used humor and shared experiences to make connections with other educators online. Social media can be welcoming enough for teachers to be comfortable sharing their positive and negative feelings. Teachers in Hur and Brush's (2009) case study felt that online social media communities were safe spaces where they did not have to feel alone or judged; there were other teachers who had similar experiences to offer support.

The closeness of teachers on social media was also measurable through social network analysis. Macià and Garcia (2017) observed in their mapping of the social networks teachers created on Twitter that educators' connections were tighter and more reciprocal than those of general users; they concluded that the tighter networks demonstrated that teachers were using Twitter for social networking, not just to collect resources.

The literature shows teachers utilizing social media as more than a database to search for new lesson ideas or strategies. This is reflective of Trust et al.'s (2016) whole teacher approach to PD—an approach that acknowledges teachers' range of needs, including social and emotional support and not just pedagogical support—the role of social media in teachers' learning experiences is not just as a library of resources, but also as a space for socializing and community.

1.5.2.3   Professional Development: Networking to Meet Diverse Needs

Many of the reviewed studies highlighted how social media offered unique affordances as platforms for professional development experiences that are not available through other forms of PD. Prestridge (2019) highlights this by summarizing how teachers differentiated between professional *development* and professional *learn-*

ing. In their interviews, the teachers described professional development as train-ings that were "formal," "agenda-based," and "content-driven" (p. 149). Comparatively, professional learning was associated with the activities teachers participated in on social media that were informal, active learning opportunities that could be individualized to meet each teacher's needs. Across the studies, teachers appreciated that social media (1) allowed for highly individualized and accessible PD, (2) facilitated interactive learning experiences, (3) exposed teachers to both diverse, global perspectives and to other like-minded educators, and (4) acted as a filter to assist teachers in curating the most current and best quality resources and practices. Nineteen articles ($n = 19$) showcased examples of how PD on social media enriched teachers' learning experiences beyond the capabilities of traditional training formats.

Some of the most cited advantages to pursuing PD on social media are the high levels of personalization available to teachers. Social media PD spaces are managed and controlled *by* educators *for* educators (Britt & Paulus, 2016; Davis, 2015; Trust, 2017a). By being able to self-select and access the content when needed, teachers are able to get the PD they need efficiently (e.g., Greenhalgh & Koehler, 2017; Rosell-Aguilar, 2018; Ross et al., 2015). In particular, the concise format and ease of use associated with social media were mentioned as appealing features that drew teachers to social media as a source of PD (Davis, 2015; Greenhalgh & Koehler, 2017). In Davis' (2015) interviews with teachers about their experiences with #Edchat, the real-time flow of conversation and the brevity of tweets made the chat accessible and a valuable resource for teachers looking to meet specific needs.

The personalization that makes PD on social media relevant and adaptive to teachers' unique interests stems from the inherent design of social media as interac-tive spaces where teachers are drawn into conversations and communities that foster participatory learning, defined as teachers' active involvement in their own learning processes. As Krutka and Carpenter (2016) concluded in their survey of teachers' uses of Twitter, "Instead of PD being another thing done to teachers (e.g., standards, testing), many participants described Twitter as a positive, creative, and emancipat-ing space for professionals" (p. 51). All teachers can be active learners on social media platforms, affording an equitable level of participation not always achievable in off-line PD settings where teachers typically take turns contributing to discus-sions and there is not always time for everyone to be heard. Online, teachers can contribute at their own pace and on their own time.

Although participatory learning can be easily facilitated on social media and is recognized as beneficial to teachers who utilize these platforms (e.g., Noble et al., 2016), promoting consistent participation was noted as a challenge in these spaces. The low numbers of active contributors on Twitter exemplify this (Gao & Li, 2017; Greenhalgh & Koehler, 2017). Additionally, Rutherford's (2010) examination of teachers' use of a Facebook group suggested challenges in getting sustained and universal participation from group members. That is, the teachers in the Facebook group largely posted ideas as individuals but did not generally engage in back-and-forth conversation with the posts of other teacher-participants (Rutherford, 2010).

While active learning seems to be a valuable affordance of PD through social media, the extent to which teachers are actually leveraging this affordance for their professional learning remains unclear. Moreover, the more passive means through which teachers may participate and derive value from social networking (e.g., observing, liking or favoriting, sharing, etc.) are largely unexplored phenomena in the research literature as mentioned in Sect. 1.5.1.1.

Regardless of teachers' active or passive participatory habits on social media, the networking features of social media connect teachers to a potentially diverse community, bringing together isolated teachers with like-minded peers and allowing for a range of perspectives to be considered. As Trust et al. (2016) found in their survey of teachers regarding their professional learning networks (PLN), teachers utilized social media to build more expansive and richer networks with peers than would be possible in only face-to-face settings. Similarly, Noble et al. (2016) argued that global connections through social media can diversify the types of conversations and experiences that teachers can leverage for educational purposes.

Digital communication on social media not only facilitates networks that would not be possible off-line but also enhances access to and engagement in other forms of PD, such as conferences (Greenhow, Gleason, & Willet, 2019; Greenhow, Li, & Mai, 2019; Li & Greenhow, 2015). Visser et al. (2014) in their survey of teachers' use of Twitter noted that the platform can be a supplement to in-person conference experiences (e.g., backchannel conversations) or an alternative space in which to virtually attend conference events. In this way, social media can help teachers find communities of support from like-minded educators and opportunities to expand their exposure to diverse perspectives that might otherwise be too geographically distant to have materialized off-line.

The reviewed literature emphasized the depth and breadth of networks that teachers can build on social media as unique features of these platforms that educators seem to value. That said, similar to the questions we raise above regarding teachers' actual levels of active participation, the research is ambivalent about whether and how networking features and network diversity actually impact teachers' learning experiences (Krutka & Carpenter, 2016).

In other ways, however, social media does directly empower teachers to control their learning; not only can teachers choose the content of the PD on social media, but the platforms can act as personalized filters for identifying new resources. Several studies noted how educators considered social media, and Twitter in particular, to be reliable sources of up-to-date, current best practices (Carpenter & Krutka, 2015; Krutka & Carpenter, 2016; Trust et al., 2016). In Holmes et al.'s (2013) analysis of tweets from 30 teachers with an influential presence on Twitter, 34% of all tweets provided links to educational resources; the authors concluded that Twitter was used to highlight relevant content for teachers that might otherwise have been difficult to discover through internet searches.

Similarly, two studies reported that teachers valued using Edmodo to find new, curated resources (Trust, 2016, 2017a). On Edmodo, teachers found new strategies and content recommended by other teachers. Posts sharing information and comments added by other users allowed teachers to browse discussion threads as if they were reviews of shared materials. Social media can support teachers' curation of trusted resources without the hassle of finding and vetting ideas drawn from general internet searches.

Whether helping teachers to find personalized support, to engage in participatory learning spaces, to expand and diversify their professional networks, or to identify the most effective new resources, social media were used by educators across the reviewed literature to fill PD gaps and make professional learning highly relevant to all teachers, despite their unique, individual needs.

1.5.2.4 Social Networking in the Classroom: For Student Learning

In the classroom, teachers found that social media could be effective for promoting student learning. In particular, social media facilitated collaboration in new and engaging ways. The interactive nature of social media can promote effective discussion and group work (e.g., Bartow, 2014; Jiang et al., 2018; Veira et al., 2014). Students learn together and from each other; they all have an equal opportunity to contribute online, rather than having to take turns speaking in class, and the written record of online discussions acts as a resource that students can return to later (Blonder & Rap, 2017; Thibaut, 2015). For example, in Vazquez-Cano's (2012) study of 15 teachers' use of Twitter in the classroom, students completed a variety of interactive assignments on Twitter, such as tweeting summaries of readings, practicing grammar, playing word games, writing poetry, and co-authoring stories with their peers. Schwarz and Caduri (2016) interviewed five teachers whose use of social media in their classrooms was exemplary; these teachers used social media for purposes such as role-playing activities and creating a flipped classroom environment.

Moreover, social media can become a community space that extends the classroom (Vivitsou et al., 2014), allowing students to connect and interact beyond what can be accomplished in off-line classrooms in schools. Lantz-Andersson (2018) similarly found that social media can help students connect their in-classroom learning with their out-of-classroom experiences and practices. In her study of two Facebook groups of English-language learners, the students played with language, moving between formal (i.e., classroom assignment) and informal (i.e., casual conversation) framings; the Facebook group setting facilitated more authentic, open-ended interaction that is ideal for language learning. In Shin (2018) a bilingual elementary school student found his writing practices validated through the collaborative aspects of using social media for classroom assignments. His interest and motivation increased as he strove to connect and engage with his peer audience.

Although the reviewed literature showcased social media as a pathway for teachers to make global connections and incorporate diverse perspectives into their PD experiences, teachers' use of social media to globally connect their students was noticeably absent from the studies. In fact, Krutka and Carpenter (2016) openly discuss how utilizing social media to build students' global citizenship is an underused affordance of Twitter. Students could engage with global peers and social activism on Twitter, but the researchers did not find that this was common practice among the teachers surveyed.

Only one study in our dataset described an example of students' making global interactions through social media. Carpenter and Justice (2017) surveyed teachers who used social media as part of the Global Read Aloud (GRA) project. Classrooms around the world shared a common reading of a text that connected students across cultural differences. Teachers in the study reflected on the challenges of the GRA, citing difficulty implementing social media effectively as one factor. In response, Carpenter and Justice (2017) suggest that teachers would benefit from additional trainings or supports to improve future iterations.

Social media have been established as effective tools to facilitate collaborative learning that can extend beyond school walls (Manca & Ranieri, 2013, 2016), yet teachers often limit the networks available for students' collaboration to their immediate classmates.

1.5.2.5 Social Networking in the Classroom: For Student Reminders

In addition to students' use of social media for classroom activities, several studies described how teachers utilized social media for administrative tasks: to make announcements or distribute class materials (Schwarz & Caduri, 2016; Van Vooren & Bess, 2013; Vivitsou et al., 2014). For instance, one teacher in Schwarz and Caduri's study (2016) used a Facebook group to share class reminders and post assignments.

The potential effectiveness of social media as a resource to help students track and manage their responsibilities is captured uniquely in Van Vooren and Bess' (2013) experimental design study. The teacher in the treatment group sent out tweets to remind eighth grade students of homework and deadlines, a support that led to those students scoring higher on standardized tests than the control group of students who did not participate on Twitter. Social media can function not only as a space for increased student participation and collaboration, but also as an organizational tool that teachers can use to provide students the structure they need to manage their own learning.

1.5.2.6 Social Networking in the Classroom: For Student-Teacher Connections

Ten of the reviewed studies ($n = 10$) addressed how social media can be used to connect students with teachers. Although increasing connectedness is recognized as a useful affordance of social media, concerns over privacy and appropriate practices to maintain professional rapport mediate how teachers approach this application of various platforms (Krutka et al., 2019). Some studies conclude that teacher-student communications through social media are underutilized (Carpenter & Krutka, 2014; Krutka & Carpenter, 2016), while others emphasize caution when attempting such interactions (Krutka et al., 2019; Kuo et al., 2017; Sumuer et al., 2014; Vivitsou et al., 2014).

Social media can be an effective tool to help students reach out to teachers when they need additional academic support (Forkosh-Baruch et al., 2015; Krutka & Carpenter, 2016), but making such connections can come at a cost to privacy. In Forkosh-Baruch et al.'s (2015) survey of teachers' and students' communication habits on Facebook, friending between teachers and students was found to support positive rapport and meaningful relationships; however, students also expressed reluctance to become privy to teachers' personal lives. They preferred to maintain their own privacy, and overall the authors found that teachers and students did not often communicate on the platform. Similarly, Visser et al. (2014) reported that only 20% of surveyed teachers interacted with students on Twitter, and Carpenter and Green (2017) noted only 10.8% of teachers surveyed contacted students and parents using Voxer.

In Asterhan and Rosenberg's (2015) survey of teachers about connecting with students on Facebook, closed groups were the most recommended form of communication to students because they can be made secure (i.e., not open to the public), and teachers and students do not have to friend each other to participate. Half of the teachers also recommended contacting students through separate, professional Facebook accounts so that teachers' personal lives could be kept private without having to sacrifice their online availability to students. Of the 74% of teachers who had experience befriending students on Facebook, 62% recommended against it.

Teachers seem to want to meaningfully connect with students online, yet identifying best practices for such relationships is complex and problematic. In Forkosh-Baruch and Hershkovitz's (2018) study, only 20% of the surveyed teachers indicated being friends with students on Facebook, but an additional 40% of participants indicated a willingness to try it. Only 20% of the teachers thought Facebook was not suitable for learning, but in practice, few teachers reported utilizing the platform with their classes. The literature suggests that while there may be benefits to connecting with students on social media, educators are reluctant to embrace this affordance, perhaps because of the difficulty in defining appropriate teacher-student relationship dynamics that can move between formal classroom settings and informal social media spaces. Ultimately, maintaining privacy and professional boundaries remain priorities for teachers.

1.5.3   Views of Learning on Social Networking Platforms

Defining and measuring learning on social media is a recurring challenge in the educational research literature. Few studies employ a pre- and post-test design or control and treatment groups to track the effects of educational networking on participant performance (Van Vooren & Bess, 2013; Vazquez-Cano, 2012; Wang et al., 2014). More frequently, teachers or students are asked to self-report their own learning gains or reflect on their motivation or engagement as indicators of impact on

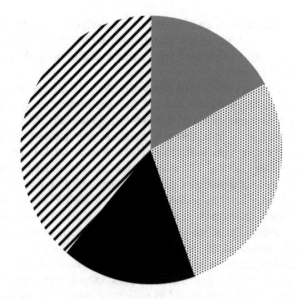

■ Quantitative performance measure: Students

∷ Qualitative engagement & motivation results: Students

■ Qualitative attitude & confidence results: Teachers

◢ Qualitative change in practice results: Teachers

Fig. 1.2 How learning was measured

learning. See Fig. 1.2 for a summary of how learning was measured ($n = 18$) across our corpus. Ten studies discussed teacher learning, and eight studies discussed student learning. In the sections below, we describe how student learning (Sect. 1.5.3.1) and teacher learning (Sect. 1.5.3.2) were depicted in the reviewed studies.

1.5.3.1   Student Learning

Student learning was discussed in the literature through either changes in assessment scores or through teachers' perceptions of students' attitudes. In our dataset, three studies measured student learning through grades and test scores. Van Vooren and Bess (2013) found that students in the treatment group who received the Twitter reminders scored higher on the class' standardized tests than those who received reminders only in the classroom. Similarly, Vazquez-Cano (2012) used pre- and post-tests to determine that linguistic competency test scores improved after the use of Twitter in the classroom. In the third study that measured student learning quantitatively, Wang et al. (2014) tracked the impact of a formal PD program on students' assessment outcomes using a quasi-experimental design. One of the two cohorts' content performance test scores was significantly higher than the control groups' scores, and both cohorts of treatment groups scored higher on the ICT skills survey than did the control groups.

Five additional articles considered changes in students' engagement and motivation rather than performance goals directly. In these studies, teachers reported the perceived impacts of social media on their students' learning. Teachers in Schwarz and Caduri's (2016) study indicated that students were actively engaged in their learning processes through platforms like Facebook and Google+. Thibaut's (2015) work with elementary school classrooms showed how having students create Edmodo profiles increased students' motivation and collaboration practices. Trust et al.'s (2016) survey of teachers added that because of their PLNs, teachers felt they were able to not only increase student engagement but also improve students' attitudes towards learning. Quantitative measures of engagement or motivation were not utilized in the reviewed studies, but qualitatively, teachers confirmed that students found social media to be an extension of the classroom (Bartow, 2014; Shin, 2018), a space where students could actively participate in their learning.

In general, social media was found to positively impact student learning; however, further investigation is necessary. Limited quantitative work and a reliance on teachers' self-reported perceptions present narrow views into how and to what extent students are learning on social media. Determining best practices for classroom implementation of social media will require insight from continued work in this area.

1.5.3.2 Teacher Learning

Teacher learning on social media, rather than being measured with quantitative tools (e.g., a pre- and post-test for a PD experience), was more often discussed in terms of how teachers' practices had changed. For example, Noble et al. (2016) found that because teachers gained confidence and were prompted to critically reflect on their teaching while on Twitter, they were inspired to make changes to their pedagogy. Rosell-Aguilar (2018) found that 87.5% of the surveyed teachers thought their teaching had improved through the use of the Twitter hashtag #MFLtwitterati. Similarly, 64% of teachers surveyed by Trust (2017a) reported that PD on Edmodo changed the way they taught by motivating them to want to improve, empowering them to self-evaluate, and inspiring them to embrace innovation and be creative. Following formal PD on how to integrate social media in the classroom, results from both Wang et al. (2014) and Blonder et al. (2013) showed that teachers had increased their effective use of technological tools and social media platforms with their students.

Furthermore, the community and connectedness facilitated through social media was specifically referenced as an affordance of social networking that impacted teacher learning. In Trust et al.'s (2016) survey, 98% of teachers indicated they changed their practice because of the PLNs they built through social media.

Goodyear et al. (2014) used social media to connect and follow up with teachers' post-PD training, and teachers offered positive reinforcements to their peers when seeing posts about each other's progress online. The added supports through social media helped teachers maintain their confidence in making pedagogical changes.

In addition to direct changes in practice, teacher learning was also described in a few studies as impacts on educators' attitudes and confidence towards their teaching. Teachers in Carpenter and Green's (2017) study reported that using Voxer led to positive changes in their attitudes and dispositions in the classroom. Similarly, Bicen and Uzunboylu (2013) found that after the formal PD in a virtual Facebook classroom, teachers had more positive opinions about how they could use Facebook with their students. Trust et al. (2016) noted that the PLNs teachers established through social media improved not only their attitudes and confidence but shaped their professional identities; they described themselves using terms such as "life-long learners" and "teacher leaders" (p. 26).

Although heavily reliant on limited self-reported data, the extant literature suggests teacher learning through social networking experiences positively influences both the pedagogy and mindsets educators bring to the classroom. Further work in this area offers a rich direction for future research as the field moves to more explicitly define *how* to successfully implement social networking in educational contexts.

1.6 Discussion: Analysis and Findings

The three main objectives we set out to address in this chapter were: (1) to present the state of the art in social networking in K-12 educational research from 2004 up to 2019; (2) to advance understanding of the knowledge base on educational networking beyond higher education settings; and (3) to evaluate and synthesize trends in the research arising from these evolving social phenomena. In this section, we turn to more closely evaluate the results above in order to illuminate educational networking, a field which combines social networking functionalities with educational content, aims, and applications.

To assist readers (e.g., academics, practitioners, and graduate students) in gaining insights and conducting research that establishes educational networking approaches towards social learning in their own contexts, we elaborate on the following themes in the sections below. First our review of over a decade of educational research suggests that there is more to learn regarding social networking platforms and how their features, functionalities, and underlying architectures impact educators' intentional uses or avoidance of them (Sect. 1.6.1). Second, definitions and measurements of learning through educational networking are underexplored terrain and ripe territory for future research (Sect. 1.6.2). Third, we are only in the beginning stages of understanding meaningful participation in educational networking, including how its dimensions may challenge traditional conceptions of effective participation (Sect. 1.6.3).

1.6.1 Views of Learning on Social Networking Platforms

Few articles from this literature review offered insights into educators' uses of particular features and functionalities within their chosen platforms, such as which features are used for which educational purposes or which features are intentionally not used and why. When framing our findings on which social media features teachers utilize, we had limited data to summarize. Research investigating how platforms' unique architectures and specific functions shape their possible applications in education is greatly lacking.

Indeed, of the range of features we profiled in Sect. 1.5.1 regarding the three most prevalent social media in the research literature (i.e., Facebook, Twitter, and Edmodo), only a handful (e.g., Facebook groups, tweets) have been thoroughly explored in educational contexts. Moreover, the most-studied features, (e.g., group posts, tweets) enable users to create their own digital content, a very active form of online participation. Other types of more passive, observational online participation (e.g., liking, retweeting, mentioning, viewing or observing others' posts) may be just as important for understanding educational networking practices, but the literature has yet to explore this possibility fully. Future research should strive to provide more comprehensive and in-depth views of how the features, architectures, and functionalities of platforms influence educational networking practices.

1.6.2 What Is Learning Through Educational Networking?

Through the course of this review, we identified the types of teacher and student learning taking place on social media and found that while social media was generally determined to positively influence teachers' and students' learning experiences, learning was mostly understood subjectively through teachers' self-reported experiences. With regard to student learning in particular, our dataset confirmed previous findings that social media can be an asset for teachers wanting to increase various types of student interaction, including collaboration, participation, critical thinking, and peer support (Ajjan & Hartshorne, 2008; DiVall & Kirwin, 2012; Mazman & Usluel, 2010). However, we also uncovered a continued lack of clarity in defining and measuring student learning that has previously been called out in past reviews (e.g., Gao et al., 2012; Manca & Ranieri, 2016; Tang & Hew, 2017).

Learning as measured by quantitative performance scores was absent from the literature with few exceptions (Van Vooren & Bess, 2013; Vazquez-Cano, 2012; Wang et al., 2014). In our review, and related literature reviews, much of the discussion surrounding learning through social networking encompassed self-reported reflections on changes in qualitative, attitudinal traits, such as perceived motivation, engagement, or enjoyment (e.g., Aydin, 2012; Tang & Hew, 2017). While future research can certainly expand in this direction and work to overcome the limitations of self-report data (Tang & Hew, 2017), we also want to call attention to the possible indicators of learning less frequently measured in this body of work: impacts on teachers' and students' social and emotional well-being.

It is clear that educators are accessing social media for community and social support, yet benefits from these affordances are not necessarily tied to specific changes in teachers' pedagogy. How does overcoming isolation change educators' perceptions of themselves and thereby the atmosphere they can create for students in the classroom? Beyond teachers' qualitative reports, more measures that consider self-efficacy (Blonder et al., 2013; Blonder & Rap, 2017), motivation, creativity, or teachers' feelings of excitement or inspiration could help more fully describe the types of influences social media have on educators.

Similarly, we found that teachers' use of social media in the classroom increases students' collaboration, but more can be done to measure the impacts of this affordance. How do students' attitudes and motivations change as a result of collaborating on social media? More attention to social media's influences on students' self-efficacy, agency, and the perceived value they place on their classwork could reflect indicators of learning and learning processes that cannot be understood through assessment scores.

Given the extensive literature showing teachers' use of social media for community and to facilitate meaningful student interactions, we note a disconnect between what is measured as social media's impacts on learning and what teachers and students report as the learning benefits afforded on social networking platforms. As the integration of educational networking both amplifies and disrupts conventional notions of learning objectives and valued outcomes (Salomon & Almog, 1998; Salomon & Perkins, 2005), the field needs more expansive and multidimensional definitions, exemplars, and assessments of learning (e.g., disciplinary, cognitive, emotional, social) in all its varied forms.

1.6.3 Understanding Meaningful Participation

Themes in the types of teacher participation on social media were repeatedly noted in the reviewed studies. While some articles emphasized the deeper connections and collaborations born from social media (Macià & Garcia, 2017; Noble et al., 2016), others noted more superficial participation (e.g., teachers contributing only one tweet during an #Edchat without deeply engaging with peers or a low incidence of "active" contributors to the Twitter stream) (Gao & Li, 2017; Greenhalgh & Koehler, 2017). Regardless of the frequency and depth of interactions, teachers still reported finding participation on social media meaningful. For teachers who needed emotional support and friendship through networking, social media were effective. For teachers who needed quick, just-in-time, specific advice that met immediate needs without overburdening or overwhelming them, social media were also effective. We argue that educational networking problematizes how we define meaningful participation in online spaces and in teacher professional development, in particular. Traditionally, high-quality teacher professional development online and off-line has emphasized teachers' continuous engagement in group learning, a departure from what is typically afforded in formal PD courses (Dede, 2006); however, our findings reveal that participating in a sustained professional development cohort may not be the only type of professional participation that teachers can benefit from, especially with respect to educational networking.

The metaphor of the learning ecology (Barron, 2004, 2006) may be most useful in helping conceptualize, study, and implement teachers' professional learning and development with, of, and through educational networking (Salomon & Perkins, 2005). Bronfenbrenner (1976) first theorized an ecology of education in which learning occurs at the individual level in an environment (e.g., home, work) and in the interconnections that exist between different environments. Building on this idea in the age of the internet and social media, Barron (2004), conceived of a learning ecology as "the set of contexts found in physical or virtual spaces that provide opportunities for learning" (p. 195). Barron argued that the internet offers learners agency as people create learning opportunities for themselves given time, resources, and interest.

Others have used learning ecology concepts to problematize the framing of social media within "formal" or "informal" learning—or, for our purposes, the framing of formal or informal professional development (Greenhow, Li, & Mai, 2019)—arguing that such terms are contested in the literature (Greenhow & Lewin, 2016); boundaries between learning activities and contexts are slippery and porous, and social media enable "seamless" integration across learning situations (Dabbagh & Kitsantas, 2012; Greenhow et al., 2009; Greenhow & Lewin, 2016). Thus, we argue for a more productive framing of teachers' educational networking situated in this broader ecology of learning, wherein teacher-learners create, manipulate, and move through intersecting "spheres of activity" (Barron, 2004, p. 5) as these meet their needs.

In the learning ecology, educators can find their own balance of formal and informal supports and active and observant participation, all of which can be valued and meaningful for learning. Thus far, the field has given precedence to studying active contributors and content creators on social media; however, this small group is not representative of how most teachers participate online. Recognizing, understanding, and valuing the experiences and the learning of "observers" (as opposed to the negative connotation of the label: "lurkers") is critical to truly demystifying the role social media play in teachers' professional learning and classroom experiences.

1.7 Conclusions

In this chapter we described a systematic review of over 15 years of educational research literature investigating teachers' use of social media, spanning from 2004 to 2019. Our results revealed that substantial work has been done to explore teachers' applications of social media. For instance, teachers' adoption of Twitter for professional development purposes was one of the most studied topics. The networking benefits of platforms like Twitter increased teachers' access to valuable resources and advice and support from colleagues. Teachers created their own learning spaces and experiences on social media, ensuring educators could craft meaningful professional development tailored to their needs. Studies noted that using social media for PD led to changes in teachers' practices and increases in teachers' motivation and inspiration.

In the classroom, teachers' use of social media was less studied, but promising findings suggest social media can be an effective tool to support students' learning. Social media facilitates peer-to-peer communication and connection, creating spaces for students to authentically share and collaborate. This trend echoes the findings of related literature reviews, such as Manca and Ranieri (2016), in which the use of Facebook was found to promote peer learning and out-of-class discussion.

Although changes in performance were studied quantitatively in only a few of the reviewed articles, multiple studies described qualitative indicators of increased motivation and engagement in students. Past reviews confirm this finding, but also point to the need to further understand how social networking changes learning in potentially negative ways (e.g., Aydin, 2012; Hew, 2011). For example, Aydin (2012) concludes that Facebook, while a space that inspires connection between students, can also facilitate cyberbullying or lead to invasions of privacy. Literature from our review addressing the use of social media for student-teacher communication similarly reports mixed findings that both laud the potential of using social networking to build strong relationships between students and teachers, but also warn against losing clear boundaries of professionalism.

Across our corpus, studies neglected to investigate whether or not teachers utilized the full range of social media features available to them. In particular, the impact of using features that complement observation-based participation (e.g., liking, retweeting, following) is not well understood. The field has focused largely on exploring social media participation that involves creating and contributing content. More attention to the range of features and participation types is necessary to fully illuminate teachers' experiences and activities in these online spaces. Related assertions from past reviews (Manca & Ranieri, 2013, 2016; Rodríguez-Hoyos et al., 2015) that the pedagogical potential of social media has not been well-defined nor accessed supports our conclusions: lack of understanding of the intricate social networking practices that take place online is likely impacting how we perceive their transfer to educational contexts.

Furthermore, our literature review revealed that conceptualizing and measuring the learning and participation that takes place through social media is challenging; meaningful learning and participation on social media often diverges from what is traditionally valued in off-line PD or classroom activities (i.e., sustained participation in a group or training and quantitatively measurable improvements in performance). For example, the reviewed articles did not highlight measures of changes in teachers' and students' attitudes and engagement, yet these benefits were some of the most oft cited in qualitative findings (e.g., the self-reported emotional and attitudinal impacts social media made on teachers or students and the emotional supports teachers sought through social media). Integrating ecological perspectives

(Barron, 2004) that account for teachers' unique movements within and across learning environments and contexts may better represent teachers' situated learning and the social affordances of social media. Performance indicators reveal one type of growth and achievement, but these may not be aligned with all the benefits teachers seek and gain through social media.

While educational networking has emerged as a burgeoning field with potential to support teachers in their professional development and classroom practices, more research is needed that examines how the informal affordances of social networking blend with what is often more formally structured educational content. The personalization and control users have over content and connections through social networking allow teachers the freedom to find and use online resources efficiently and as needed, granting them increased agency and ultimately expanding the possible applications of educational content. Educational networking offers new directions for teacher growth and support: potential that is evident in the literature, but not yet fully understood and achieved.

References

Ajjan, H., & Hartshorne, R. (2008). Investigating faculty decisions to adopt Web 2.0 technologies: Theory and empirical tests. *The Internet and Higher Education, 11*, 71–30.

Alias, N., Sabdan, M. S., Aziz, K. A., Mohammed, M., Hamidon, I. S., & Jomhari, N. (2013). Research trends and issues in the studies of Twitter: A content analysis of publications in selected journals (2007-2012). *Procedia-Social and Behavioral Sciences, 103*, 773–780.

Asterhan, C. S. C., & Rosenberg, H. (2015). The promise, reality and dilemmas of secondary school teacher–student interactions in Facebook: The teacher perspective. *Computers & Education, 85*, 134–148. https://doi.org/10.1016/j.compedu.2015.02.003

Aydin, S. (2012). A review of research on Facebook as an educational environment. *Educational Technology Research and Development, 60*, 1093–1106.

Barron, B. (2004). Learning ecologies for technological fluency: Gender and experience differences. *Journal of Educational Computing Research, 31*(1), 1–36.

Barron, B. (2006). Interest and self-sustained learning as catalysts of development: A learning ecologies perspective. *Human Development, 49*, 193–224.

Bartow, S. M. (2014). Teaching with social media: Disrupting present day public education. *Educational Studies, 50*(1), 36–64. https://doi.org/10.1080/00131946.2013.866954

Bicen, H., & Uzunboylu, H. (2013). The use of social networking sites in education: A case study of Facebook. *Journal of Universal Computer Science, 19*, 658–671.

Blonder, R., Jonatan, M., Bar-Dov, Z., Benny, N., Rap, S., & Sakhnini, S. (2013). Can you tube it? Providing chemistry teachers with technological tools and enhancing their self-efficacy beliefs. *Chemistry Education Research and Practice, 14*(3), 269–285. https://doi.org/10.1039/C3RP00001J

Blonder, R., & Rap, S. (2017). I like Facebook: Exploring Israeli high school chemistry teachers' TPACK and self-efficacy beliefs. *Education and Information Technologies, 22*, 697–724. https://doi.org/10.1007/s10639-015-9384-6

Boyd, D., Golder, S., & Lotan, G. (2010, January). Tweet, tweet, retweet: Conversational aspects of retweeting on twitter. In *System Sciences (HICSS) 2010 43rd Hawaii International Conference* (pp. 1–10). IEEE.

Britt, V. G., & Paulus, T. (2016). "Beyond the four walls of my building": A case study of #Edchat as a community of practice. *American Journal of Distance Education, 30*(1), 48–59. https://doi.org/10.1080/08923647.2016.1119609

Bronfenbrenner, U. (1976). The experimental ecology of education. *Educational Researcher, 5*(9), 5–15.

Buettner, R. (2013). The utilization of Twitter in lectures. *GI-Jahrestagung*, 244–254.

Carpenter, J. P., & Green, T. D. (2017). Mobile instant messaging for professional learning: Educators' perspectives on and uses of Voxer. *Teaching and Teacher Education, 68*, 53–67. https://doi.org/10.1016/j.tate.2017.08.008

Carpenter, J. P., & Justice, J. E. (2017). Can technology support teaching for global readiness? The case of the Global Read Aloud. *LEARNing Landscapes, 11*(1), 65–85.

Carpenter, J. P., & Krutka, D. G. (2014). How and why educators use Twitter: A survey of the field. *Journal of Research on Technology in Education, 46*, 414–434. https://doi.org/10.1080/15391523.2014.925701

Carpenter, J. P., & Krutka, D. G. (2015). Engagement through microblogging: Educator professional development via Twitter. *Professional Development in Education, 41*, 707–728. https://doi.org/10.1080/19415257.2014.939294

Cinkara, E., & Arslan, F. Y. (2017). Content analysis of a Facebook group as a form of mentoring for EFL teachers. *English Language Teaching, 10*(3), 40–53. https://doi.org/10.5539/elt.v10n3p40

Dabbagh, N., & Kitsantas, A. (2012). Personal Learning Environments, social media, and self-regulated learning: A natural formula for connecting formal and informal learning. *Internet and Higher Education, 15*, 3–8.

Davis, K. (2015). Teachers' perceptions of Twitter for professional development. *Disability and Rehabilitation, 37*, 1551–1558. https://doi.org/10.3109/09638288.2015.1052576

Dede, C. (2006). *Online professional development for teachers: Emerging models and methods.* Cambridge, MA: Harvard Education Press.

DiVall, M. V., & Kirwin, J. L. (2012). Using Facebook to facilitate course-related discussion between students and faculty members. *American Journal of Pharmaceutical Education, 76*(2), 1–5.

Ebner, M., Lienhardt, C., Rohs, M., & Meyer, I. (2010). Microblogs in higher education – A chance to facilitate informal and process-oriented learning? *Computers & Education, 55*(1), 92–100.

Ellison, N. B., & Boyd, D. (2013). Sociality through social network sites. In W. H. Dutton (Ed.), *The Oxford handbook of internet studies* (pp. 151–172). Oxford: Oxford University Press.

Forkosh-Baruch, A., & Hershkovitz, A. (2018). Broadening communication yet holding back: Teachers' perceptions of their relationship with students in the SNS-era. *Education and Information Technologies, 23*, 725–740. https://doi.org/10.1007/s10639-017-9632-z

Forkosh-Baruch, A., Hershkovitz, A., & Ang, R. P. (2015). Teacher-student relationship and SNS-mediated communication: Perceptions of both role-players. *Interdisciplinary Journal of E-Skills and Lifelong Learning, 11*, 273–289.

Forkosh-Baruch, A., & Hershovitz, A. (2012). A case study of Israeli higher-education institutes sharing scholarly information with the community via social networks. *The Internet and Higher Education, 15*, 58–68. https://doi.org/10.1016/j.iheduc.2011.08.003

Gao, F., Juo, T., & Zhang, K. (2012). Tweeting for learning: A critical analysis of research on microblogging in education published in 2008–2011. *British Journal of Educational Technology, 43*, 783–801. https://doi.org/10.1111/j.1467-8535.2012.01357.x

Gao, F., & Li, L. (2017). Examining a one-hour synchronous chat in a microblogging-based professional development community: Microblogging-based professional development community. *British Journal of Educational Technology, 48*, 332–347. https://doi.org/10.1111/bjet.12384

Goodyear, V. A., Casey, A., & Kirk, D. (2014). Tweet me, message me, like me: Using social media to facilitate pedagogical change within an emerging community of practice. *Sport, Education and Society, 19*, 927–943. https://doi.org/10.1080/13573322.2013.858624

Greenhalgh, S. P., & Koehler, M. J. (2017). 28 days later: Twitter hashtags as "just in time" teacher professional development. *TechTrends, 61*(3), 273–281.

Greenhow, C., & Askari, A. (2017). Learning and teaching with social network sites: A decade of research in K-12 related education. *Education and Information Technologies, 22*(2), 623–645.

Greenhow, C., Cho, V., Dennen, V., & Fishman, B. (2019). Education and social media: Research directions to guide a growing field. *Teachers College Record, 121*(14). Retrieved from: https://www.tcrecord.org/Content.asp?ContentId=23050

Greenhow, C., Galvin, S., Brandon, D., & Askari, E. (in press). Fifteen years of research on K-12 education and social media: Insights on the state of the field. *Teachers College Record.*

Greenhow, C., Gleason, B., & Willet, K. B. S. (2019). Social scholars: Learning through tweeting in the academic conference backchannel. *British Journal of Educational Technology, 50,* 1656–1672. https://doi.org/10.1111/bjet.12817

Greenhow, C., & Lewin, C. (2016). Social media and education: Reconceptualizing the boundaries of formal and informal learning. *Learning, Media and Technology, 41*(1), 6–30. https://doi-org.proxy1.cl.msu.edu/10.1080/17439884.2015.1064954

Greenhow, C., Li, J., & Mai, M. (2019). Social scholarship revisited: Changing scholarly practices in the age of social media. *British Journal of Educational Technology, 50*(3), 987–1004. https://doi.org/10.1111/bjet.12772

Greenhow, C., Robelia, E., & Hughes, J. (2009). Web 2.0 and classroom research: What path should we take now? *Educational Researcher, 38*(4), 246–259.

Hew, K. F. (2011). Students' and teachers' use of Facebook. *Computers in Human Behavior, 27,* 662–676.

Holmes, K., Preston, G., Shaw, K., & Buchanan, R. (2013). "Follow" me: Networked professional learning for teachers. *Australian Journal of Teacher Education (Online), 38*(12), 55–65. https://doi.org/10.14221/ajte.2013v38n12.4

Hunter, L. J., & Hall, C. M. (2018). A survey of K-12 teachers' utilization of social networks as a professional resource. *Education and Information Technologies, 23,* 633–658. https://doi.org/10.1007/s10639-017-9627-9

Hur, J. W., & Brush, T. A. (2009). Teacher participation in online communities: Why do teachers want to participate in self-generated online communities of K-12 teachers? *Journal of Research on Technology in Education, 41*(3), 279–303. https://doi.org/10.1080/15391523.2009.10782532

Jiang, H., Tang, M., Peng, X., & Liu, X. (2018). Learning design and technology through social networks for high school students in China. *International Journal of Technology and Design Education, 28,* 189–206.

Junco, R., & Cotton, S. R. (2013). No A 4 U: The relationship between multitasking and academic performance. *Computers & Education, 59,* 505–514.

Kelly, N., & Antonio, A. (2016). Teacher peer support in social network sites. *Teaching and Teacher Education, 56,* 138–149. https://doi.org/10.1016/j.tate.2016.02.007

Kerr, S. L., & Schmeichel, M. J. (2018). Teacher Twitter chats: Gender differences in participants' contributions. *Journal of Research on Technology in Education, 50,* 241–252. https://doi.org/10.1080/15391523.2018.1458260

Kirschner, A. P., & Karpinski, A. C. (2010). Facebook and academic performance. *Computers in Human Behavior, 26,* 1237–1245.

Krutka, D., Manca, S., Galvin, S., Greenhow, C., Koehler, M., & Askari, E. (2019). Teaching "against" social media: Confronting problems of profit in the curriculum. *Teachers College Record, 121*(14). Retrieved from https://www.tcrecord.org/Content.asp?contentid=23046

Krutka, D. G., & Carpenter, J. P. (2016). Participatory learning through social media: How and why social studies educators use Twitter. *Contemporary Issues in Technology and Teacher Education, 16*(1), 38–59.

Kuo, F. W., Cheng, W., & Yang, S. C. (2017). A study of friending willingness on SNSs: Secondary school teachers' perspectives. *Computers & Education, 108,* 30–42. https://doi.org/10.1016/j.compedu.2017.01.010

Lantz-Andersson, A. (2018). Language play in a second language: Social media as contexts for emerging Sociopragmatic competence. *Education and Information Technologies, 23*, 705–724. https://doi.org/10.1007/s10639-017-9631-0

Li, J., & Greenhow, C. (2015). Scholars and social media: tweeting in the conference backchannel for professional learning. *Educacional Media International, 52*(1), 1–14. https://doi.org. proxy1.cl.msu.edu/10.1080/09523987.2015.1005426

Macià, M., & Garcia, I. (2017). Properties of teacher networks in Twitter: Are they related to community-based peer production? *The International Review of Research in Open and Distributed Learning, 18*(1), 110–140.

Madden, M. (2012). Privacy management on social media sites. *Pew Internet Report*, 1–20.

Manca, S., & Ranieri, M. (2013). Is it a tool suitable for learning? A critical review of the literature on Facebook as a technology-enhanced learning environment. *Journal of Computer Assisted Learning, 29*, 487–504.

Manca, S., & Ranieri, M. (2016). Is Facebook still a suitable technology-enhanced learning environment? An updated critical review of the literature from 2012 to 2015. *Journal of Computer Assisted Learning, 32*, 503–528. https://doi-org.proxy2.cl.msu.edu/10.1111/jcal.12154

Mazman, S. G., & Usluel, Y. K. (2010). Modeling educational uses of Facebook. *Computers in Education, 55*, 444–453.

Moher, D., Liberati, A., Tetzlaff, J., Altman, D. G., & The PRISMA Group. (2009). Preferred reporting items for systematic reviews and meta-analyses: The PRISMA statement. *Public Library of Science Medicine, 6*(7), e1000097.

Noble, A., McQuillan, P., & Littenberg-Tobias, J. (2016). "A lifelong classroom": Social Studies educators' engagement with professional learning networks on Twitter. *Journal of Technology and Teacher Education, 24*(2), 187–213.

Obar, J. A., & Wildman, S. (2015). Social media definition and the governance challenge: An introduction to the special issue. *Telecommunications Policy, 39*, 745–750.

Prestridge, S. (2019). Categorising teachers' use of social media for their professional learning: A self-generating professional learning paradigm. *Computers & Education, 129*, 143–158.

Rashid, R. (2017). Dialogic reflection for professional development through conversations on a social networking site. *Reflective Practice, 19*(1), 105–117. https://doi.org/10.1080/1262394 3.2017.1379385

Rehm, M., Manca, S., Brandon, C., Greenhow, C. (2019). Beyond disciplinary boundaries: Mapping educational science in the discourse on social media. *Teachers College Record, 121*(14). Available online first at: https://www.tcrecord.org ID Number: 23050, Date Accessed: 9/3/2019

Reich, J., Willet, J., & Murnane, R. J. (2012). The state of wiki usage in U.S. K-12 schools: Leveraging web 2.0 data warehouses to assess quality and equity in online learning environments. *Educational Researcher, 41*(1), 7–15.

Risser, H. S., & Waddell, G. (2018). Beyond the backchannel: Tweeting patterns after two educational conferences. *Educational Media International, 55*(3), 199–212. https://doi.org/10.1080 /09523987.2018.1512449

Robson, J. (2017). Performance, structure and ideal identity: Reconceptualising teachers' engagement in online social spaces: Teachers' engagement in online social spaces. *British Journal of Educational Technology, 49*, 439–450. https://doi.org/10.1111/bjet.12551

Rodesiler, L. (2014). Weaving contexts of participation online: The digital tapestry of secondary English teachers. *Contemporary Issues in Technology and Teacher Education, 14*(2), 72–100.

Rodesiler, L. (2015). The nature of selected English teachers' online participation. *Journal of Adolescent & Adult Literacy, 59*(1), 31–40. https://doi.org/10.1002/jaal.427

Rodesiler, L., & Pace, B. G. (2015). English teachers' online participation as professional development: A narrative study. *English Education, 47*, 347–378.

Rodríguez-Hoyos, C., Salmón, I. H., & Fernández-Díaz, E. (2015). Research on SNS and education: The state of the art and its challenges. *Australasian Journal of Educational Technology, 31*(1), 100–111.

Rosell-Aguilar, F. (2018). Twitter: A professional development and community of practice tool for teachers. *Journal of Interactive Media in Education, 1*(6), 1–12. https://doi.org/10.5334/jime.452

Ross, C., Maninger, R., LaPrairie, K., & Sullivan, S. (2015). The use of Twitter in the creation of educational professional learning opportunities. *Administrative Issues Journal Education Practice and Research, 5*(1), 55–76. https://doi.org/10.5929/2015.5.1.7

Rutherford, C. (2010). Facebook as a source of informal teacher professional development. *In Education, 16*(1), 60–74.

Salomon, G., & Almog, T. (1998). Educational psychology and technology: A matter of reciprocal relations. *Teachers College Record, 100*(2), 222–241.

Salomon, G., & Perkins, D. (2005). Do technologies make us smarter? Intellectual amplification with, of and through technology. In R. J. Sternberg & D. D. Preiss (Eds.), *Intelligence and technology: The impact of tools on the nature and development of human abilities* (pp. 71–86). Mahwah, NJ: Lawrence Erlbaum Associates.

Schwarz, B., & Caduri, G. (2016). Novelties in the use of social networks by leading teachers in their classes. *Computers & Education, 102*, 35–51. https://doi.org/10.1016/j.compedu.2016.07.002

Shin, D. (2018). Social media & English learners' academic literacy development. *21st Century Learning & Multicultural Education, 25*(2), 13–16.

Statista. (2017). *Number of social media users worldwide from 2010–2021 (in billions)*. Retrieved from https://www.statista.com/statistics/278414/number-of-worldwide-social-network-users/

Sumuer, E., Esfer, S., & Yildirim, S. (2014). Teachers' Facebook use: Their use habits, intensity, self-disclosure, privacy settings, and activities on Facebook. *Educational Studies, 40*, 537–553. https://doi.org/10.1080/03055698.2014.952713

Tang, Y., & Hew, K. F. (2017). Using Twitter for education. Beneficial or a waste of time? *Computers & Education, 106*, 97–118.

Thibaut, P. (2015). Social network sites with learning purposes: Exploring new spaces for literacy and learning in the primary classroom. *Australian Journal of Language and Literacy, 38*(2), 83–94.

Trust, T. (2015). Deconstructing an online community of practice: Teachers' actions in the Edmodo math subject community. *Journal of Digital Learning in Teacher Education, 31*(2), 73–81. https://doi.org/10.1080/21532974.2015.1011293

Trust, T. (2016). New model of teacher learning in an online network. *Journal of Research on Technology in Education, 48*(4), 290–305. https://doi.org/10.1080/15391523.2016.1215169

Trust, T. (2017a). Motivation, empowerment, and innovation: Teachers' beliefs about how participating in the Edmodo math subject community shapes teaching and learning. *Journal of Research on Technology in Education, 49*(1–2), 16–30. https://doi.org/10.1080/15391523.2017.1291317

Trust, T. (2017b). Using cultural historical activity theory to examine how teachers seek and share knowledge in a peer-to-peer professional development network. *Australasian Journal of Educational Technology, 33*(1), 98–113.

Trust, T., Krutka, D. G., & Carpenter, J. P. (2016). "Together we are better": Professional learning networks for teachers. *Computers & Education, 102*, 15–34. https://doi.org/10.1016/j.compedu.2016.06.007

Van Vooren, C., & Bess, C. (2013). Teacher tweets improve achievement for eighth grade science students. *Journal of Education, Informatics & Cybernetics, 11*(1), 33–36.

Vazquez-Cano, E. (2012). Mobile learning with Twitter to improve linguistic competence at secondary schools. *The New Educational Review, 29*(3), 134–147.

Veira, A., Leacock, C., & Warrican, S. (2014). Learning outside the walls of the classroom: Engaging the digital natives. *Australasian Journal of Educational Technology, 30*(2), 227–244.

Visser, R. D., Evering, L. C., & Barrett, D. E. (2014). #TwitterforTeachers: The implications of Twitter as a self-directed professional development tool for K-12 teachers. *Journal of Research on Technology in Education, 46*, 396–413.

Vivitsou, M., Tirri, K., & Kynäslahti, H. (2014). Social media in pedagogical context: A study on a Finnish and a Greek teacher's metaphors. *International Journal of Online Pedagogy and Course Design, 4*(2), 1–18.

Wang, S.-K., Hsu, H.-Y., Reeves, T. C., & Coster, D. C. (2014). Professional development to enhance teachers' practices in using information and communication technologies (ICTs) as cognitive tools: Lessons learned from a design-based research study. *Computers & Education, 79*, 101–115. https://doi.org/10.1016/j.compedu.2014.07.006

Wesely, P. M. (2013). Investigating the community of practice of world language educators on Twitter. *Journal of Teacher Education, 64*(4), 305–318.

Wilson, R. E., Gosling, S. D., & Graham, L. T. (2012). A review of Facebook research in the social sciences. *Perspectives on Psychological Science, 7*(3), 203–220.

Xing, W., & Gao, F. (2018). Exploring the relationship between online discourse and commitment in Twitter professional learning communities. *Computers & Education, 126*, 388–389.

Zhang, S., Liu, Q., Chen, W., Wang, Q., & Huang, Z. (2017). Interactive networks and social knowledge construction behavioral patterns in primary school teachers' online collaborative learning activities. *Computers & Education, 104*, 1–17. https://doi.org/10.1016/j.compedu.2016.10.011

Chapter 2
Reviewing Mixed Methods Approaches Using Social Network Analysis for Learning and Education

Dominik Froehlich, Martin Rehm, and Bart Rienties

Abstract Across the globe researchers are using social network analysis (SNA) to better understand the visible and invisible relations between people. While substantial progress has been made in the last 20 years in terms of quantitative modelling and processing techniques of SNA, there is an increased call for SNA researchers to embrace and mix methods developed in qualitative research to understand the what, how, and why questions of social network relations. In this chapter, we will reflect on our experiences with our latest edited book called *Mixed Methods Approaches to Social Network Analysis for Learning and Education*, which contained contributions from 20+ authors. We will first review the empirical literature of mixed methods social network analysis (MMSNA) by conducting a systematic literature review. Secondly, by using two case studies from our own practice, we will critically reflect on how we have used MMSNA approaches. Finally, we will discuss the potential limitations of MMSNA approaches, in particular given the complexities of mastering two ontologically different methods.

Keywords Social network analysis · Mixed method · MMSNA · Systematic review

D. Froehlich
University of Vienna, Vienna, Austria
e-mail: dominik.froehlich@univie.ac.at

M. Rehm
Pädagogische Hochschule Weingarten, Weingarten, Germany
e-mail: rehmm@ph-weingarten.de

B. Rienties (✉)
Open University, Milton Keynes, UK
e-mail: bart.rienties@open.ac.uk

© Springer Nature Switzerland AG 2020
A. Peña-Ayala (ed.), *Educational Networking*, Lecture Notes in Social Networks, https://doi.org/10.1007/978-3-030-29973-6_2

43

Abbreviations

AD	Academic development
AMOT	Amotivated students
CET	Cognitive Evaluation Theory
CSCL	Computer-supported collaborative learning
EMER	External motivation to external regulation
EMID	External motivation to identified regulation
EMIN	External motivation to introjected regulation
IMES	Intrinsic motivation to experience stimulation
IMTA	Intrinsic motivation to accomplish
IMTK	Intrinsic motivation to know
MM	Mixed methods
MMSNA	Mixed methods social network analysis
MRQAP	Multiple regressions quadratic assignment procedure
PBL	Problem-based learning
SDT	Self-Determination Theory
SNA	Social network analysis
VLE	Virtual learning environment

2.1 Introduction

Social network theory postulates that individuals' behaviour can be predicted by the underlying network structure of relations in which they are embedded. Social network analysis (SNA) investigates these structures and helps to determine and understand social interactions between individuals (e.g. workers, managers, students, cohorts) and in contrast to main disciplines in the social science research tradition expands the focus to also include larger entities (e.g. groups, communities) (Scott, 2012; Wassermann & Faust, 1994). Over the last two decades, a wealth of mostly quantitative research (i.e. systematic empirical investigation of observable phenomena via statistical, mathematical, or computational techniques using numbers) in social science and education in particular has shown that social networks and ties of individuals to others can have a significant influence on a myriad of different aspects, including attitudes, behaviours, and cognition of individuals, groups, and even wider society (e.g. Borgatti, Mehra, Brass, & Labianca, 2009; Cela, Sicilia, & Sánchez, 2015; Coburn & Russell, 2008; Daly & Finnigan, 2011; Hommes et al., 2014; Moolenaar, 2012; Rienties & Tempelaar, 2018).

SNA researchers typically distinguish between two types of data, namely, objective and subjective SNA data (Hanneman & Riddle, 2005; Scott, 2012). Objective data refers to the actual trace data of interactions between individuals nodes, such as email conversations (McCallum, Wang, & Corrada-Emmanuel, 2007), discussion threads (De Laat, Lally, Lipponen, & Simons, 2007; Peña-Ayala, Cárdenas-Robledo, & Sossa, 2017), or Wikipedia feeds (Rehm, Littlejohn, & Rienties, 2018). This type of "objective" data sheds light on who central actors might be within a network, or whether subgroups might be present within a larger network structure.

Similarly, a large number of SNA researchers have used so-called subjective (or self-reported) SNA data approaches (Coburn & Russell, 2008; Daly & Finnigan, 2011; Rienties, Johan, & Jindal-Snape, 2015), whereby in a closed (i.e. a clearly delineated group of participants: department, village, classroom) or open network approach participants are asked SNA questions like "who are your friends", "who do you go for advice", or "from whom have you learned in the last 4 weeks". For example, in our own research, we found that how students form friendship and learning relations significantly influenced how they maintained learning relations over time, which significantly impacted on their academic grades and long-term academic performance (Rienties & Tempelaar, 2018). Similarly, in a medical programme Hommes et al. (2012) found that although motivation and academic integration significantly predicted academic performance of first-year medical students, the largest predictor for performance were the respective ties with whom students were learning from.

As argued by Froehlich, Rehm, and Rienties (2019), "quantitative" SNA data might under- or overestimate complex and dynamic underlying network structures (De Laat & Lally, 2004; Jindal-Snape & Rienties, 2016; Rienties et al., 2015; Rienties & Hosein, 2015). There may be several methodological and ontological issues with collecting such subjective and objective SNA data (Moolenaar, 2012; Scott, 2012). More specifically, subjective data might be subject to issues of recall (Neal, 2008), remembering social interactions, and the type of information shared.

Several researchers (Cela et al., 2015; Crossley, 2010; Dado & Bodemer, 2017; Froehlich & Brouwer, Forthcoming; Rienties & Hosein, 2015) have urged SNA researchers to embrace and mix methods developed in qualitative research (i.e. a scientific method of observation to gather non-numerical data) to understand the what, how, and why questions of social network relations. For example, if "objective" SNA data identified that a person was central in an email network, this does not automatically imply that this person is also central in the respective company (e.g. an administrator standardly included in conversation). Indeed, our own work using SNA data from a range of specific historical Wikipedia pages indicated were mostly uploaded by a couple of dedicated people (Rehm et al., 2018), but in hindsight these were primarily people who had support roles, rather than the actual historians who just sent the information to the administrators.

In a case study reported by Froehlich, Mamas, and Schneider (2019), a similar phenomenon was observed when studying advice networks: newcomers appeared to be quite central in these networks. This, however, does not mean that they play an actual central role in the respective networks. Instead, it is an artefact produced by the newcomers' rather unfocused attempts in seeking contact to many people in the early days within a new organizational environment.

Therefore, a number of researchers have started to explore whether integrating SNA approaches with other (quantitative and qualitative) approaches could help to increase our complex understandings of social networks (Cela et al., 2015; Dado & Bodemer, 2017; Froehlich & Bohle Carbonell, 2020; Froehlich, Rehm, et al., 2019), while at the same time providing a potentially more robust, reliable, and interlinked approach. In our forthcoming book, we argued that an increasing number of social network studies make use of mixed methods (MM) to generate their findings (Froehlich, Rehm, et al., 2019). This surge in recent MMSNA research is based upon the realization that quantitative (or formal) and qualitative SNA each have their very own sets of strengths and weaknesses (Crossley, 2010). For example, qualitative SNA (e.g. asking people in an interview with whom they have worked intensively in the last month and what their lived experiences were working with these people) often lacks an overview of the structural properties of a network. However, this often is a central piece of relational thinking. For example, it would be rather difficult to ask all employees of a large enterprise a sociocentric question via interviews and build a coherent network structure.

In contrast, as argued by Froehlich and Brouwer (Forthcoming), quantitative SNA often may be "too abstract" to consider what is actually exchanged between dyads (e.g. working on a joined research project, sharing a new idea of a potential patent). Furthermore, quantitative SNA may fail to account for any fluctuations of the dyads' relationships over time (i.e. "I worked very intensively together with Jennifer 2 months ago, but now she has moved to a different project, but I am still occasionally going for a beer to discuss how our project is going, and picking her brain on my new project"). Hence, mixed methods social network analysis (MMSNA) can be particularly relevant in unveiling (social) complexities within organizations, and educational research in particular. For example, Froehlich and Gegenfurtner (2019) referred to MMSNA as being useful to measure the transfer of training. Additionally, Froehlich and Bohle Carbonell (2020) proposed MMSNA to investigate issues around team and group learning (Rienties & Tempelaar, 2018).

According to Froehlich, Rehm, et al. (2019), MMSNA may be formally defined as "any SNA study that draws from both qualitative and quantitative data, or uses qualitative and quantitative methods of analysis, and thoughtfully integrates the different research strands with each other". Here, we define mixing as the combination of types of data being used. Both qualitative and quantitative data need to be incorporated in a study. Alternatively, we can consider whether the methods being used may be more quantitatively or qualitatively oriented (Hesse-Biber, 2010). Finally, we may look at the "mixing" itself: Are different strands of research integrated in a thoughtful, purposeful manner (Schoonenboom, Johnson, & Froehlich, 2018)? While these considerations are helpful, they only provide information on the nature and the potential of mixing methods.

However, the formal definition of MMSNA by Froehlich, Rehm, et al. (2019) may tell us little about the actual form that MMSNA studies may take. For example, when would a study using social network concepts (e.g. cohesion, number of links) be classified as an SNA study, or MMSNA study, or just a qualitative or qualitative study? Does a MMSNA study require both quantitative (e.g. SNA graphs) and qualitative SNA data (e.g. ethnographical mapping of a network in a company), or can a MMSNA study also contain multiple qualitative data, multiple quantitative data, or a combination of the two? Therefore, in this chapter we will first review 44 studies that have used MMSNA in the field of learning and education. Afterwards, we will provide two studies with practical applications from our own practice of how one could potentially mix SNA with other methods (and vice versa).

2.2 Methods

In this chapter, we will first provide a systematic literature review of the MMSNA literature. While there are a range of studies that have systematically reviewed SNA studies (Borgatti et al., 2009; Borgatti & Halgin, 2011; Carpenter, Li, & Jiang, 2012), and in education (Golonka, Bowles, Frank, Richardson, & Freynik, 2014; McConnell, Hodgson, & Dirckinck-Holmfeld, 2012; Van den Bossche & Segers, 2013; Vera & Schupp, 2006), and CSCL and SNA in particular (Cela et al., 2015; Dado & Bodemer, 2017), there is to the best of our knowledge no systematic literature review of MMSNA. Afterwards, using our experiences when writing the MMSNA book (Froehlich, Rehm, et al., 2019), we will critically reflect on two exemplars of MMSNA. We are not claiming that these exemplars are the best in MMSNA, but are mere reflections of how we as authors in SNA and mixed methods research have tried, and often failed, to combine MMSNA. We hope that by sharing our lessons learned, it will inspire others to think about adopting similar approaches.

2.2.1 Study 1: Systematic Literature Review

As a first part of this chapter, we conducted a systematic literature review of MMSNA studies that was originally published by Froehlich (2019). Texts were retrieved from the Education Resources Information Center using a predefined set of keywords. This set was developed and refined in multiple rounds of test searches and in consultation with MM researchers. Given that this study focuses on methodological approaches rather than any knowledge generated thereby, the procedures of back-tracking and forward-tracking to find additional MMSNA studies seemed unwarranted. Also, further databases were not queried, given that the focus of this chapter is the presentation of a novel approach to analysing texts and methods, and not a systematic literature review. Throughout the process, the search strategy was discussed with experts in the fields of literature reviews and MM.

2.2.1.1 Data Collection

Texts were retrieved from the Education Resources Information Center using a predefined set of keywords. The data set was the repeatedly refined in consultation with MM researchers. As search terms, Froehlich (2019) used two blocks of keywords: (1) First is the application of social network analysis through the search terms (social network analys* OR SNA or social network* or network analys*). (2)

Second is searching for MM studies only by applying the following search terms: (mixed method∗ OR MMMR OR multiple method∗) OR (qualitative OR unstructured interview∗ OR open interview∗ OR semi-structured interview∗ OR focus group∗ OR grounded theory OR grounded theories OR ethnograph∗ OR etnograf∗ OR ethnograf∗ OR phenomelogic∗ OR hermeneutic∗ OR life history∗ OR life stor∗ OR participant observation∗ OR open interview OR thematic analyses OR content analyses OR observational methods OR constant comparative method OR field notes OR field study OR audio recording) AND (quant∗). For time period, Froehlich (2019) opted to search for articles and studies between 1996 and 2018. The two blocks were connected using an "AND" operator.

The search yielded 657 results. The title and abstracts of all found publications were screened before the full texts were read. Of those screened, 159 papers were excluded because they did not fit the field of learning and instruction; 52 papers were excluded because they were not empirical; 346 papers were excluded because they did not use social network analysis; 21 papers were excluded because they did not include MM (i.e. either quantitative or qualitative data or quantitative or qualitative methods of analysis); and 35 articles were excluded because they did not meet any of the other inclusion criteria (e.g. the full text of five texts could not be retrieved). In the end, 44 texts were selected to be included in the study. For further details of the search strategy, and inclusion and exclusion criteria, we refer to Froehlich (2019). Eventually, 44 texts formed the basis for Study 1.

2.2.1.2 Data Analysis

First, the methods used in each of the selected articles were coded. Second, we applied SNA to investigate this network of methods. For the selected articles, the 32 methods used were coded for data collection (e.g. semi-structured interviews or sociometric surveys) and analysis (e.g. linear regression analyses or multiple regression quadratic assignment procedures). The temporal order of the methods used (e.g. sequential or parallel designs) was then used to build a relational data set.

Quantitative SNA (Wassermann & Faust, 1994) was used to analyse the relational data set created. While this method was originally used to depict relationships between human beings (Freeman, 2003), as also highlighted by Rehm et al. (2018), these procedures can also be applied to uncover the relations between concepts or, as in our case, research methods. The 32 identified methods were the nodes used, while the ties represent the sequence of their usage. Note that the network is also weighted, as the ties were aggregated (e.g. if two studies were coded to show a sequence of X to Y, this would result in a weight of "two"). While 287 relationships were coded, only 98 ties were present in the aggregated, weighted graph. The network data was cleaned and prepared using the igraph package (Csardi & Nepusz, 2006) for R (R Core Team, 2014); Gephi (Bastian, Heymann, & Jacomy, 2009) was used to visualize the network and calculate basic key metrics.

2.2.2   Study 2: Objective SNA with Content Analysis

In the second part of this chapter, we explore the use of two exemplars of use of MMSNA approaches. In Study 2, we applied an objective SNA with content analysis in a context of an online summer course in economics in the Netherlands. Study 2 was chosen to illustrate how one quantitative SNA approach could be combined with a more qualitative approach of content analysis. In Study 3 we explored a different MMSNA approach using an open-SNA approach with qualitative follow-up analysis.

2.2.2.1   Context and Setting

In order to provide a critical perspective how MMSNA could be used in practice, in the second part of the chapter, we explored two practical MMSNA studies. In Study 2 we used an exemplar of an online summer course followed by 82 participants at a business school in the Netherlands in 2005/2006 (Rienties, 2010, 2019; Rienties, Tempelaar, Giesbers, Segers, & Gijselaers, 2014; Rienties, Tempelaar, Van den Bossche, Gijselaers, & Segers, 2009) to explore how objective SNA could be combined with content analysis of actual discourse. This summer course is part of a wider summer course programme that has been offered since 2004 to over a thousand students and has been fully integrated in the admission and application processes of the respective business school (See Rienties et al., 2009; Rienties, Giesbers, et al., 2012; Rienties, Tempelaar, Waterval, Rehm, & Gijselaers, 2006; Tempelaar et al., 2011). All students who subscribed to the International Business and Economics Bachelor programme were informed of the opportunity to participate in this online summer course via a letter and email with information about the course and a link to a prior knowledge test. Based on their score on the prior knowledge test, students could decide to voluntarily enrol. Students with a low prior knowledge score who did not enrol received a follow-up e-mail recommending enrolment.

The primary aim of this course was to bridge the gap in economics prior knowledge with the requirements for the degree programme (Giesbers, Rienties, Tempelaar, & Gijselaers, 2013, 2014; Rienties et al., 2006, 2009; Rienties, Giesbers, et al., 2012; Tempelaar et al., 2011). This online course was given over a period of 6 weeks in which learners were assumed to work for 10–15 hours per week. The participants never met face to face before or during the course and had to learn economics using the virtual learning environment "on-the-fly"; that is, learners had to learn how to use the VLE and PBL learning phases while undertaking the programme. In particular, the aim of this research was to determine how students' motivation might have been related to their participation in online discussion fora. Moreover, we also investigated whether collaboration within these fora might have led to a higher cognitive discourse among students.

2.2.2.2 Instruments Used

First, we collected *objective SNA data* from the discussion forums and calculated both Freeman's degree of centrality (Wassermann & Faust, 1994), as well as ego network density. Second, we employed *content analysis* in order to analyse what students were contributing to the fora and possibly contributing to our understanding of underlying processes of learning and knowledge construction (Rehm, Gijselaers, & Segers, 2015; Rienties et al., 2009; Rienties, Giesbers, et al., 2012). The content analysis was based on an instrument developed by Veerman and Veldhuis-Diermanse (2001), which distinguished between non-task-related and task-related discourse activity. Departing from this data, we linked individuals' degree centrality with the cognitive level of their discourse. The content analysis of Veerman and Veldhuis-Diermanse (2001) has been used and validated by other researchers (e.g. Rehm et al., 2015; Rienties et al., 2009; Rienties, Giesbers, et al., 2012; Schellens & Valcke, 2005). When comparing various content analysis schemes, Schellens and Valcke (2005) conclude that the Veerman and Veldhuis-Diermanse's (2001) scheme is particularly suited for analysing knowledge construction among novice students. Veerman and Veldhuis-Diermanse (2001) make a distinction between the so-called non-task-related (1, planning; 2, technical; 3, social; 4, nonsense) and task-related discourse activity (5, facts, 6, experience/opinion; 7, theoretical ideas; 8, explication; 9, evaluation). Three independent coders coded all messages. The network data was cleaned and prepared using UCINET (Borgatti, Everett, & Freeman, 2002); Netdraw (Borgatti, 2002) was used to visualize the network and calculate basic key metrics.

Finally, in order for us to draw conclusions on the impact of motivation on the discussion fora, we used the *Academic Motivation Scale* by Vallerand et al. (1992), which builds on the well-established Self-Determination Theory (SDT). According to Ryan and Deci (2000, p. 56), intrinsic motivation is "… a critical element in cognitive, social, and physical development because it is through acting on one's inherent interests that one grows in knowledge and skills". In a sub-theory of SDT, Cognitive Evaluation Theory (CET), social and environmental factors play an important role in determining what facilitates and what hinders intrinsic motivation. More specific, in SDT, feelings of competence and social relatedness in combination with a sense of autonomy (defined as basic psychological needs) are important facilitators for intrinsic motivation to occur, to maintain, and to enhance.

Externally motivated learning refers to learning that is a means to an end and not engaged for its own sake. In contrast to classical theories of motivation that regard extrinsic motivation as a single construct, SDT proposes that extrinsic motivation is a construct with different facets that vary greatly with the degree to which the learner is autonomous (Deci & Ryan, 1985; Ryan & Deci, 2000). That is, besides intrinsic motivation and amotivation, SDT distinguishes four different forms of extrinsic motivation that constitute a motivational continuum reflecting an increasing degree of self-determined behaviour, namely, external regulation, introjection, identification, and integration (Ryan & Deci, 2000).

2.2.3   Study 3: Open-SNA Approach with Follow-Up In-Class Discussion

2.2.3.1   Context and Setting

The way teachers build relations and network with fellow colleagues and people outside their teaching and learning environment has been found to substantially influence their academic development (AD). For example, research in the context of primary school teachers in Portugal, the Netherlands, and the USA have provided robust and reliable evidence that social networks have a strong impact on trust (Coburn & Russell, 2008), collective efficacy (Moolenaar, Sleegers, & Daly, 2012), sharing of lesson materials (de Lima, 2007), teacher involvement in shared decision-making (Daly, Moolenaar, Bolivar, & Burke, 2010; de Lima, 2007), and schools' innovative climate (Daly et al., 2010; Daly & Finnigan, 2010). Furthermore, there is an emerging body of research that has found that social networks of teachers are also important in secondary and higher education (Rienties & Kinchin, 2014; Roxå & Mårtensson, 2009; Thomas, Tuytens, Devos, Kelchtermans, & Vanderlinde, 2019; Van Waes, Van de Bossche, Moolenaar, De Maeyer, & Van Petegem, 2015).

For example, Roxå and Mårtensson (2009) found that most academics discussed their teaching experience and reflections with a limited number of (mutually trusted) colleagues with whom they reciprocate the sharing of each other's knowledge in a private setting. Using longitudinal modelling, Van Waes et al. (2015) found that most academics maintained primarily relations with colleagues, while only incidental relations were built within the AD. In contrast, Rienties and Kinchin (2014) found that academics maintained on average four links with other academics in an interdisciplinary AD programme. Furthermore, academics maintained on average three contacts outside their AD to discuss their teaching practice. Therefore, Rienties and Kinchin (2014) argue that these interactions may have an impact on AD within and beyond the classroom.

Therefore, in Study 3 we reviewed the use of so-called open-SNA approaches with follow-up qualitative approaches among 114 academics from four faculties from a UK university participated in an 18-month AD programme (Rienties & Hosein, 2015; Rienties & Kinchin, 2014) in 2011–2012. In contrast to traditional, workshop-based AD accredited programmes often taken by early-career academics in the UK, where participants typically follow a "pre-described" programme with a bi-weekly 2-hour session on topic A, B, and C (Parsons, Hill, Holland, & Willis, 2012), this AD programme used a distinct approach starting from the academics' daily practice and reflected on the educational problems academics may face (Rienties & Kinchin, 2014). During the first module, participants worked together on these educational problems in small groups consisting of four or five members, using principles of inquiry-based learning (Rienties & Hosein, 2015; Rienties & Kinchin, 2014). The meeting times and setting for each small group were negotiated with the tutor. As a primary learning objective, participants were expected to develop greater understanding of their role as an academic within the learning environment.

With an estimated workload of 150 hours per module, the majority of hours were self-study, as only five face-to-face meetings of 2–3 hours with an academic developer were arranged per module. During the third and fourth module, participants conducted an individual piece of action research within their own teaching practice.

As indicated by Rienties and Hosein (2015), participants were from 23 different departments, primarily from business (14%), engineering, hospitality and tourism (both 11%), mathematics (7%), and psychology and biosciences (both 6%). The majority of participants were within the first year and half of their contract at the university as it is a contractual obligation to follow the AD programme. This meant the participants were not familiar with most of the other participants except for those perhaps in their own respective discipline.

2.2.3.2 Instruments Used

After the initial 9 months, we conducted both a closed-network analysis in combination with an open-network approach. First, we used a closed-network analysis technique (Daly et al., 2010; Rienties & Kinchin, 2014) after participants had worked together for 9 months, whereby lists with names of the 54 and 60 participants were provided. Participants answered three social network questions, namely, "In the AD programme, I have learned from …," "I have worked a lot with …," and "I am friends with …" in a checkbox manner.

Second, we asked participants using an open-network approach (Daly et al., 2010; Rienties & Kinchin, 2014) the following: "In addition to members of the [AD] programme, we are interested to know with whom you discuss your learning and teaching issues (e.g. how to prepare for a lecture, how to create an assessment, how to provide feedback). This could for example be with a colleague, a friend, family, or partner who is not following the [AD] programme." Participants were asked the name of each network contact, the frequency of contact (as proxy for strength of tie), the type of relation, and where each contact works. A response rate of 88% was established for the open- and closed-SNA questions. Overall, the response rate for both types of data was 88%. The network data was cleaned and prepared using UCINET (Borgatti et al., 2002); Netdraw (Borgatti, 2002) was used to visualize the network and calculate basic key metrics.

2.2.3.3 Qualitative Follow-Up Reflection Exercise

In order to gain more insights the complex nature of the underlying relationships, we triangulated the SNA data (Crossley, 2010; Daly & Finnigan, 2010). One month after the SNA questionnaire was distributed, we presented the results in the form of three social network graphs (i.e. learning and friendship network of AD, external network) during one of the four face-to-face sessions, which were attended by 45 and 32 participants, respectively. This took the form of showing a set of social network graphs, which participants were asked to individually reflect on. As indicated by Rienties and Hosein (2015), participants were asked to reflect individually on the social network graphs for about 10 minutes using predefined questions (e.g. What is the first thing that comes to mind when looking at these networks? Why do you choose these persons to talk to (and not others)? To what extent is it challenging to work with people from different disciplines? In hindsight, would you have chosen the same group members?). Afterwards, the exercise was repeated in pairs and then as a facilitated discussion within the entire group. The verbal responses of 77 participants were recorded and transcribed, and 37 out of 77 (48%) participants who attended the follow-up session were willing to share these reflections (Rienties & Hosein, 2015).

2.3 Results

In order to explore the affordances and limitations of MMSNA approaches, we first explored what other researchers have published in this field using a systematic literature review in Study 1. Afterwards, we used two own exemplars (Rienties, 2019) of how we used MMSNA ourselves to explore what we have learned using MMSNA in practice.

2.3.1 Study 1: Systematic Literature Review

As indicated by Froehlich (2019), we were able to identify 44 studies that met our search criteria, which used in total 32 research methods. In Fig. 2.1 the quantitatively oriented methods are coloured white, while the qualitatively oriented methods are coloured grey. As visible in Fig. 2.1, network surveys are being used as the most prominent type of data collection of the investigated MMSNA studies.

One often used MMSNA research design included the sequential use of network surveys that are being analysed quantitatively, followed by a qualitative exploration using semi-structured interviews and some type of qualitative analysis (e.g. thematic analysis). For example, this approach was followed by Pifer (2011) when researching intersectionality in institutional contexts. Also, Rienties et al. (2015) made use of this approach for studying intercultural learning relations, whereby international students made conscious and deliberate decisions how to network with other international and host-national students. From the interviews with five case study participants, it became clear that cultural sensitivity, motivation to do well, willingness to share with others, and respecting others were crucial elements why some international students became cross-cultural bridge builders, while others did not (Rienties et al., 2015).

Parallel research designs were found in the data, too. For instance, the dissertation of Hiltz-Hymes (2011) used interviews and network surveys (and the associated subsequent analyses) in parallel. Similarly, Sarazin (2019) used a mix of

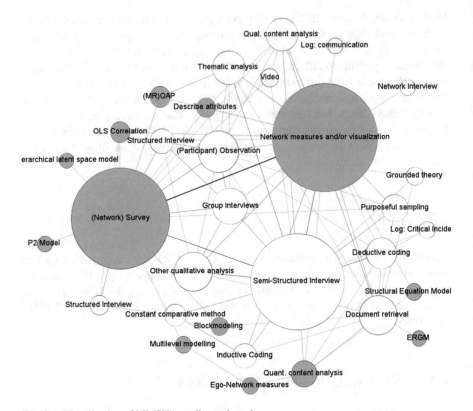

Fig. 2.1 Visualization of MMSNA studies reviewed

interviews, notes, and ethnographic approaches as well as SNA to explore how school children in a disadvantaged background made use of their formal and informal friendship networks. This way, the findings of thematic analyses can be compared and triangulated with basic network measures and the results of an MRQAP analysis. In a longitudinal study of 10 beginning teachers in a range of primary schools, Thomas et al. (2019) explored how these teachers used their social networks within schools over time and how they reflected on their experiences using three separate interviews.

The weighted nodal degrees (*d*) showed a clear separation between methods being used, including basic network metrics (*d* = 72), sociometric surveys (*d* = 67), and semi-structured interviews (*d* = 63), as featured, among others, in Froehlich, Mamas, et al. (2019). Overall, our systematic review of 44 MMSNA studies identified a diverse, widespread, and rich practice of using different MMSNA approaches. The generated map of methods allows for the differentiation of often-used combinations of methods and the "paths" less travelled by MMSNA researchers. These paths visualized in Fig. 2.1 are important starting points for researchers who are considering to potentially use MMSNA to make sense of the world. In other words, MMSNA researchers may use Fig. 2.1 to navigate the complex field of methods within SNA. Given that MM in general, and MMSNA in particular, draw their concepts and methods from a range of disciplines and ways of thinking, this is a daunting task and deserves attention (Frels, Newman, & Newman, 2015). MMSNA researchers need to be able to apply an array of both quantitative and qualitative methods of data collection and data analysis and take care of the sound integration of both. While some archetypes of MM have been developed in social science research about this approach (e.g. parallel designs, sequential designs, etc.), these archetypical solutions may be too abstract for the novice MM researcher.

While some approaches were more commonly combined than others, it would be impossible to argue that the best approach to answer a particular research question is by combining method X with method Y. Nonetheless, we argue that Fig. 2.1 may be a useful tool for MMSNA researchers by highlighting common combinations of approaches. In the remainder of this chapter, we will provide two example studies that have combined different quantitative and qualitative approaches to objective (Study 2) and subjective (Study 3) SNA data.

2.3.2 Study 2: Objective SNA with Content Analysis

As illustrated in Fig. 2.2, we extracted interaction data from participants who used discussion forums to discuss complex problems in economics. The discussion board was an integrated tool developed by Maastricht University, which provided a more scaffolded approach of interaction in comparison to the standard discussion forum tool in Blackboard at the time (Rehm et al., 2015; Rienties et al., 2006, 2009; Rienties, Giesbers, et al., 2012). These discussion messages were afterwards coded by the three independent coders, which assigned the respective Veerman and Veldhuis-Diermanse (2001) codes to these messages. As these quantitative SNA data and qualitative data were collected simultaneously, we argue in line with Fig. 2.1 that this is a parallel design. To illustrate the power of mixing objective SNA data with this content analysis data in understanding the interaction of contributions of individuals, the social network of all discourse activity (Fig. 2.2) as well as only higher cognitive discourse (Fig. 2.3) of one team are illustrated.

First, using the Netdraw visualization tool in UCINET (Borgatti, 2002), the social networks visualized with whom individuals were communicating and whether the connection was bilateral. The social networks illustrate to whom a learner (i.e. a red node) was communicating with and what the direction of communication was. For example, Peter and Caroline had a so-called reciprocal link on the right side of Fig. 2.3, as they reacted both to each other's contribution, and the arrow goes in both directions (Hanneman & Riddle, 2005; Scott, 2012). All learners, except Michael, were connected in discourse in this online summer course. However, Peter and Caroline did not have any direct link when looking at higher cognitive discourse in. Moreover, as shown in Fig. 2.4, we were able to show that their interaction was primarily on a lower cognitive level.

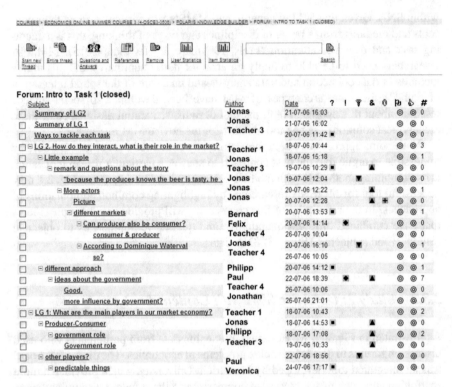

Fig. 2.2 Discussion forum interactions in online summer course

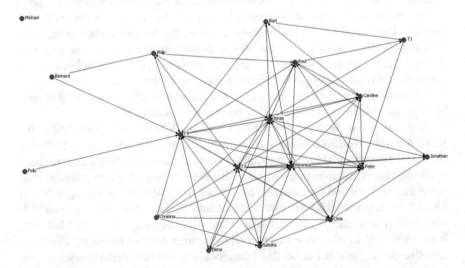

Fig. 2.3 Social network of all discourse activity. (Source: Rienties et al. (2009))

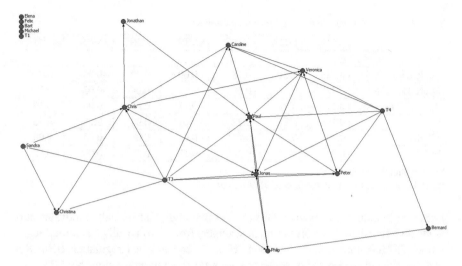

Fig. 2.4 Social network of all discourse activity. (Source: Rienties et al. (2009))

Second, some learners, like Veronica and Jonas, were central in Fig. 2.3, while others, like Jonas and Paul but not Veronica, were relatively central in the higher cognitive network of Fig. 2.4. In other words, not every learner (e.g. Veronica) who was central in the overall network (Fig. 2.3) was also central in the higher cognitive network (Fig. 2.4). Other learners (e.g. Paul) who were not necessarily central in the overall network became a central contributor to higher cognitive discourse. Indeed, as found by Rienties et al. (2009) on average in the course, the learners contributed 25.64 (SD = 28.07) messages and there are substantial differences among individuals with respect to the amount of discourse as assessed by a chi-square test (χ^2 ($df = 81$; $N = 82$) 2258.17, $p < 0.001$). In addition, if we distinguish between task- and non-task-related discourse, again significant differences are found.

Furthermore, several learners, including Bart, Elena, Felix, and Michael, were not present in conversations on a higher cognitive discourse. This does not mean that they were not learning from these higher cognitive discussion forum messages as they could read and "digest" these interactions, but at least they were not actively contributing to these discourses. Hence, by integrating content analysis with SNA, we were able to distinguish multilayered interaction patterns among learners based upon the type of discourse.

By using an integrated social network analysis, a detailed picture of the role of motivation on learning interaction processes can be established when comparing Fig. 2.3 with Fig. 2.4. All three aspects of intrinsic motivation (intrinsic motivation to know (IMTK); intrinsic motivation to accomplish (IMTA); intrinsic motivation to experience stimulation (IMES)) were positively correlated with the three centrality measures from our social network analysis (size, task-related size, and higher cognitive size): see Table 2.1.

What is particularly interesting is that those correlations for task-related and higher-cognitive-related discourse become larger relative to overall discourse for

Table 2.1 Centrality, ego density, and academic motivation

	IMTK	IMTA	IMES	EMID	EMIN	EMER	AMOT
Social network analysis							
Reply degree	0.23[a]	0.21	0.18	0.00	0.00	−0.03	0.05
Reply TR degree	0.27[a]	0.21	0.18	0.01	−0.04	−0.07	−0.12
Reply HC degree	0.27[a]	0.24[a]	0.20	0.11	0.06	0.05	−0.16
Size	0.24[a]	0.21	0.22	−0.01	0.04	0.05	−0.02
TR size	0.29[a]	0.26[a]	0.23[a]	−0.02	0.05	−0.03	−0.08
HC size	0.29[a]	0.29[a]	0.24[a]	0.13	0.05	0.04	−0.12

Source: Rienties et al. (2009)

[a]Correlation is significant at the 0.05 level (2-tailed)

intrinsically motivated students. This implies that highly intrinsically motivated students distinguished themselves (much stronger) from extrinsically (identified regulation (EMID), introjected regulation (EMIN), and external regulation (EMER)) and amotivated students (i.e. students without a specific motivation, AMOT), also with respect to their position in the network. Especially students with high levels of intrinsic motivation to know are central contributors to overall discourse (Reply Degree) (IMTK $r = 0.24$, $p < 0.05$). Those students are also more central in task-related discourse (IMTK $r = 0.27$, $p < 0.05$) and in contributions of higher cognitive discourse (IMTK $r = 0.27$; IMTA $r = 0.24$, $p < 0.05$).

This implies that highly intrinsically motivated learners both show up in the centre of the network but also on the outer fringe, but then as learners who have above average connections to other learners. Students who are highly extrinsically motivated do not distinguish from the average student in our online setting. The number of links of highly extrinsically motivated learners is on average. Amotivation demonstrates a negative but non-significant relationship with higher cognitive centrality ($r = −0.16$, n.s.) and higher cognitive size ($r = −0.12$, n.s.).

In other words, the results from the "objective" parallel MMSNA Study 2 show that individuals' motivation can affect their participation in online discussion fora (Rienties et al., 2009). Some learners became active contributors to discourse, while other learners contributed only a limited amount to discourse. Although these results have already been found in other studies, this study was the first to empirically investigate how motivation can explain differences in quantity and quality of individuals' online contributions. Hence, by employing MMSNA, we were able to unveil complex, and often hidden, social interactions between learners, that perhaps may not have been visible when just looking quantitatively at the number of posted messages, or qualitatively just on what participants were posting.

2.3.2.1 Lessons Learned

Study 2 is one of the most cited studies in CSCL using SNA (Cela et al., 2015; Dado & Bodemer, 2017) and has encouraged researchers to look beyond pure SNA data to aim to understand what participants are actually talking about, and why (Bogler,

Caspi, & Roccas, 2013; De Laat & Schreurs, 2013; Giesbers et al., 2013, 2014; Kirschner & Erkens, 2013). Yet, the content analysis constituted a substantial amount of work. Consequently, one can wonder whether a simple closed-network survey at the end of the module with a question like "from whom have you learned the most about economics during this online summer course" might have been easier. Moreover, follow-up interviews, incorporating the SNA results, might have provided an additional, interesting perspective.

Another reflection is that increasingly students will be using other social media technologies to interact with each other. When we conducted these studies in 2005/2006, Facebook and WhatsApp did not exist, and many students did not have smartphones (Clough, Jones, McAndrew, & Scanlon, 2008; Dillenbourg et al., 2011). Therefore, in order to share experiences and learn from each other, they had to connect to the university's virtual learning environment (VLE). With the rise of Web 2.0 tools like Facebook (Ellison, Steinfield, & Lampe, 2007; Madge, Meek, Wellens, & Hooley, 2009; Rienties, Tempelaar, Pinckaers, Giesbers, & Lichel, 2012; Sharples et al., 2013; Sharples et al., 2016), Twitter (Colleoni, Rozza, & Arvidsson, 2014; Rehm & Cornelissen, 2019; Rehm & Notten, 2016), WhatsApp (Madge et al., 2019), Wikipedia (De Wever, Van Keer, Schellens, & Valcke, 2011; Klein & Konieczny, 2015; Rehm et al., 2018; Sharples et al., 2015), and YouTube (Duffy, 2008; Holmes, Clark, Burt, & Rienties, 2013; T. Jones & Cuthrell, 2011), obviously students will increasingly learn outside the formal boundaries of the institutional VLE (Johnson et al., 2016; Law & Jelfs, 2016; Okada & Moreira, 2008; Roberts, Daly, Held, & Lyle, 2017; Watson, Watson, & Reigeluth, 2015). Several authors refer to this notion as ubiquitous learning (Cárdenas-Robledo & Peña-Ayala, 2018; Lally, Sharples, Tracy, Bertram, & Masters, 2012; Zhang, 2015), whereby students and staff have access to technology and information 24/7. For example, a wealth of studies on use of Facebook (Kirschner, 2015; Kirschner & Karpinski, 2010; Manca & Ranieri, 2013) have found that many students spent more time on Facebook than they do in the institution's VLE.

Given the omnipresence of knowledge and information on the web (Bråten, Strømsø, & Salmerón, 2011; Ferguson et al., 2017; Knight et al., 2017; Sharples et al., 2014) and the ubiquitous presence of technology for learning (Cárdenas-Robledo & Peña-Ayala, 2018; Lally et al., 2012; Zhang, 2015), one wonders whether conducting the same study again with similar results would be found. One way to measure these informal networks is to extract and link Web 2.0 data with the VLE data using learning analytics approaches (Peña-Ayala, 2018; Peña-Ayala et al., 2017; Tempelaar, Rienties, & Giesbers, 2015; Tempelaar, Rienties, Mittelmeier, & Nguyen, 2018). However, the linking of various data sources might be technically complex, given that learners might use different user profiles and have lack of interoperability between these systems (Conde, García, Rodríguez-Conde, Alier, & García-Holgado, 2014; Ferguson et al., 2016), and perhaps more importantly the interconnections between these systems might lead to strong ethical issues and undesired consequences (Boyd & Crawford, 2012; Buchanan, 2011; Hoel, Griffiths, & Chen, 2017; Korir, Mittelmeier, & Rienties, 2019; Lally et al., 2012; Prinsloo & Slade, 2017). For example, the Open University UK has specifically excluded Web

2.0 tools like Facebook and Twitter as safe spaces for students, whereby learning analytics models will not take into consideration what students are saying or doing in these spaces (Open University UK, 2014). Other institutions do take into consideration how students are sharing their concerns and issues in Web 2.0 tools like Facebook in order to support their well-being (Ferguson et al., 2016; Hartrey, Denieffe, & Wells, 2017; E. Jones, Samra, & Lucassen, 2018; Piper & Emmanuel, 2019).

One way to address how knowledge and information is shared in these informal and social media spaces could be to combine objective SNA data extracted from the VLE with self-reported SNA data from students. To a certain degree, it does not really matter if students are sharing nodes, information, or ideas (Hommes et al., 2012, 2014) using the VLE, or using informal social media tools, as long as they are able to work effectively together. These informal data points could be collated by researchers using self-reported closed- or open-SNA approaches. Researchers will need to balance when, what, and how often participants are asked to complete these self-reported instruments and whether it may be more appropriate to use qualitative follow-up approaches to explore the lived experiences of these students.

2.3.3 Study 3: Open-SNA Approach with Follow-Up In-Class Discussion

In our Study 3 of 114 academics working in an academic development (AD) programme, we employed a sequential design whereby we first collected quantitative self-reported SNA data with follow-up in-class discussions of the results by participants 1 month after the quantitative data was collected. The results indicated that most academics developed cohesive links, either within their own assigned group or within the wider AD programme after 9 months. As illustrated in Table 2.2, on average, participants developed 3.09 friendship relations within the AD programme; 1.07 friendships were based upon the initial group division during the first

Table 2.2 Social ties within and outside groups/academic development

SNA metric	M	SD	Range
Friendship within group	1.07	1.23	(0–4)
Friendship within AD	3.09	1.97	(0–10)
Work relations within group	2.37	1.29	(0–4)
Work relations within AD	4.84	2.22	(1–11)
Learning within group	2.30	1.36	(0–4)
Learning within AD	4.84	2.43	(0–15)
Learning outside AD	3.17	2.31	(0–10)
Same discipline	2.39	2.06	(0–8)
Externals outside the institution	0.94	1.11	(0–4)

Source: Rienties and Hosein (2015)

module, when participants were assigned to work in smaller groups of 4–5 partici-pants; and 2.02 of their self-reported friendship relations were based outside their first group and can be characterized as informal ties. Participants had on average 4.84 learning and teaching relations within the AD programme, of which 2.30 were based upon the initial group division.

Second, as illustrated in Fig. 2.5, while some academics were centrally posi-tioned, others were less connected and positioned towards the fringes of the net-work. The colour and shape of the node represents the respective faculty of each participant, while the number of the node referred to the respective group an aca-demic was enrolled in the AD programme. Note that the visualization software tool Netdraw (Borgatti, 2002) positions the academics at random across the X- and Y-axis based upon the (perceived) social interactions between academics, whereby academics who share similar connections are positioned more closely together

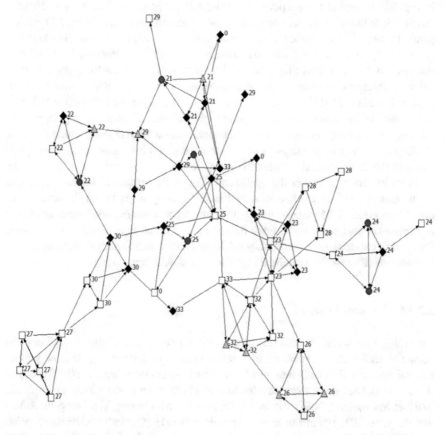

Fig. 2.5 Learning and teaching network after 9 months. Note: Blue circle, Arts and Social Science; white square, STEM; black diamond, business studies; grey triangle, health and social care. Note that the numbers refer to the respective group division during the first module. (Source: Rienties and Hosein (2015))

(Rienties & Kinchin, 2014). Being on the left of the graph is not necessarily better or worse than being on the right, top, or bottom, but academics with similar connections are positioned closer together.

As illustrated by links between nodes with different colours and shapes in Fig. 2.5, most academics developed a range of interdisciplinary learning relations with other academics. For example, on the left of Fig. 2.5, four members (each from a different faculty) of group 22 were connected to each other, while on the left of Fig. 2.5, five members of group 35 (three from arts and social science faculty, two from other faculties) were connected to each other. MMSNA researchers could use these SNA visualizations to identify how cohesive the learning climates within groups are. For example, it would perhaps be interesting to know why members of group 29 were often not connected to their group.

As illustrated in Fig. 2.6, the learning and teaching network of the two cohorts and their external network contacts were rather diverse. The grey and black colour in Fig. 2.6 indicated the participants in the AD programmes (Rienties & Hosein, 2015), while the white nodes were the so-called external relations of the AD participants. In total, 293 network contacts outside the AD were used to discuss learning and teaching issues with the 114 AD participants (Rienties & Hosein, 2015). 92 participants (81%) indicated they discussed their learning and teaching practice with external relations. Although the visualization is complex, these open-network approaches allow MMSNA researchers to map and analyse the informal social relations outside the academic development programme. Figure 2.6 highlights the intensity of the usage of the informal network outside the AD programme for most participants, whereby participants of the second cohort also learned from the experiences of the first cohort (Rienties & Hosein, 2015).

From the open coding of the qualitative data of the completed forms, transcripts of the discussions, and reflections by the facilitators, three broad thematic areas could be identified. More specifically, the data suggest professional, emotional, and academic support as being relevant for the case at hand. Participants were in need for sparing partners, with whom they could talk about their feelings, challenges, and frustrations (Rienties & Hosein, 2015; Rienties & Kinchin, 2014).

2.3.3.1 Lessons Learned

Extending beyond the boundaries of the closed networks and also including open networks definitely contributed to a more refined understanding of how people learned within and outside their academic development programme. Whether or not these external relations were developed using formal networks (Moolenaar, 2012; Roxå & Mårtensson, 2009; Van Waes, De Maeyer, Moolenaar, Van Petegem, & Van den Bossche, 2018) or informal social media networks (Cárdenas-Robledo & Peña-Ayala, 2018; Rehm et al., 2018; Rehm & Cornelissen, 2019; Rehm & Notten, 2016; Rienties & Kinchin, 2014), many academics indicated that these networks were very useful "to keep them sane". Furthermore, organizing a reflection meeting 1 month after the SNA data certainly added another relevant dimension to the

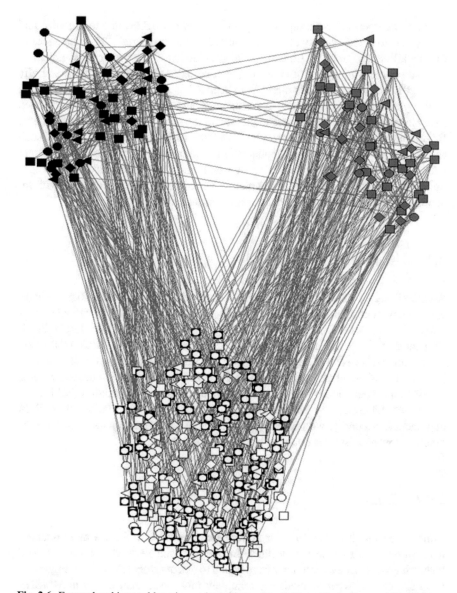

Fig. 2.6 External and internal learning and teaching network. Note: Circle, Arts and Humanities; square, STEM; diamond, business; triangle, health and social care. Circle in box = external outside university. Black = first AD cohort. Grey = second AD cohort. (Source: Rienties and Hosein (2015))

question of why and how participants networked in and outside the formal programme. Nonetheless, while the open-SNA approach, as well as the data triangulation, can be considered as a methodological improvement, we would like to highlight two points of caution.

First, the open-network approach quickly became troublesome to organize and codify, due to a wide variety of possible answers (e.g. "my wife", "my colleagues in my department", "John"). Linked to the previous study, whether it is ethical to involve people who were not consulted, can be debated (Korir et al., 2019; Lally et al., 2012). Second, the data might have been subject to a biased sample. What we mean by this is that participants were academics and researchers themselves, which likely influenced their willingness to partake in the study and learn from the results. It remains to be seen if this approach also works with other target audiences, such as younger participants or participants who cannot be easily followed over a longer period of time. However, the research by Rehm and Cornelissen (2019) does highlight that when done carefully, triangulating large open (Twitter) networks with in-depth interviews could work under certain circumstances.

2.4 Discussion

An increasing number of researchers are using and integrating other social science approaches in conjunction with social network analysis (SNA), as indicated by a range of reviews of SNA studies (Borgatti et al., 2009; Borgatti & Halgin, 2011; Carpenter et al., 2012), and in education (Golonka et al., 2014; McConnell et al., 2012; Van den Bossche & Segers, 2013; Vera & Schupp, 2006), and CSCL and SNA in particular (Cela et al., 2015; Dado & Bodemer, 2017). In this chapter, we first provided a systematic review of mixed method social network studies (MMSNA), where we identified 44 studies that mixed qualitative approaches with SNA approaches. Secondly, we provided two MMSNA study exemplars from our own practice how such studies could be applied and evaluated.

2.4.1   Study 1

Building on Froehlich (2019), our systematic review allowed for a more nuanced description of MM research designs. Applying social network thinking to MM research gives us a new dictionary from which to discuss research designs. Figure 2.1 is an estimate of what methods are relatively more used than others in MMSNA designs. These methods of data collection may be interpreted as methods that integrate previous analyses, where one strand of research merges into another. In other words, the generated map of methods allows for the differentiation of often-used combinations of methods and the "paths" less travelled. This is important to inspire new ways of making sense of the world.

Given the complexity of adopting and mixing the various quantitative and qualitative approaches in MMSNA in various sequential cycles (e.g. first quantitative self-report SNA, follow-up semi-structured interviews, alternatively start with semi-structured interviews and afterwards longitudinal self-reported SNA), theoretically there could be a near-infinite number of potential combinations of implementing MMSNA. As a result, providing a clear 1–2–3 step guideline of implementing MMSNA may be undesirable or perhaps unrealistic. Nonetheless, one approach that could bear fruit is to learn from researchers who have already implemented a particular MMSNA approach, and learn, build on their lessons learned. In other words, we encourage researchers and practitioners to use Fig. 2.1 to find whether there are potentially compatible designs, read up on these respective studies as identified by Froehlich (2019), and explore how these researchers have implemented their respective designs.

2.4.2   Study 2

Whatever MMSNA method one adopts, we have to be mindful that the interpretation of these networks and related data are just our unique interpretations of reality, which are strongly influenced by our own lenses and perspectives (Hesse-Biber, 2010; Kadushin, 2005; Korir et al., 2019; Prinsloo & Slade, 2017; Sarazin, 2019; Schoonenboom et al., 2018). As highlighted from Study 2, we were able to empirically determine that intrinsically motivated students were more likely to contribute to higher cognitive discourse than extrinsically motivated students (Rienties et al., 2009, Rienties, Giesbers et al., 2012). Nonetheless, there could be a myriad of reasons why these results could be different in another contexts, or even when the study would be replicated in 2020. For example, in many discussion forums, participation is rather unequal and dominated by certain participants (Rehm et al., 2015, 2018). Indeed other researchers have found that gender (Bevelander & Page, 2011), ethnicity (Jindal-Snape & Rienties, 2016; Rienties et al., 2015), seniority (Rehm et al., 2015), and technology access (Gemmell & Harrison, 2017) could be mediating factors in engagement, explaining a divergence of practice.

Furthermore, given the ubiquitous presence of technology (Cárdenas-Robledo & Peña-Ayala, 2018; Sharples et al., 2015) and the opportunities of learners to use both formal and informal learning tools (Clough et al., 2008; Duffy, 2008; Rehm & Notten, 2016), one wonders how MMSNA researchers need to effectively balance and integrate these different potentially rich data sources. As highlighted from our Study 2, there could be several ethical issues when combining different data sets. Furthermore, the lens of Study 2 was focused on Self-Determination Theory (Deci & Ryan, 1985; Ryan & Deci, 2000; Vallerand et al., 1992) and motivation, while perhaps intersectionality with other factors like gender or ethnicity could be an important explanatory factor.

2.4.3   Study 3

In Study 3 we identified that many participants actively developed learning links within their academic development (AD) programme, while at the same time maintaining many links with "external" people outside the AD programme (Rienties & Hosein, 2015; Rienties & Kinchin, 2014). For many participants these relations were maintained to keep them "sane" and to rapport and reflect (up)on their professional, emotional, and academic support. As the lens of the research team was embedded into the AD programme, any other lenses in terms of intersectionality, like gender (Bevelander & Page, 2011), seniority (Rehm et al., 2015), or disciplinary differences (Rienties & Tempelaar, 2018), were not explored explicitly. As a result, researchers with different research questions and/or approaches could potentially find different patterns and results when conducting MMSNA.

Starting-out researchers may use Fig. 2.1. to navigate the complex field of methods within SNA. Given that MM in general, and MMSNA in particular, draw their concepts and methods from a range of disciplines and ways of thinking, this is a daunting task and deserves attention (Frels et al., 2015). MM researchers need to be able to apply an array of both quantitative and qualitative methods of data collection and data analysis and take care of the sound integration of both. While some archetypes of MM have been developed in general research about this approach (e.g. parallel designs, sequential designs, etc.), these archetypical solutions may be too abstract for the novice MM researcher. Furthermore, not all of these archetypes are applicable to our domain of interest, which is education and learning research.

The generated map of MMSNA is also a tool that contributes to the objective of making MMSNA more accessible for researchers less experienced in MM. This is achieved by moving away from typologies derived from theory that "have become almost too refined" (Bryman, 2006) and hence may be overly complex for novice researchers. Instead, as also evidenced in Study 2 and Study 3, we took an approach that is close to research practice (following Bryman, 2006) and that offers an innovative and intuitive way to understand the mixing of methods in the field of education and learning research.

2.5   Conclusions

First, the literature review indicates that the reporting practices of MMSNA studies may be improved. This is understandable, since the very notion of MMSNA is very new and it so far even lacked a coherent definition. Second, the sequencing of methods was often described in very imprecise terms, too. For example, Pifer (2011) writes that she "completed a social network analysis, followed by in-depth one-on-one interviews"—however, this leaves out the step of quantitative analysis happening between these two data collection procedures that would tie the different strands of research together.

Third, the primarily lesson we learned from writing the MMSNA book (Froehlich, Rehm, et al., 2019) is to continuously engage the stakeholders in the design, implementation, and evaluation of the research (De Laat & Lally, 2004; Hommes et al., 2014). By discussing initial trends and visualizations of SNA with the rich narratives from the qualitative approaches, some interesting and unexpected findings emerged over time. At the same time, while great care was taken to code data with multiple coders, each with their own unique lenses, a major limitation of all three studies was that the interpretations of these findings were primarily reliant on the respective research teams. Future research should aim to not only triangulate quantitative SNA with other quantitative and qualitative data but also to go back to the main stakeholders to determine whether (or not) the "emergent" findings resonate with the actual lived experiences of students, teachers, and senior management.

Fourth, as highlighted by the three studies, there could be strong ethical concerns when gathering diverse and rich MMSNA data from various data sources. For example, participants who agree to participate in a piece of research in most social science approaches are able to do so in an anonymous and confidential format (Borgatti & Molina, 2003; Conway, 2014; Kadushin, 2005). Yet this is not the case in SNA, and as argued by Korir et al. (2019) in particular in MMSNA contexts, there might even be stronger ethical concerns when researchers are able to link social media profiles of learners with learning processes and outcomes (Korir et al., 2019; Rehm & Cornelissen, 2019; Schoonenboom et al., 2018). As described in a review of ethics in SNA studies by Borgatti and Molina (2003, p. 338), "the most obvious difference is that in a network study, anonymity at the data collection stage is not possible … [as] the researcher must know who the respondent was to record a link from that respondent to the persons with whom the respondent indicates having relationships". Yet, confidentiality of respondents' identities are considered to be an essential part of ethics and privacy in social science research (Boyd & Crawford, 2012; Buchanan, 2011; Hoel et al., 2017; Korir et al., 2019; Lally et al., 2012; Prinsloo & Slade, 2017) and (MM)SNA research in particular (Kadushin, 2005; Korir et al., 2019). Therefore, as argued by Korir et al. (2019), strong procedures and policies are needed in SNA studies to ensure that, during the data processing phase, the researcher acts as an independent gatekeeper between respondents and potential end users of social network data (i.e. peers within the social network, managers making promotion and firing decisions, teachers addressing potential dropout issues, other researchers).

For example, in Study 3 we informed participants at the start of the study of the main purpose of the study (i.e. informed consent) and the options not to participate (i.e. right to be excluded). Practically, participants could hand in their responses anonymously if they desired by putting their responses in an envelope. Furthermore, if participants did not want to share their responses at all, this was also feasible. Finally, any concerns by participants could be shared with the main researcher or by contacting an independent third party by the information sheet provided separately in the informed consent and subsequent "report" shared with participants. In other words, in line with common ethics practices, we encourage MMSNA researchers to be transparent and open about the purpose of the MMSNA study to all recruits and participants. This is particularly important given the opportunities to connect and triangulate data of people who have opted not to join the research (Borgatti & Molina, 2003; Crossley, 2010; Korir et al., 2019).

A particular concern for MMSNA research is the inclusion of students who have not indicated consent to participate in a piece of research. In SNA it is often still possible to generate a social network profile of a non-respondent based upon the perceived relations provided by others in the network (Conway, 2014; Kadushin, 2005), and in MMSNA in particular, it can be relatively straightforward to identify a particular person when triangulating different forms of data, either from interviews (Froehlich & Brouwer, Forthcoming; Rienties, 2019; Sarazin, 2019; Schoonenboom et al., 2018; Thomas et al., 2019) or from social media data. Of course this leads to substantial and difficult ethical dilemmas for researchers to address, irrespective of whether particular student made a conscious (e.g. refusal to participate, fear of being marginalized) or unconscious decision (e.g. incomplete submission, forgot to indicate the respondent's name, forgot to participate in online survey) to (not) participate. Therefore, Korir et al. (2019) encourage MMSNA researchers and practitioners should take into consideration the following three questions:

1. Are you going to report about the non-respondents?
2. Are you going to use non-respondents' data in any way?
3. In this case while the participants do own their own perceptions, is it ethical to make any statements/inferences about non-respondents?

Overall, we hope that our chapter has contributed to an overview of the challenges and affordances of using MMSNA. When combining and integrating different forms of data and artefacts, there are several complex issues that researchers, practitioners, and managers need to take into consideration. As highlighted by Winston Churchill, success consists of going from failure to failure without loss of enthusiasm.

References

Bastian, M., Heymann, S., & Jacomy, M. (2009). *Gephi: An open source software for exploring and manipulating networks.* Paper presented at the Third international AAAI conference on weblogs and social media.

Bevelander, D., & Page, M. J. (2011). Ms. Trust: Gender, networks and trust—Implications for management and education. *Academy of Management Learning & Education, 10*(4), 623–642. https://doi.org/10.5465/amle.2009.0138

Bogler, R., Caspi, A., & Roccas, S. (2013). Transformational and passive leadership: An initial investigation of university instructors as leaders in a virtual learning environment. *Educational Management Administration & Leadership, 41*(3), 372–392. https://doi.org/10.1177/1741143212474805

Borgatti, S. P. (2002). NetDraw: Graph visualization software.

Borgatti, S. P., Everett, M. G., & Freeman, L. C. (2002). *Ucinet for windows: Software for social network analysis.* Harvard, MA: Analytic Technologies.

Borgatti, S. P., & Halgin, D. (2011). On network theory. *Organization Science, 22*(5), 1168–1181. https://doi.org/10.1287/orsc.1100.064

Borgatti, S. P., Mehra, A., Brass, D. J., & Labianca, G. (2009). Network analysis in the social sciences. *Science, 323*(5916), 892–895. https://doi.org/10.1126/science.1165821

Borgatti, S. P., & Molina, J. L. (2003). Ethical and strategic issues in organizational social network analysis. *The Journal of Applied Behavioral Science, 39*(3), 337–349. https://doi.org/10.1177/0021886303258111

Boyd, D., & Crawford, K. (2012). Critical questions for big data: Provocations for a cultural, technological, and scholarly phenomenon. *Information, Communication & Society, 15*(5), 662–679. https://doi.org/10.1080/1369118X.2012.678878green

Bråten, I., Strømsø, H. I., & Salmerón, L. (2011). Trust and mistrust when students read multiple information sources about climate change. *Learning and Instruction, 21*(2), 180–192. https://doi.org/10.1016/j.learninstruc.2010.02.002

Bryman, A. (2006). Integrating quantitative and qualitative research: how is it done? *Qualitative Research, 6*(1), 97–13. https://doi.org/10.1177/1468794106058877

Buchanan, E. A. (2011). Internet research ethics: Past, present, and future. In M. Consalvo, C. Ess, & R. Burnett (Eds.), *The handbook of internet studies* (pp. 83–108). Somerset: Wiley.

Cárdenas-Robledo, L. A., & Peña-Ayala, A. (2018). Ubiquitous learning: A systematic review. *Telematics and Informatics, 35*(5), 1097–1132. https://doi.org/10.1016/j.tele.2018.01.009

Carpenter, M. A., Li, M., & Jiang, H. (2012). Social network research in organizational contexts: A systematic review of methodological issues and choices. *Journal of Management, 38*(4), 1328–1361. https://doi.org/10.1177/0149206312440119

Cela, K. L., Sicilia, M. Á., & Sánchez, S. (2015). Social network analysis in E-learning environments: A preliminary systematic review. *Educational Psychology Review, 21*(1), 219–246. https://doi.org/10.1007/s10648-014-9276-0

Clough, G., Jones, A. C., McAndrew, P., & Scanlon, E. (2008). Informal learning with PDAs and smartphones. *Journal of Computer Assisted Learning, 24*(5), 359–371. https://doi.org/10.1111/j.1365-2729.2007.00268.x

Coburn, C. E., & Russell, J. L. (2008). District policy and teachers' social networks. *Educational Evaluation and Policy Analysis, 30*(3), 203–235. https://doi.org/10.3102/0162373708321829

Colleoni, E., Rozza, A., & Arvidsson, A. (2014). Echo chamber or public sphere? Predicting political orientation and measuring political homophily in Twitter using big data. *Journal of Communication, 64*(2), 317–332. https://doi.org/10.1111/jcom.12084

Conde, M. A., García, F., Rodríguez-Conde, M. J., Alier, M., & García-Holgado, A. (2014). Perceived openness of Learning Management Systems by students and teachers in education and technology courses. *Computers in Human Behavior, 31*, 517–526. https://doi.org/10.1016/j.chb.2013.05.023

Conway, S. (2014). A cautionary note on data inputs and visual outputs in social network analysis. *British Journal of Management, 25*(1), 102–117. https://doi.org/10.1111/j.1467-8551.2012.00835.x

Crossley, N. (2010). The social world of the network. Combining qualitative and quantitative elements in social network analysis. *Sociologica, 4*(1), 1–34. https://doi.org/10.2383/32049

Csardi, G., & Nepusz, T. (2006). The igraph software package for complex network research. *InterJournal, Complex Systems, 1695*(5), 1–9.

de Lima, J. Á. (2007). Teachers' professional development in departmentalised, loosely coupled organisations: Lessons for school improvement from a case study of two curriculum departments. *School Effectiveness and School Improvement, 18*(3), 273–301. https://doi.org/10.1080/09243450701434156

Dado, M., & Bodemer, D. (2017). A review of methodological applications of social network analysis in computer-supported collaborative learning. *Educational Research Review, 22*(Supplement C), 159–180. https://doi.org/10.1016/j.edurev.2017.08.005

Daly, A. J., & Finnigan, K. S. (2010). A bridge between worlds: Understanding network structure to understand change strategy. *Journal of Educational Change, 11*(2), 111–138. https://doi.org/10.1007/s10833-009-9102-5

Daly, A. J., & Finnigan, K. S. (2011). The ebb and flow of social network ties between district leaders under high-stakes accountability. *American Educational Research Journal, 48*(1), 39–79. https://doi.org/10.3102/0002831210368990

Daly, A. J., Moolenaar, N. M., Bolivar, J. M., & Burke, P. (2010). Relationships in reform: The role of teachers' social networks. *Journal of Educational Administration, 48*(3), 359–391. https://doi.org/10.1108/09578231011041062

De Laat, M., & Lally, V. (2004). It's not so easy: Researching the complexity of emergent participant roles and awareness in asynchronous networked learning discussions. *Journal of Computer Assisted Learning, 20*(3), 165–171. https://doi.org/10.1111/j.1365-2729.2004.00085.x

De Laat, M., Lally, V., Lipponen, L., & Simons, R.-J. (2007). Investigating patterns of interaction in networked learning and computer-supported collaborative learning: A role for Social Network Analysis. *International Journal of Computer-Supported Collaborative Learning, 2*, 87–103. https://doi.org/10.1007/s11412-007-9006-4

De Laat, M., & Schreurs, B. (2013). Visualizing informal professional development networks: Building a case for learning analytics in the workplace. *American Behavioral Scientist, 57*(10), 1421–1438. https://doi.org/10.1177/0002764213479364

De Wever, B., Van Keer, H., Schellens, T., & Valcke, M. (2011). Assessing collaboration in a wiki: The reliability of university students' peer assessment. *The Internet and Higher Education, 14*(4), 201–206. https://doi.org/10.1016/j.iheduc.2011.07.003

Deci, E. L., & Ryan, R. M. (1985). *Intrinsic motivation and self-determination in human behaviour*. New York: Plenum.

Dillenbourg, P., Sharples, M., Fisher, F., Kollar, I., Tchounikine, P., Dimitriadis, Y., et al. (2011). *Trends in orchestration. Second research & technology scouting report*. Retrieved from: http://hal.archives-ouvertes.fr/hal-00722475/

Duffy, P. (2008). Engaging the YouTube Google-eyed generation: Strategies for using web 2.0 in teaching and learning. *Electronic Journal of E-learning, 6*(2), 119–130. http://www.ejel.org/issue/download.html?idArticle=64

Ellison, N. B., Steinfield, C., & Lampe, C. (2007). The benefits of Facebook "friends": Social capital and college students' use of online social network sites. *Journal of Computer-Mediated Communication, 12*(4), 1143–1168.

Ferguson, R., Barzilai, S., Ben-Zvi, D., Chinn, C. A., Herodotou, C., Hod, Y., et al. (2017). *Innovating pedagogy 2017: Open University innovation report 6*. Milton Keynes: The Open University.

Ferguson, R., Brasher, A., Cooper, A., Hillaire, G., Mittelmeier, J., Rienties, B., et al. (2016). *Research evidence of the use of learning analytics; implications for education policy*. Retrieved from Luxembourg: https://ec.europa.eu/jrc/en/publication/eur-scientific-and-technical-research-reports/research-evidence-use-learning-analytics-implications-education-policy

Freeman, L. (2003). *The development of social network analysis*. Vancouver: Empirical Press.

Frels, R. K., Newman, I., & Newman, C. (2015). Mentoring the next generation in mixed methods research. In S. Hesse-Biber & R. B. Johnson (Eds.), *The Oxford handbook of multimethod and mixed methods research inquiry* (pp. 333–353). Oxford: Oxford Internet Institute.

Froehlich, D. (2019). Mapping mixed methods approaches to social network analysis in learning and education. In D. Froehlich, M. Rehm, & B. Rienties (Eds.), *Mixed methods approaches to social network analysis*. London: Routledge.

Froehlich, D., & Bohle Carbonell, K. (2020). Social influences on team learning. In D. Gijbels, C. Harteis, & E. Kyndt (Eds.), *Research approaches to workplace learning*. Heidelberg: Springer.

Froehlich, D., & Brouwer, J. (Forthcoming). Social network analysis as mixed analysis. In A. J. Onwuegbuzie & R. B. Johnson (Eds.), *Reviewer's guide for mixed methods research analysis*. London: Routledge.

Froehlich, D., & Gegenfurtner, A. (2019). Social support in transitioning from training to the workplace: A social network perspective. In H. Fasching (Ed.), *Beziehungen in pädagogischen Arbeitsfeldern [Relations in pedagogical work]*. Bad Heilbrunn: Klinkhardt.

Froehlich, D., Mamas, X., & Schneider, Y. (2019). Mapping mixed methods approaches to social network analysis in learning and education. In D. Froehlich, M. Rehm, & B. Rienties (Eds.), *Mixed methods approaches to social network analysis*. London: Routledge.

Froehlich, D., Rehm, M., & Rienties, B. (2019). *Mixed methods approaches to social network analysis*. London: Routledge.

Gemmell, I., & Harrison, R. (2017). A comparison between national and transnational students' access of online learning support materials and experience of technical difficulties on a fully online distance learning master of public health programme. Open Learning: *The Journal of Open, Distance and e-Learning, 32*(1), 66–80. https://doi.org/10.1080/02680513.2016.1253463

Giesbers, B., Rienties, B., Tempelaar, D. T., & Gijselaers, W. H. (2013). Investigating the relations between motivation, tool use, participation, and performance in an e-learning course using web-videoconferencing. *Computers in Human Behavior, 29*(1), 285–292. https://doi.org/10.1016/j.chb.2012.09.005

Giesbers, B., Rienties, B., Tempelaar, D. T., & Gijselaers, W. H. (2014). A dynamic analysis of the interplay between asynchronous and synchronous communication in online learning: The impact of motivation. *Journal of Computer Assisted Learning, 30*(1), 30–50. https://doi.org/10.1111/jcal.12020

Golonka, E. M., Bowles, A. R., Frank, V. M., Richardson, D. L., & Freynik, S. (2014). Technologies for foreign language learning: A review of technology types and their effectiveness. *Computer Assisted Language Learning, 27*(1), 70–105. https://doi.org/10.1080/09588221.2012.700315

Hanneman, R. A., & Riddle, M. (2005). *Introduction to social network methods*. Riverside, CA: University of California.

Hartrey, L., Denieffe, S., & Wells, J. S. G. (2017). A systematic review of barriers and supports to the participation of students with mental health difficulties in higher education. *Mental Health & Prevention, 6*, 26–43. https://doi.org/10.1016/j.mhp.2017.03.002

Hesse-Biber, S. (2010). Qualitative approaches to mixed methods practice. *Qualitative Inquiry, 16*(6), 455–468. https://doi.org/10.1177/1077800410364611

Hiltz-Hymes, C. E. (2011). *The role of emotional contagion and flooding in the group process of children exposed to domestic violence*. Barbara, CA: Fielding Graduate University.

Hoel, T., Griffiths, D., & Chen, W. (2017). *The influence of data protection and privacy frameworks on the design of learning analytics systems*. Paper presented at the Proceedings of the Seventh International Learning Analytics & Knowledge Conference, Vancouver, Canada.

Holmes, V., Clark, W., Burt, P., & Rienties, B. (2013). Engaging teachers (and students) with media streaming technology, the case of Box of Broadcasts. In L. Wankel & P. Blessinger (Eds.), *Increasing student engagement and retention using mobile applications: Smartphones, Skype and texting technologies* (Vol. 6D, pp. 211–240). Bingley, UK: Emerald Publishing Group.

Hommes, J., Arah, O. A., de Grave, W., Bos, G., Schuwirth, L., & Scherpbier, A. (2014). Medical students perceive better group learning processes when large classes are made to seem small. *PLoS One, 9*(4), e93328. https://doi.org/10.1371/journal.pone.0093328

Hommes, J., Rienties, B., de Grave, W., Bos, G., Schuwirth, L., & Scherpbier, A. (2012). Visualising the invisible: A network approach to reveal the informal social side of student learning. *Advances in Health Sciences Education, 17*(5), 743–757. https://doi.org/10.1007/s10459-012-9349-0

Jindal-Snape, D., & Rienties, B. (Eds.). (2016). *Multi-dimensional transitions of international students to higher education*. London: Routledge.

Johnson, L., Adams Becker, S., Cummins, M., Estrada, V., Freeman, A., & Hall, C. (2016). *NMC horizon report: 2016 higher education edition*. Retrieved from Austin, Texas: http://cdn.nmc.org/media/2016-nmc-horizon-report-he-EN.pdf

Jones, E., Samra, R., & Lucassen, M. (2018). The world at their fingertips? The mental wellbeing of online distance-based law students. *The Law Teacher*, 1–21. https://doi.org/10.1080/03069400.2018.1488910

Jones, T., & Cuthrell, K. (2011). YouTube: Educational potentials and pitfalls. *Computers in the Schools, 28*(1), 75–85. https://doi.org/10.1080/07380569.2011.553149

Kadushin, C. (2005). Who benefits from network analysis: Ethics of social network research. *Social Networks, 27*(2), 139–153. https://doi.org/10.1016/j.socnet.2005.01.005

Kirschner, P. A. (2015). Facebook as learning platform: Argumentation superhighway or dead-end street? *Computers in Human Behavior, 53*, 621–625. https://doi.org/10.1016/j.chb.2015.03.011

Kirschner, P. A., & Erkens, G. (2013). Toward a framework for CSCL research. *Educational Psychologist, 48*(1), 1–8. https://doi.org/10.1080/00461520.2012.750227

Kirschner, P. A., & Karpinski, A. C. (2010). Facebook® and academic performance. *Computers in Human Behavior, 26*(6), 1237–1245. https://doi.org/10.1016/j.chb.2010.03.024

Klein, M., & Konieczny, P. (2015). *Wikipedia in the world of global gender inequality indices: What the biography gender gap is measuring*. Paper presented at the Proceedings of the 11th International Symposium on Open Collaboration, San Francisco, CA.

Knight, S., Rienties, B., Littleton, K., Tempelaar, D. T., Mitsui, M., & Shah, C. (2017). The orchestration of a collaborative information seeking learning task. *Information Retrieval Journal, 20*(5), 480–505. https://doi.org/10.1007/s10791-017-9304-z

Korir, M., Mittelmeier, J., & Rienties, B. (2019). Is mixed methods social network analysis ethical? In D. Froehlich, M. Rehm, & B. Rienties (Eds.), *Mixed methods approaches to social network analysis*. London: Routledge.

Lally, V., Sharples, M., Tracy, F., Bertram, N., & Masters, S. (2012). Researching the ethical dimensions of mobile, ubiquitous and immersive technology enhanced learning (MUITEL): A thematic review and dialogue. *Interactive Learning Environments, 20*(3), 217–238. https://doi.org/10.1080/10494820.2011.607829

Law, P., & Jelfs, A. (2016). Ten years of open practice: A reflection on the impact of OpenLearn. *Open Praxis, 8*(2), 7. https://doi.org/10.5944/openpraxis.8.2.283

Madge, C., Breines, M., Beatrice Dalu, M. T., Gunter, A., Mittelmeier, J., Prinsloo, P., et al. (2019). WhatsApp use among African international distance education (IDE) students: Transferring, translating and transforming educational experiences. *Learning, Media and Technology, 44*(3), 267–282.

Madge, C., Meek, J., Wellens, J., & Hooley, T. (2009). Facebook, social integration and informal learning at university: "It is more for socialising and talking to friends about work than for actually doing work". *Learning, Media and Technology, 34*(2), 141–155. https://doi.org/10.1080/17439880902923606

Manca, S., & Ranieri, M. (2013). Is it a tool suitable for learning? A critical review of the literature on Facebook as a technology-enhanced learning environment. *Journal of Computer Assisted Learning, 29*(6), 487–504. https://doi.org/10.1111/jcal.12007

McCallum, A., Wang, X., & Corrada-Emmanuel, A. (2007). Topic and role discovery in social networks with experiments on enron and academic email. *Journal of Artificial Intelligence Research, 30*, 249–272.

McConnell, D., Hodgson, V., & Dirckinck-Holmfeld, L. (2012). Networked learning: A brief history and new trends. In L. Dirckinck-Holmfeld, V. Hodgson, & D. McConnell (Eds.), *Exploring the theory, pedagogy and practice of networked learning* (pp. 3–24). New York: Springer.

Moolenaar, N. M. (2012). A social network perspective on teacher collaboration in schools: Theory, methodology, and applications. *American Journal of Education, 119*(1), 7–39. https://doi.org/10.1086/667715

Moolenaar, N. M., Sleegers, P. J. C., & Daly, A. J. (2012). Teaming up: Linking collaboration networks, collective efficacy, and student achievement. *Teaching and Teacher Education, 28*(2), 251–262. https://doi.org/10.1016/j.tate.2011.10.001

Neal, J. W. (2008). "Kracking" the missing data problem: Applying Krackhardt's cognitive social structures to school-based social networks. *Sociology of Education, 81*(2), 140–162. https://doi.org/10.1177/003804070808100202

Okada, A., & Moreira, P. (2008). *Enhancing informal learning through videoconferencing and knowledge maps*. Paper presented at the EDEN 2008, Universidade Aberta, Lisbon.

Open University UK. (2014). *Ethical use of student data for learning analytics policy*. Retrieved from: http://www.open.ac.uk/students/charter/essential-documents/ethical-use-student-data-learning-analytics-policy

Parsons, D., Hill, I., Holland, J., & Willis, D. (2012). *Impact of teaching development programmes in higher education*. Retrieved from York: http://www.heacademy.ac.uk/assets/documents/research/HEA_Impact_Teaching_Development_Prog.pdf

Peña-Ayala, A. (2018). Learning analytics: A glance of evolution, status, and trends according to a proposed taxonomy. *Wiley Interdisciplinary Reviews: Data Mining and Knowledge Discovery, 8*(3), e1243. https://doi.org/10.1002/widm.1243

Peña-Ayala, A., Cárdenas-Robledo, L. A., & Sossa, H. (2017). A landscape of learning analytics: An exercise to highlight the nature of an emergent field. In A. Peña-Ayala (Ed.), *Learning analytics: Fundaments, applications, and trends: A view of the current state of the art to enhance e-learning* (pp. 65–112). Cham: Springer International Publishing.

Pifer, M. J. (2011). Intersectionality in context: A mixed-methods approach to researching the faculty experience. *New Directions for Institutional Research, 2011*(151), 27–44. https://doi.org/10.1002/ir.397

Piper, R., & Emmanuel, T. (2019). *Co-producing mental health strategies with students: A guide for the higher education sector*. Retrieved from London: https://www.studentminds.org.uk/co-productionguide.html

Prinsloo, P., & Slade, S. (2017). Ethics and learning analytics: Charting the (un)charted. In C. Lang, G. Siemens, A. F. Wise, & D. Gasevic (Eds.), *Handbook of learning analytics* (pp. 49–57). Society for Learning Analytics Research. https://solaresearch.org/wp-content/uploads/2017/05/hla17.pdf

R Core Team. (2014). *R: A language and environment for statistical computing*. Retrieved from: http://www.r-project.org

Rehm, M., & Cornelissen, F. (2019). Power to the people?! Twitter discussions on (educational) policy processes. In D. Froehlich, M. Rehm, & B. Rienties (Eds.), *Mixed methods approaches to social network analysis*. London: Routledge.

Rehm, M., Gijselaers, W., & Segers, M. (2015). The impact of hierarchical positions on communities of learning. *International Journal of Computer-Supported Collaborative Learning, 10*(2), 117–138. https://doi.org/10.1007/s11412-014-9205-8

Rehm, M., Littlejohn, A., & Rienties, B. (2018). Does a formal wiki event contribute to the formation of a network of practice? A social capital perspective on the potential for informal learning. *Interactive Learning Environments, 26*(3), 308–319. https://doi.org/10.1080/10494820.2017.1324495

Rehm, M., & Notten, A. (2016). Twitter as an informal learning space for teachers!? The role of social capital in Twitter conversations among teachers. *Teaching and Teacher Education, 60*, 215–223. https://doi.org/10.1016/j.tate.2016.08.015

Rienties, B. (2010). *Understanding social interaction in Computer-Supported Collaborative learning: the role of motivation on social interaction* (Ph.D. manuscript, Océ Business Services, Maastricht) (978-90-5681-328-4, 841).

Rienties, B. (2019). Powers and limitations of MMSNA: Experiences from the field of education. In D. Froehlich, M. Rehm, & B. Rienties (Eds.), *Mixed methods approaches to social network analysis*. London: Routledge.

Rienties, B., Giesbers, B., Tempelaar, D. T., Lygo-Baker, S., Segers, M., & Gijselaers, W. H. (2012). The role of scaffolding and motivation in CSCL. *Computers & Education, 59*(3), 893–906. https://doi.org/10.1016/j.compedu.2012.04.010

Rienties, B., & Hosein, A. (2015). Unpacking (in)formal learning in an academic development programme: A mixed method social network perspective. *International Journal of Academic Development, 20*(2), 163–177. https://doi.org/10.1080/1360144X.2015.1029928

Rienties, B., Johan, N., & Jindal-Snape, D. (2015). Bridge building potential in cross-cultural learning: A mixed method study. *Asia Pacific Education Review, 16*, 37–48. https://doi.org/10.1007/s12564-014-9352-7

Rienties, B., & Kinchin, I. M. (2014). Understanding (in)formal learning in an academic development programme: A social network perspective. *Teaching and Teacher Education, 39*, 123–135. https://doi.org/10.1016/j.tate.2014.01.004

Rienties, B., & Tempelaar, D. T. (2018). Turning groups inside out: A social network perspective. *Journal of the Learning Sciences, 27*(4), 550–579. https://doi.org/10.1080/10508406.2017.13 98652

Rienties, B., Tempelaar, D. T., Giesbers, B., Segers, M., & Gijselaers, W. H. (2014). A dynamic analysis of social interaction in Computer Mediated Communication; a preference for autonomous learning. *Interactive Learning Environments, 22*(5), 631–648. https://doi.org/10.1080/1 0494820.2012.707127

Rienties, B., Tempelaar, D. T., Pinckaers, M., Giesbers, B., & Lichel, L. (2012). The diverging effects of social network sites on receiving job information for students and professionals. In E. Coakes (Ed.), *Technological change and societal growth: Analyzing the future* (pp. 202–217). Hershey, PA: IGI Global.

Rienties, B., Tempelaar, D. T., Van den Bossche, P., Gijselaers, W. H., & Segers, M. (2009). The role of academic motivation in Computer-Supported Collaborative Learning. *Computers in Human Behavior, 25*(6), 1195–1206. https://doi.org/10.1016/j.chb.2009.05.012

Rienties, B., Tempelaar, D. T., Waterval, D., Rehm, M., & Gijselaers, W. H. (2006). Remedial online teaching on a summer course. *Industry and Higher Education, 20*(5), 327–336. https://doi.org/10.5367/000000006778702300

Roberts, C., Daly, M., Held, F., & Lyle, D. (2017). Social learning in a longitudinal integrated clinical placement. *Advances in Health Sciences Education, 22*(4), 1011–1029. https://doi.org/10.1007/s10459-016-9740-3

Roxå, T., & Mårtensson, K. (2009). Significant conversations and significant networks – exploring the backstage of the teaching arena. *Studies in Higher Education, 34*(5), 547–559. https://doi.org/10.1080/03075070802597200

Ryan, R. M., & Deci, E. L. (2000). Intrinsic and extrinsic motivations: Classic definitions and new directions. *Contemporary Educational Psychology, 25*(1), 54–67.

Sarazin, M. (2019). Ethnographic mixed methods social network analysis studies: Opportunities and challenges. In D. Froehlich, M. Rehm, & B. Rienties (Eds.), *Mixed methods approaches to social network analysis*. London: Routledge.

Schellens, T., & Valcke, M. (2005). Collaborative learning in asynchronous discussion groups: What about the impact on cognitive processing? *Computers in Human Behavior, 21*(6), 957–975.

Schoonenboom, J., Johnson, R. B., & Froehlich, D. (2018). Combining multiple purposes of mixing within a mixed methods research design. *International Journal of Multiple Research Approaches, 10*(1), 271–282.

Scott, J. P. (2012). *Social network analysis: A handbook* (3rd ed.). London: Sage Publications Ltd..

Sharples, M., Adams, A., Alozie, N., Ferguson, F., FitzGerald, E., Gaved, M., et al. (2015). *Innovating pedagogy 2015*. Retrieved from Milton Keynes: http://proxima.iet.open.ac.uk/public/innovating_pedagogy_2015.pdf

Sharples, M., Adams, A., Ferguson, R., Gaved, M., McAndrew, P., Rienties, B., et al. (2014). *Innovating pedagogy 2014*. Retrieved from Milton Keynes: http://www.open.ac.uk/iet/main/files/iet-web/file/ecms/web-content/Innovating_Pedagogy_2014.pdf

Sharples, M., de Roock, R., Ferguson, R., Gaved, M., Herodotou, C., Koh, E., et al. (2016). *Innovating pedagogy 2016: Open University innovation report 5*. Retrieved from Milton Keynes: http://www.open.ac.uk/blogs/innovating/

Sharples, M., McAndrew, P., Weller, M., Ferguson, R., FitzGerald, E., Hirst, T., & Gaved, M. (2013). *Innovating pedagogy 2013*. Retrieved from Milton Keynes. https://iet.open.ac.uk/file/innovating-pedagogy-2013.pdf

Tempelaar, D. T., Rienties, B., & Giesbers, B. (2015). In search for the most informative data for feedback generation: Learning analytics in a data-rich context. *Computers in Human Behavior, 47*, 157–167. https://doi.org/10.1016/j.chb.2014.05.038

Tempelaar, D. T., Rienties, B., Kaper, W., Giesbers, B., Van Gastel, L., Van de Vrie, E., et al. (2011). Effectiviteit van facultatief aansluitonderwijs wiskunde in de transitie van voortgezet naar hoger onderwijs (effectiveness of voluntary remedial education in mathematics to facilitate the transition from secondary to higher education). *Pedagogische Studiën, 88*(4), 231–248.

Tempelaar, D. T., Rienties, B., Mittelmeier, J., & Nguyen, Q. (2018). Student profiling in a dispositional learning analytics application using formative assessment. *Computers in Human Behavior, 78*, 408–420. https://doi.org/10.1016/j.chb.2017.08.010

Thomas, L., Tuytens, M., Devos, G., Kelchtermans, G., & Vanderlinde, R. (2019). Unpacking beginning teachers' collegial network structure: A mixed-method social network study. In D. Froehlich, M. Rehm, & B. Rienties (Eds.), *Mixed methods approaches to social network analysis*. London: Routledge.

Vallerand, R. J., Pelletier, L. G., Blais, M. R., Brière, N. M., Senécal, C., & Vallières, E. F. (1992). The academic motivation scale: A measure of intrinsic, extrinsic, and amotivation in education. *Educational and Psychological Measurement, 52*, 1003–1017.

Van den Bossche, P., & Segers, M. (2013). Transfer of training: Adding insight through social network analysis. *Educational Research Review, 8*, 37–47. https://doi.org/10.1016/j.edurev.2012.08.002

Van Waes, S., De Maeyer, S., Moolenaar, N. M., Van Petegem, P., & Van den Bossche, P. (2018). Strengthening networks: A social network intervention among higher education teachers. *Learning and Instruction, 53*, 34–49. https://doi.org/10.1016/j.learninstruc.2017.07.005

Van Waes, S., Van de Bossche, P., Moolenaar, N. M., De Maeyer, S., & Van Petegem, P. (2015). Know-who? Linking faculty's networks to stages of instructional development. *Higher Education, 70*(5), 807–826. https://doi.org/10.1007/s10734-015-9868-8

Veerman, A. L., & Veldhuis-Diermanse, E. (2001). Collaborative learning through computer-mediated communication in academic education. In P. Dillenbourg, A. Eurelings, & K. Hakkarainen (Eds.), *European perspectives on computer-supported collaborative learning: Proceedings of the 1st European conference on computer-supported collaborative learning* (pp. 625–632). Maastricht: University of Maastricht.

Vera, E. R., & Schupp, T. (2006). Network analysis in comparative social sciences. *Comparative Education, 42*(3), 405–429. https://doi.org/10.1080/03050060600876723

Wassermann, S., & Faust, K. (1994). *Social network analysis: Methods and applications*. Cambridge: Cambridge University Press.

Watson, W. R., Watson, S. L., & Reigeluth, C. M. (2015). Education 3.0: Breaking the mold with technology. *Interactive Learning Environments, 23*(3), 323–343. https://doi.org/10.1080/10494820.2013.764322

Zhang, M. (2015). Internet use that reproduces educational inequalities: Evidence from big data. *Computers & Education, 86*, 212–223. https://doi.org/10.1016/j.compedu.2015.08.007

Chapter 3
Educational Networking: A Glimpse at Emergent Field

Alejandro Peña-Ayala ⓘD

Abstract As a sample of the evolution that characterizes twenty-first century education, diverse Technology-Enhanced Teaching and Learning (TETL) approaches emerge, where *educational networking* (EN) is relevant. In this context, Web 2.0 represents a suitable environment to foster the creation, usage, and growth of online social networks (OSNs) that enable users to meaningfully connect, create, contribute, and collaborate in pursuit of common affairs, such as education. Hence, EN represents more than a platform for advertisement, breaking news, message dissemination, and content sharing, as it merges OSN functionalities with learning resources and its tools suitable for online teaching to instill *social learning for all*. Although EN is an incipient field in vigorous progress, it represents a new fashion of e-learning settings that gregariously transforms and spreads education worldwide. This is why a review of the evolution of EN labor and recent achievements leads to appreciate its nature, development, impact, and trends to consider its exploitation as part of an extended and novel academic offer. Thus, in this chapter a sample of works published since 2015 up to date is analyzed to shape a landscape of the endeavors performed in the EN arena. Therefore, a taxonomy for EN is designed to report a summary of the EN labor, whose features are outlined through a pattern to depict EN applications. The work reveals that most of the publications study how OSNs affect students' learning, attitudes, and behavior, and general-purpose OSNs represent the most demanded platforms to foster EN.

Keywords Educational networking · Social network · Online social network · Social network sites · Social learning

A. Peña-Ayala (✉)
WOLNM, Artificial Intelligence on Education Lab, Mexico City, Mexico

Instituto Politécnico Nacional, Escuela Superior de Ingeniería Mecánica y Eléctrica, Zacatenco, Mexico City, Mexico
e-mail: apenaa@ipn.mx

© Springer Nature Switzerland AG 2020
A. Peña-Ayala (ed.), *Educational Networking*, Lecture Notes in Social Networks, https://doi.org/10.1007/978-3-030-29973-6_3

Abbreviations

EN   Educational networking
HE   Higher education
HS   High school
ICT   Information communication technology
OSN   Online social network
PDENA   Pattern to Describe Educational Networking Applications
SN   Social network
SNA   Social network analysis
SNS   Social network site
STEM   Science, Technology, Engineering, and Mathematics
TEN   Taxonomy for educational networking
TETL   Technology-Enhanced Teaching and Learning

3.1   Introduction

It is well known that investing in education contributes to the common good of society, nurturing economic prosperity, supporting family stability, and leading individual's personal growth (Pellegrino & Hilton, 2013). However, nowadays education is even more valuable before the globalization phenomenon, science development, economic interdependence, industrial automation of human labor, emergence of diverse community issues, intromission of information communication technology (ICT) in human life, and the worldwide spread of vast information and social media at real time that people produce, access, interpret, and respond.

Those trends have prompted rulers, policymakers, education reformers, businesspersons, and social leaders to claim that classic curriculums, teaching practices, learning habits, cognitive skills, inner attitudes, academic settings, and educational content and resources are no longer suitable for enabling the new generations of students to deal with and overcome such universal tendencies (Jerald, 2009).

Before the mentioned demands, academic institutions are compelled to conceive, design, develop, and deploy a novel model of education coined as *twenty-first century education*. In this sense, Jerald (2009) proposes four axes to sketch such a model: educational attainment, foundational knowledge, practical literacies, and broader competences. As for the last axis, it lays on *twenty-first century skills* that concern creativity, productivity, communication, collaboration, digital literacy, citizenship, problem solving, and critical thinking (Voogt & Roblin, 2012). All of these are essential so that individuals effectively behave in school, workplace, and society (Siddiq, Gochyyev, & Wilson, 2017), where ICTs expand their capabilities to achieve human duties (Dede, 2010) as well as to challenge people to efficiently use them.

In addition to disciplines such as artificial intelligence and data science, ICTs provide the logistic and material bench to nurture diverse TETL settings in innovative and transformative ways. Technology affords diverse opportunities that extend teaching and learning processes, offering improved possibilities for knowledge and skills transference and acquisition (Goodyear & Retalis, 2010). Essentially, TETL boosts (1) digital delivery of course content and learning resources; (2) access, communication, collaboration, and reflection; and (3) data tracking from students' interaction with e-learning systems that is automatically learned and interpreted to provide adaptive and personalized support for students (Davies, Mullan, & Feldman, 2017).

As an essential resource to foment the development of TETL approaches, *Web 2.0* constitutes an essential facility, a term that by the way was coined during a session between Tim O'Reilly and MediaLive International (O'Reilly, 2007) to mean: "The network as platform, spanning all connected devices, where applications are those that make the most of the intrinsic advantages of that platform...." Web 2.0 improves the classic one-way information flow (e.g., reading, receiving, and researching) to a two-way conversation cycle (e.g., creating, contributing, and collaborating) by means of some platforms (e.g., twittering, wikis, and blogs), whose sites' primary content is provided by users through a given framework for their participation (Hargadon, 2010). According to O'Reilly (2007), Web 2.0 success lays on the "long tail" criterion, stated by Chris Anderson, which asserts: "Leverage customer self–service and algorithm data management to reach out the entire Web... to the long tail and not just the head." Web 2.0 offers interactive and collaborative functionalities for participation, harnessing collective intelligence, and rich user experiences (Chen, 2011). Web 2.0 fosters users to easily create and share multimedia content as well as contributes to global conversations among communities (Drexler, Baralt, & Dawson, 2008). Web 2.0 raises a social constructivist approach for e-learning and facilitates inquiry practice, individual expression, and literacy.

In another vein, the world society continues its evolution and historical advances, looking for better welfare levels and solutions for complex challenges (e.g., pollution, migration, climate change, etc.). It is thought the aforesaid progress relies on community work, where as a manifestation of human gregarious sense, webs of social relations and associations persistently emerge with diverse purposes, evolve during the time, collapse, and vanish after a while. In this regard, John Arundel Barnes labeled the concept of *social network* (SN) in 1995 to describe patterns of social relationships that are not easily explained by classic social units such as extended families or work groups (Heaney & Israel, 2008). A SN represents ties between people that may or may not offer social support and that may serve functions other than providing support. In other words, a SN is a set of socially networked members connected by one or several relations, whose patterns are object of study (Marin & Wellman, 2011), where *social capital* is an emergent property that depicts some resources and norms that arise from SNs (Ferlander, 2007). According to Lin (2001), social capital is "An investment in social relations with expected returns in the marketplace", as well as "A social asset by virtue of actors' connections and access to resources in the group they pertain."

In this context, ICTs contribute to boost the appearance, organization, and expansion of SNs through a *computer-mediated communication*, where an emergent vehicle is known as *OSN* that virtualizes the webs of social relations. Specially, in the Web 2.0, OSN relies on an ideal bench to interactively spread data, communication, and social media among worldwide communities of users. An OSN enables users to connect and communicate with each other, where the *network externalities* trait reveals the utility of a platform to a user depends on the number of platform users among the user's connections (Lőrincz, Koltai, Győr, & Takács, 2019). Lőrincz et al. (2019) define two relationships between OSNs and social capital: (1) *bridging social capital* that resumes factors correlated with access to opportunities, resources, and information through social links and (2) *bonding social capital* that gathers the benefits from trust, intimacy, and cooperation in stronger, dense, and close relationships. In short, an OSN fosters users to stand by in relatively close social contact with others through the use of a *social network site* (SNS) (Cain, 2008).

Concerning a SNS, it is a relationship facilitator Web 2.0 platform that supports persons to build connections with others. SNSs provide web-based services that further users to build a profile within a bounded system, organize a list of several users with whom they share a connection, and view and traverse their list of connections and those made by others within the site (Boyd & Ellison, 2008). The core of a SNS is composed of visible profiles and networks of friends, who are also users of the site. A process stated by Harrison and Thomas (2009) for being a member of an OSN is explained as follows: (1) user signs up to a SNS, (2) user creates a personal profile, (3) user is then regarded as a member, and (4) user creates his/her OSN by means of a list of friends, who are also users of the site that is organized through consensus and mutual recognition (i.e., user invites known people who accept or decline), a unidirectional process (i.e., people who follow a given user), or community based on a particular topic (i.e., a user creates a group within a SNS, where people interested in such a topic join that group). Hence, the single term *network*, as part of the composed words SN, OSN, and SNS, denotes that users are mainly interested in making *connections* with familiar people, whereas *networking* as a member of those terms highlights that users are mostly concerned in setting new links with strangers who they expect to meet. In consequence, when OSN members are primarily communicating with familiar individuals keeping their existing social connections alive, rather than for making new ones, then the distinctive term is *network* (e.g., as it is stated for the SN, OSN, SNS acronyms). Otherwise, the *networking* term is used to emphasize that individuals are mainly interested in meeting new people so that the corresponding terms are social networking, online social networking, and social networking site. Thus, in the remainder of this chapter, the already declared acronyms are used, unless the purpose is to seek and initiate new human affairs.

On the other hand, the usage of SNSs in education represents a shift toward social and community-based web applications that cultivate and sustain discipline-specific SNs (Brady, Holcomb, & Smith, 2010). In this sense, EN is the use of social networking technologies for educational purposes (Goldfarb, Pregibon,

Shrem, & Zyko, 2011). Even though the utilization of SNSs for education faces diverse claims, such as the fear students are distracted when accessing their SNSs during school and could be object of negative biases and effects exerted by wrong friends (Nee, 2014), privacy and safety concerns (Holcomb, Brady, Smith, & Bethany, 2010), and foment of antisocial and unproductive behavior. However, there is a growing evidence that these SNSs improve technology proficiency, enhance social skills, and support learning (Goldfarb et al., 2011), as well as EN impacts on students' creativity, interaction, motivation, engagement, retention, and satisfaction facilitating communication and the expression of diverse views. In resume, online social networking educational labor is defined as the use of suitable social technologies that facilitate a range of teaching and learning tasks in collaborative settings (Hamid, Waycott, Kurnia, & Chang, 2015).

Since privacy, legal, and safety issues, as well as students' misleading behavior and learning achievements, prevent policymakers and educators from adopting EN initiatives, education systems strive to overcome such issues to use SNSs for teaching and learning purposes (Goldfarb et al., 2011). Even more, academic institutions resist change and are slowly adopting new technologies (Drexler et al., 2008). In consequence, a problem for disseminating EN in academic institutions arises: *How to encourage academic provosts, officers, and staff to pragmatically consider the development and use of SNSs to enhance and extend current academic offer?*

In addition, the following research questions are made: (1) How has EN evolved? (2) What are the key lines that distinguish EN labor? (3) What are the relevant traits that characterize EN research? (4) What about the recent labor carried out in EN? (5) Which are the main affairs considered by recent EN works and how are they approached? (6) Which are the expectations concerning the situation and development of EN?

With the aim of contributing to solve the problem for spreading EN and answer the prior six research inquiries, the following solution is given: *to tailor a landscape of the EN field*, which embraces four constructs: (1) the design of a taxonomy that organizes the work lines, (2) the characterization of the related works through a set of relevant features, (3) the statement of a profile that resumes the achieved labor reported in recent and representative related works, and (4) the analysis of the status, work lines, and outcomes. The purpose is to provide a briefing of the accomplished labor and a report of relevant experiences that offer an empirical and valuable reference worthy to be considered for investing in EN.

In this concern, previous reviews focus on specific features of the EN labor, such as the following: Guraya (2016) gathers ten articles published between 2004 and 2014 to identify the medical students' extent of usage of SNSs for educational affairs and examine utility, quality, strength, target, and setting of the evidence, finding that 75% of the respondents in those studies admitted using EN. In some sense, another review is edited by Issa, Isaias, and Kommers (2015), where dozens of authors shape a national and international scenery about the use of social networking as a tool on the education of countries from diverse regions in addition to propose a Social Networking and Education Model to further and deploy EN. With the aim of answering whether SNS use is beneficial or harmful for academic performance, Liu, Kirschner, and Karpinski (2017) develop a meta-analysis of 24 studies disclosed up to 2016, producing two findings, a negative correlation between SNS use and grade point average and a positive correlation for SNS use and language test. As for one popular SNS, Tang and Hew (2017) examine 51 empirical studies, divulged between 2006 and 2015, concerning the use of Twitter in teaching and learning duties (e.g., ways in which Twitter is employed, the impacts on interactions, bias on students' learning results) to determine whether its exploitation favors students, giving as a result that Twitter is mostly used for communication and assessment affairs, whereas Nagle (2018) reviews the literature about how social media is spread through Twitter in teacher education, discussing ethical implications of using Twitter as a pedagogical tool and proposing policies to define critical social media literacy.

In contrast to the introduced related works, this review pursues as the main objective: *to provide a briefing of recent studies, applications, and proposals that report the achieved labor and identify main outcomes to unveil the benefits and constraints produced by the implementation of EN in academic settings.* Moreover, four specific goals are also considered: (1) gather representative publications that offer a broad landscape of the arena, (2) discover relevant lines of work that label the essence of the reported articles, (3) design a pattern that depicts key features of the presented research, and (4) critically analyze the pros and cons of the field, and propose how they could be enhanced and solved, respectively.

As a sample of the results achieved in this work, the following outcomes stand out: firstly, an updated and wide review of the EN recent labor is outlined; secondly, a repository of 92 relevant EN-related works has been gathered and organized; thirdly, a functional viewpoint to classify the essence of the EN work is proposed; fourthly, a set of features to be instantiated with the particular values of each collected article to reveal a standard profile; and fifthly, an evaluation of advantages and limitations that represent the application of EN to extend the academic offer.

The remainder of this chapter is organized as follows: the underlying elements of this work are introduced in Sect. 3.2, where the method designed to develop the review is illustrated and a chronicle of the EN evolution is traced, including the identification of the material gathered to shape the glimpse and the description of a proposal for organizing and characterizing the related works by means of a taxonomy and a pattern, respectively. In Sect. 3.3, a landscape of the EN field is outlined through a summary of the 92 collected articles that are classified according to ten cases of SNSs that are distinguished in the suggested taxonomy, which are edited as subsections. An overall view of the already introduced EN labor is pictured in Sect. 3.4 through annual frequencies with their corresponding interpretation, as well as the discussion of the EN field (i.e., composed of four viewpoints: strengths, weaknesses, opportunities, and threats) and the responses given for the research problem and questions. Afterward, the conclusions are expressed in Sect. 3.5 by means of the acknowledgment of achievements and constraints, in addition to the identification of future work. Lastly, in Appendix, the record of features that characterize the 92 EN-related works is edited in Table 3.8 according to the order in which works are introduced in Sect. 3.3.

3.2 Review Baseline

The development of this work follows the steps defined for a proposed method that is oriented to produce reviews of a given domain. Thus, the first topic edited in this section corresponds to the overall description of such a method. As a consequence of its application, the second subject recognizes the background of the EN field, and the third identifies the material gathered to generate the account, while the last themes respond to the needs of proposing a taxonomy that classifies the EN labor achieved up to date and a pattern to characterize the features that depict the works.

3.2.1 Method

In relation to the method used to produce the present review, it embraces three steps that pursue a specific goal through the accomplishment of several tasks as is sketched in Fig. 3.1. In such an illustration, the three steps correspond to *gathering, compilation,* and *evaluation,* whereas the tasks that compose each step are sequentially edited inside the shapes, whose workflow is drawn by means of arrows that identify sequential and cyclical flows.

3.2.2 Background

With the goal of tracing the *evolution of EN*, some related works are chronologically stated to highlight pioneer and relevant achievements, and key terms are identified with their respective concepts, giving as a result a proper definition of the EN concept. Therefore, the first work is a dossier of 23 *computer networks* (e.g., ARPANET Net and California State College Network) that higher education (HE) institutions built for academic aims (Chambers & Poore, 1975). In this report, "*computer networking*" labels: The efforts to share computer power between HE schools that own such facilities, as well as with those which hold little or lack those resources. Later, "*networking*" emerges to mean: Educational affairs where based on acquired skills people and resources strive through intentional acts, while stakeholders collaborate, rather than compete, under a self-directed learning guide (Fonte & Davis, 1979).

As one of the earliest *international networking services*, it corresponds to the use of the NASA's ATS-1 satellite that provides educational and social service networking to Pacific island nations through the transmission of voice and data (Lange et al., 1984). Afterward, the *computer-mediated communication* paradigm occurs to bring students together as they collaborate with peers and adults in diverse sectors (Riel, 1990). Riel (1990) compiles studies that assert the positive effect of *electronic networking* on students' reading and writing skills and their aim in meaningful educational tasks. Moreover, McAnge Jr. (1990) edits a survey

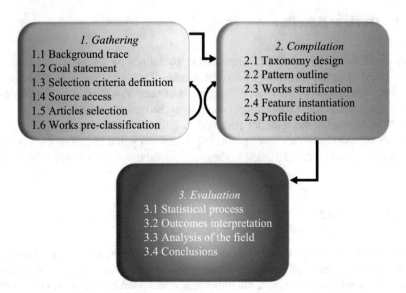

Fig. 3.1 Method applied to develop the review of the educational networking field

about the status of *computer networking in K–12 education* and how network linkages impact students' actions, regional networks, networking initiatives, and public access networks.

Afterward, the use of *wide area networks* enabled students and teachers of K–12 schools to access educational resources, communicate with peers abroad, share information, and work together (Eurich-Fulcer & Schofield, 1995). In this sense, when local area networks and wide area networks link organizations, people, and machines, they become *computer-supported social networks* that back virtual communities, cooperative work, and telework (Wellman et al., 1996).

Later, the *World Wide Web* flooded the international market giving as a result new ways of social media publishing, front e-learning systems, and the *first wave of SNSs*, producing as a consequence a suitable bench for the birth of EN. Related to publication, *Blogger* started in 1999 as a blog-publishing service that encourages people to keep a personal diary (Johnston, 2002). Concerning e-learning, a pioneer system fostered collaboration among students in the UK, the USA, and Ireland to analyze in real time online content of Third World following interactive learning strategies to reach a common goal (Warf, Vincent, & Purcell, 1999). As for SNSs, in 1997 *SixDegrees.com* emerged as the earliest SNS, whose principle asserts: Every human being on the face of the earth can be connected to another by approximately a chain of six degrees of levels (Ezumah, 2013; Rielgelsberger & Sasse, 2000). Regarding EN, some primary approaches deploy online discussion forums, where users type messages to convey meaning, build knowledge, and create socio-informational networks (Allan, 2004). Furthermore, early studies focus on friendly and adversarial effects of OSNs on students' performance in classroom and distance education that uses networking as an adjunct mode (Yang & Tang, 2003).

In another vein, the "dot-com" collapse in autumn of 2001 marked a turning point for the Web as a call to action such as "Web 2.0," whose definition was posted in September 2003 (O'Reilly, 2007). Since then, the Web is harnessed as a suitable platform for OSNs, among other purposes, where people dispatch, publish, and share content, while other individuals reply their respective opinion triggering collective expression (Craig, 2007). Around the occurrence of Web 2.0, the explosion of the *second wave of SNSs* is pictured by Boyd and Ellison (2008) as a timeline that shows the launch year of various SNSs as follows: *2001*, Cyworld; *2002*, Friendster and LinkedIn; *2003*, MySpace and Hi5; *2004*, Orkut, Facebook exclusive for Harvard, and Catster; *2005*, Ning, Yahoo! 360, YouTube, and Facebook for High School (HS) networks; and *2006*, Windows Live Spaces, Twitter, and Facebook for everyone.

Due to the occurrence of those SNSs, their use for EN affairs grew up as the next sequence of *basic studies* reveals: Sharples (2006) studies the support given by *MySpace* during informal learning, which facilitates people's media authoring, knowledge working, and collaboration; Mitchell and Watstein (2007) affirm how both e-learning and *Facebook* are settings that combined further students and scholars to work, collaborate, share, and plan academic duties, as well as recognize how intrinsic Facebook is among NetGens and Millennials scholars like a companion that facilitates its daily duties; Kaufman (2007) drafts a guide for research, development, and application of moving images and open content for education affairs to be deployed on *YouTube* by scholars, librarians, producers, curators, and distributors; Charlton, Marshall, and Devlin (2008) analyze how learning performance is affected through the use of *Friendster* as a SNS that furthers student interaction, community building, and social awareness affordances; Park, Heo, and Lee (2008) analyze whether *Cyworld* is a meaningful context to foment adult informal learning through communication and social tasks, in addition to study user perception of online activities in relation to learning as acquisition or as reflection; Grosseck and Holotescu (2008) explore pedagogical uses of *Twitter* as a tool in school, finding some benefits (e.g., dissemination of content, feedback to students, discussions), drawbacks (e.g., distraction, time-consuming, rudeness, addiction, etc.), and logistics (e.g., language and meaning definition, topic to support, learn self-discipline); List and Bryant (2009) report how *Ning* is used as a SNS to set up the private SN of an HS, where students post blog entries, videos, and discussions to forward the "learning of all" as a way to show students work and boost collaboration.

During the lustrum of 2010–2014, the use of OSNs raised all over the world, giving the chance to consolidate several SNSs (e.g., Facebook, Twitter, etc.), while others collapsed (e.g., Orkut in 2014), some emerged (e.g., Instagram and Pinterest in 2010), and others evolved (e.g., *Edmodo* in 2010). In this sense, Al-Ali (2014) reports the use of *Instagram* in classrooms as an authoring tool of visual content and support to socially connect learners through the "like" and "comment" functions to stimulate English learning as a second language. Concerning *Pinterest*, it enables user to tailor boards and assign pins by board of any subject with the aim for teachers to create boards that gather content of topics to discuss, while students review and save pins to their boards (Blair & Serafini, 2014). With regard to *Edmodo*, Nee (2014) finds classroom evidence, where control students were overcome by peers who learn direct, simple, and complex Biology concepts assisted by *Edmodo*, a public (i.e., Freemium account) and commercial (i.e., Premium account) SNS designed for scholars to lead *private* social interaction in classrooms that granted in 2010 parent accounts for communicating with teachers and students to have a surveillance of their children's assignments and advances.

As a consequence of the mentioned evolution, various market visions and scopes were progressively maturing for accessing, funding, and operating SNSs, including the development and spread of EN as an *education-based social networking* trend mainly through *general-purpose SNSs* (e.g., Facebook, Twitter, YouTube, etc.) and *education-oriented SNSs* (e.g., Edmodo), as the following account explains:

- *General-purpose SNSs*: There are *public*, *commercial*, and *open* SNSs that permit people and organizations to create accounts and manage their own OSNs for free, by a fee, and open source, respectively. As for public SNSs, they are classified into *popular* and *diverse* subcategories (i.e., according to the count of related works that use them in this review), including *several* (i.e., when various SNSs are considered in a given work). In this context, authors of these SNSs provide facilities, administrate the setting, and make business to obtain incomes (e.g., hosting, digital marketing, monetizing, custom domain selling, premium features, business accounts, etc.).
- *Education-oriented SNSs*: They deploy SNS functionalities (e.g., communication, networking, collaboration, content sharing, etc.) to socialize scholars' teaching and learning labor in classroom and distance settings as well as complement the scope of classic face to face and virtual educational resources and systems. Such sites can be *public* (i.e., no cost, e.g., Edmodo) and *private* (i.e., made and maintained by the own academic user; e.g., The Hive).

Once a broad scene has been outlined to depict the nature, evolution, and application of EN, an own definition is given for *educational networking* according to the incoming considerations. Firstly, this compound term appends the "*ing*" suffix to distinguish it from classic *educational networks* make-up of scholars or institutions as well as emphasizes the concept concerns to *computer-supported social networks*. However, it relies on the essence stated for the single term *network* instead of the one formerly defined for *networking*, where the aim corresponds to *connecting* with recognized persons in lieu of *captivating* unknown individuals. In this sense, several fears and claims against the use of OSNs for education affairs are confronted, such as privacy, safety, cyberbullying, harassment, and sexual predators.

Having said that, *educational networking* is a kind of e-learning setting backed by SNSs mainly oriented to facilitate online social interactions among members of an academic community, where teachers orchestrate content delivery, students perform collaborative practices to reach social learning, and parents supervise children advances through services that foster social media delivery, assignments, discussions, assessment, feedback, and security that close the system to strangers.

3.2.3 Resources

With the aim of gathering suitable resources to outline a representative landscape of the EN field, the following criteria should be met by the articles to be chosen: (1) the topic corresponds to EN or education based on SNs, OSNs, and SNSs, including *networking* versions, whose publication's title, keywords, abstract, and content concern one of these subjects, or specific instances of SNSs; (2) the reported work corresponds to studies, approaches, or logistic proposals; (3) articles published since 2015 up to date; and (4) relevant papers backed by journals indexed in the Science Citation Index Expanded™ and Social Sciences Citation Index® databases held by the Journal Citation Reports® of Clarivate Analytics©. As a result of the application of such guidelines, a sample of 92 key works is gathered with the purpose of tailoring an updated landscape of the EN arena, producing the annual frequencies edited in Table 3.1, where the quantity for 2019 corresponds up to its first half of year.

3.2.4 A Proposed Taxonomy for Educational Networking

Since EN is an incipient field, its research, development, and application already achieved can be seen from two perspectives, the SNS that is used as a platform to deploy an EN application and the kind of work published in the article. Both views compose the two categories of the proposed *taxonomy for educational networking* (TEN), which are edited at the left column of Table 3.2. Concerning the *SNSs* category, it embraces three hierarchical levels of subcategories and an additional tier to show their respective instances, which are stated in the four remaining columns of Table 3.2, while the second category is composed of just one level of subcategories that are identified in the second column.

3.2.5 A Pattern to Depict Educational Networking Applications

Inspired by the TEN, a *Pattern to Depict Educational Networking Applications* (PDENA) is proposed with the aim of providing a set of attributes–values that reveal representative traits of a specific EN work. Such characteristics are organized into 11 features, which are named, described, and instantiated by a specific value that clearly identifies a feature of the EN work as they are edited in Table 3.3.

Table 3.1 Annual frequencies of related works

Category	2015	2016	2017	2018	2019	Sum
Indexed journal	21	25	22	11	13	92

Table 3.2 Taxonomy for educational networking, which contains two categories

Category	Subcategory 1	Subcategory 2	Subcategory 3	Instances
1. Social network sites	General–purpose	Public	Popular	Facebook
				Twitter
				Wikis-Wikipedia
				YouTube
			Diverse	LinkedIn, WhatsApp, Google+, Google Documents, Blogger, Pinterest
			Several	Facebook, Twitter, etc.
		Commercial		Ning
		Open		Elgg, Story & Painting House
	Education oriented	Public		Edmodo
		Private		EO-P, SALEO, KINSHIP, The Hive, KakaoTalk
2. Kinds of works	Approach			
	Experiment			
	Logistic			
	Review			
	Study			
	Tool			

Table 3.3 Pattern to Depict Educational Networking Applications

Feature	Description	Instance
SNS	Official name	Facebook, Twitter, etc.
Work	Type of the research	Approach, study, etc.
School	Formal academic hierarchy level	Primary, K–12, HS, HE, etc.
Domain	Topic to be taught–learned	Business, language, etc.
Learning	Learning underlying paradigm	Social learning, reflection, etc.
Function	Functionality that SNS fosters	Collaboration, interaction, etc.
Effect	Expected bias that OSN exerts on users	Acceptance, performance, etc.
Setting	Academic setting enhanced by the SNS	Classroom, blended, etc.
Country	Nation where the research is made	USA, UK, Taiwan, etc.
Sample	Number of volunteers	Any quantity
Subjects	Academic-user role	Students, faculty, etc.

3.3 A Gleam of the Social Network Site Usage

Once the overview of the EN field has been stated in the Introduction, as well as the EN background traced, now a broad review of recent EN labor is outlined. Thus, according to the SNSs category of the TEN, including its three subcategories and instances, *ten cases* of SNSs are distinguished. This is why ten subsections are reserved to describe each case of SNSs, as follows: six subsections are devoted for *public general-purpose* SNSs (e.g., the first four present *popular* instances of public SNSs, and the remainder identify *diverse* and *several* SNSs), two additional subsections are oriented to present *commercial* and *open general-purpose* SNSs, and the residuary subsections offer a resume of *public* and *private education-oriented* SNSs. Together, the ten subsections shape a profile of the sample that embraces 92 articles according to the SNS they mainly use, whose order of presentation considers the year of publication and the subcategories of the TEN "kinds of works." With this in mind, a summary of EN labor is given to highlight its nature, traits, and achievements, including suitable citations, whereas the PDENA that corresponds to each of the 92 works is edited in Table 3.8 of Appendix, following similar sequence to facilitate the analysis of the features that characterize each related work.

3.3.1 Facebook-Based Education

Without doubt, Facebook is the most popular public general-purpose SNS, which gathers billions of users who promote their profile, post social media, interact with familiar people, establish new social affairs, and spread diverse messages. In this context, education has been benefited through diverse applications, studies, and outcomes that contribute to revolutionize e-learning traditional systems, as well as classic collaborative learning approaches. This is why in the present review, Facebook as a SNS for EN represents 41% of the sample! Hence, in this subsection a chronological summary of 38 works is outlined as follows.

In order to compare the impact that traditional design studio and a social media-based design exert on students' learning *habits*, *performance*, and *perceptions*, Güler (2015) develops a trial with HE students who use Facebook during the studio process in contrast with the control group that applies a traditional studio process. Both processes are evaluated from the students' and instructors' perspectives (e.g., opinions, experiences, panel evaluation of the submitted works). Findings show that ease of communication, unlimited exposure to peer advances, backtracking, and archiving capabilities attained through social media use offer advantages on student success, while, contrary to expectations, pedagogical lurking and peer plagiarism do not negatively affect learners' performance. On the other hand, Ainin, Naqshbandi, Moghavvemi, and Jaafar (2015) consider how *socialization* biases Facebook *use* and then how such a usage exerts on learners' academic *performance*. Through a survey, where 1165 HE Malaysian students participated, they found that the *sociality accepted* construct stimulates the use, and as a consequence, it sets up a positive relationship with the apprentice's academic performance. Hamid et al. (2015) explore the effects of group *discussions* through Facebook to complement learning activities at classroom in HE universities in Malaysia. They find that as a result of students' interactions with peers, collaboration is improved, lectures enhance learners' confidence, critical thinking is fostered, and students self-monitor their learning progress.

Additional research published in 2015 that concerns the *use of Facebook* for educational aims (e.g., communication, collaboration, and resource-material sharing) is related to diverse factors, such as students' engagement, communication, and instructor credibility (Imlawi, Gregg, & Karimi, 2015); academic performance (Michikyan, Subrahmanyam, & Dennis, 2015); learning from reading argumentative group discussions (Asterhan & Hever, 2015); argumentative knowledge construction (Greenhow, Gibbins, & Menzer, 2015; Puhl, Tsovaltzi, & Weinberger, 2015); students' perceptions of their faculties' credibility, professionalism, and approachability in the classroom (Sarapin & Morris, 2015); group awareness support (Tsovaltzi, Judele, Puhl, & Weinberger, 2015); pre-service teachers' pattern of usage to warn of potential threats (Sendurur, Sendurur, & Yilmaz, 2015); and teaching aid in academic courses (Miron & Ravid, 2015).

With regard to the usage of Facebook for educational affairs, a sample of related works published in 2016 is reported as follows:

- Akcaoglu and Bowman (2016) study the impact of instructor-guided SNS usage through a survey of opinions given by HE students who participate in a Facebook-based course, where most are interested in such a modality and recognize more value in course content, and feel their instructors are involved with the process and think effects on instructor credibility are clear.
- Gupta and Irwin (2016) are involved in the analysis of diverse intrusions (e.g., distractions, multitasking, and interruptions) that happen when students use Facebook to manipulate lecture interest-value and access material; hence, they study Facebook exposure, posts order, and time of interruption, giving, as a result, students claim that primary learning task is of low-interest goal-relevant, and therefore, they are more susceptible to distractions, besides intrusions negatively affecting their interest in lecture comprehension.
- Sharma, Joshi, and Sharma (2016) apply structural equation modeling to discover that resource sharing, perceived usefulness and enjoyment, collaboration, and social influence are the most relevant factors that motivate students to use Facebook in HE. Later, they feed such outcomes to a neural network-based approach to find out that relevant predictors of Facebook usage for academic aims correspond to collaboration, resource sharing, perceived enjoyment, social influence, and appreciated usefulness.
- Kaewkitipong, Chen, and Ractham (2016) examine how students are engaged in a customer relationship building by means of a walkthrough of an information system trip. The aim is for students to know and inquire about ICT services and products directly from vendors' Facebook, as a way to foster *active learning* and reinforce classroom content to improve learners' satisfaction.
- Finally, Facebook has been also related to learners' academic performance (Lambić, 2016), general use (e.g., social relations, work-related activities, and daily activities) (Manasijević, Živković, Arsić, & Milošević, 2016), generational behavior differences (Čičević, Samčović, & Nešić, 2016), and enhancing self-efficacy (Argyris & Xu, 2016).

Otherwise, with the aim of contributing to reduce *academic dishonesty* during exams made in classrooms, Topîrceanu (2017) proposes a student seating arrangement as a way to reduce copying practices. Thus, he builds an approach based on social network analysis (SNA) and genetic algorithms to examine the students' underlying friendship topology of their classmates that are recorded in Facebook. The analysis is fulfilled by means of empirical measurement of interaction between students during exam and a metric for the placement effectiveness. The case reports average improvements of ×2:8 in terms of breaking up real-world friendships and an ×3:3 reduction in terms of empirically measured student interaction. Rap and Blonder (2017) analyze the stage to which students use SNSs and specifically their attitudes to the facility of a medium for learning through Facebook. As a result, active Facebook groups consider their often use as a relevant learning experience.

As for Lantz-Andersson, Peterson, Hillman, Lundin, and Rensfeldt (2017), they study the types of online discussions that emerge in teacher Facebook group composed of 13,000 subjects, as well as their established repertoires. The outcomes reveal that requesting and giving tips, asking for and providing concrete instructional examples, and questioning and justifying the flipped classroom approach are the outstanding findings. Tsovaltzi, Judele, Puhl, and Weinberger (2017) wonder how argumentation scripts and individual preparation foster teachers' argumentation in Facebook and facilitate learning. The study asserts that individual preparation produces negative bias, while scripting argumentation exerts positive effects.

Other works printed in 2017 concern an approach to extract Facebook data to author a personalized sample data set for increasing intrinsic students' motivation (Marzo, Ardaiz, Sanz de Acedo, & Sanz de Acedo, 2017), a study of how students' personality types affect their academic performance and how the use of Facebook affects these relations (Naqshbandi, Ainin, Jaafar, & Shuib, 2017), and a prediction of the intensive Facebook usage among adolescents based on the uses and gratification theory and educational affordances (Dhir, Khalil, Lonka, & Tsai, 2017).

Lately, the development of recent research fulfilled during 2018 reports the following advances related to the usage of Facebook for educational aims:

- Datu, Yang, Valdez, and Chu (2018) examine how students' Facebook involvement is associated with academic engagement through the analysis of the relationship between usage intensity dimensions (e.g., persistence, boredom, overuse, self-expression) with engagement (e.g., agentic, behavioral, cognitive, emotional engagement), giving as a result three key correlations: self-boredom positively predicts behavioral engagement, overuse negatively predicts behavioral engagement, and expression positively predicts agentic engagement.
- Lin (2018) investigates anonymous versus identified Facebook online peer assessment through measuring anonymity's effects on the distributions of affective, cognitive, and metacognitive peer feedback. Later he examines the effects of anonymity on students' perceived learning, their beliefs of whether peer assessment was fair, and their attitudes toward the SNS. The evidence reveals that the identifiable assessment offers more affective feedback, whereas the

anonymous assessment provides more cognitive feedback, and involved subjects recognize a better learning and positive attitudes toward the system although they also perceived peer comments as being less fair than the identifiable mode.

- Keles (2018) takes into account the community of inquiry framework to investigate the reflections of the teaching process when an online learning community of prospective teachers discusses diverse topics, achieves social activities, and develops community service projects through Facebook. He finds that the SNS boosts teaching presence for instructors and students and enables them to share responsibility for the teaching endeavor as well as contributes to the social presence, social sensitivity, and awareness among prospective teachers.
- Atroszko et al. (2018) aim to provide empirical data on the validity and reliability of the Bergen Facebook Addiction Scale through a survey applied to 1157 HE students. Inclusive, they design a Potential Facebook Addiction Personality Risk Factors Model, whose traits and well-being indicators are also measured. The model shows that Facebook addiction is related to higher extraversion, loneliness, narcissism, lower self-efficacy, and social anxiety. Inclusive, addiction is also related to impoverished well-being (e.g., impaired general health, decreased sleep quality, and higher perceived stress).

Concerning fresh EN approaches based on Facebook as a SNS, some are oriented to foster communities of learning practice (Chen, Kuo, & Hsieh, 2019) and support social connectedness (Thai, Sheeran, & Cummings, 2019) and academic procrastination (Toker & Baturay, 2019), while additional labor is carried out by the following:

- Saini and Abraham (2019) assess the effectiveness of OSN instruction in contrast with conventional instruction on preservice teachers learning achievement and engagement for a pedagogy of mathematics courses. As a result, Facebook users reach a relevant learning achievement and engagement.
- Awidi, Paynter, and Vujosevic (2019) identify elements of students' learning experiences (e.g., support and motivation, participation and collaboration, etc.) to be considered for the improvement in learning design of HE courses and recognize traits of students' Facebook practice that bias their perception about overall course experience, where students feel a sense of community and are encouraged to learn, in addition to qualify the experience as satisfactory.
- Moorthy et al. (2019) include enjoyment and self-efficacy as part of the technology acceptance model to examine student's perceptions of using Facebook for learning. The results show that perceived usefulness, ease of use, enjoyment, and self-efficacy determinants have significant positive relationships with intention and behavior to use Facebook, except for perceived usefulness.
- Feng, Wong, Wong, and Hossain (2019) consider the distraction-conflict and cognitive load theories to propose a Theoretical Framework of the Factors of Facebook Usage on Academic Distraction that relates three daily frequencies of Facebook and Internet usage for entertainment and study, as well as the number of friends on Facebook that together the four measures influence on students' academic distraction, which in turn affect students' academic achievement.

Therefore, the study reveals that students with a high-daily frequency utilization of Facebook for entertainment tended to be more distracted in academic endeavors and then reach lower grade point average.

3.3.2 Twitter-Based Education

Twitter is the most used microblogging Internet platform that facilitates the broadcast of text strings and social media objects, whose communities of followers are attentive to any tweet that a given user triggers through his/her account. As a SNS, Twitter has also been used to spread messages among the academic members through mentions (e.g., @user), responses to (e.g., RT @user), or hashtags (e.g., #topic). This is why in the current review, Twitter reaches the second spot with nearly 10% of the sample as the next account of nine related works reveals.

- Aramo-Immonen, Jussila, and Huhtamäki (2015) evaluate the utilization of Twitter during a social media conference to boost co-learning virtual environments through discussions among professional attenders.
- Menkhoff, Chay, Bengtsson, Woodard, and Gan (2015) foster pedagogical tweeting to find what students are thinking during learning practices while they are in and out the classroom, all of this to encourage unwilling students to participate in discussions, question raising, and sharing advice on assignments.
- Rehm and Notten (2016) take into account the social capital theory to study underlying communication processes and products among teachers as part of their professional growth through boosting informal learning. Hence, they analyze a hashtag conversation among teachers to measure the relevance of the structural dimension of social capital. The outcome shows that participating in a hashtag chat on Twitter contributes to teachers' formation of structural social capital, facilitates the emergence of new ties and individuals' personal networks, and supports that teachers attain central spots within the conversation.

Afterward, Tang and Hew (2017) examine 51 reports of Twitter usage in teaching and learning from 2006 to 2015 to question whether this practice benefits learners. They edit a briefing composed of the profile of studies, ways of using Twitter in education, impacts on interactions, and influence on students' learning. The resume asserts that Twitter is a push technology that enables professors to send course information, give homework assignments, and inform test deadlines, concluding the SNS is suitable for peer interaction and assessment purposes.

Lackovic, Kerry, Lowe, and Lowe (2017) aim to know students' concerns about Twitter utilization during the regular lectures and seminars, such as frequency, perceptions, discouraging issues, and facilitation agents. In this setting, HE students also make questions and comments, in addition to share information and links under a hashtag. The contrasting results warn that though students consider Twitter as a SNS where their knowledge and power are subordinated to leading Twitter users of key fields, they also perceive Twitter as a career vehicle and request tutors' help to create a Twitter learning habit. To reach this goal, a pedagogical orientation toward a critical analysis of and acting upon social media information is demanded.

After, as result of a research about Twitter usage by educators that collaborate within professional learning networks, Nagle (2018) claims: Such a practice is rife with misogyny and racial violence. So he discusses ethical implications for teacher educators who use Twitter as a pedagogical tool and suggests strategies to develop critical social media literacy practices, as, for instance, fostering social media as a tool and participation examination.

In another vein, under the hypothesis "online communication attitudinal background impacts aims for communication, then it in turn influences student involvement on Twitter," Denker, Manning, Heuett, and Summers (2018) tailor a model based on the Media Choice and Polymediation Theories to identify ways students prefer to interact with teachers in and out of the classroom via Twitter. During lectures assistant professor utilized Twitter to engage with students who were motivated to voice their expressions by tagging the course's Twitter handle. Evidence shows that online communication attitudes shape students motives to communicate with their teachers using Twitter, and therefore, such reasons bias student participation.

Recently, a study of how Twitter boosts teachers' identity mined 33,184 K–12 American teachers' accounts to identify the degree to which accounts appeared to be used for personal and professional aims based on context collapse and boundary work concepts. The approach found that profiles infrequently disclosed personal information, while hashtags frequently included in teachers' tweets correspond to professional matters (Carpenter, Kimmons, Short, Clements, & Staples, 2019).

Lastly, with the aim of providing a comprehension of self-generating online professional learning, Prestridge (2019) investigates reasons to explain why teachers use Twitter to learn and seek online sources and common activities of networking on an individual basis. As a result, a Typology of Reasoning based along Self and Interactivity Continuums is conceived, where four models of Teacher Engagement Online emerge: (1) info-networker, actively networks to find interesting and valuable people to follow; (2) self-seeking contributor, posts their knowledge, ideas, or curriculum materials to their professional learning SN for self-determining reasons; (3) vocationalist, promotes the good of the profession as implicit in their own professional learning; and (4) info-consumer, seeks to find and take away new knowledge and resources from social mediated spaces.

3.3.3 Wikis and Wikipedia-Based Education

Wiki is a kind of content that is the result of the contribution given by a community of users that are interested in publishing social media of a specific topic. The content part organizes the text, multimedia objects, documents, hyperlinks, and diverse digital items that users provide and others modify for which revision history is tracked, while the comment part is where interested users may leave short messages for their collaborators. Similarly, Wikipedia is a kind of autonomous cultural informal system, where the users' contributions progressively generate the largest online encyclopedia in the world. Both, wikis and Wikipedia, offer a useful and easy tool to authoring educational collective knowledge to be distributed in academic settings, as the following five related works, which compose the 6.5% of the sample, report.

Firstly, Brailas, Koskinas, Dafermos, and Alexias (2015) consider the grounded theory to design a model for utilizing Wikipedia's community in education, which embraces five constructs: community resistance, organizing intervention, acculturation stress, community benefit, and educational profit with the aim of interpreting the acculturation of the educational group into the culture of a hosting virtual community. During academic interventions, where students write and edit Wikipedia's articles, the group's acculturation is manifested as the culture of a hosting virtual community by means of collaborative actions, disturbances, and issues resulting into the creation of a collective zone of proximal development.

Afterward, Chu, Capio, van Aalst, and Cheng (2017) evaluate how secondary school teenagers use a wiki tool, PBworks, to boost collaborative group writing through sharing information and files, exchanging comments and ideas, and thereby co-constructing the writing outcome. As a consequence, higher writing achievements are reached by those who used the tool, even though pedagogical strategies need to be known and applied by teachers to lead their students to harness the tool. What is more, Heimbuch and Bodemer (2017) study the potentials of supporting learning-related processes at the Wiki talk page level by analyzing the bias of visual controversy awareness information on content-related discussion threads, which in turn benefit learning when contradictory evidence leads Wiki discussions. Such an issue is tested as an implicit guidance that addresses students learning processes and fosters internalized knowledge representations. Results show Wiki talk page benefits users with additional structuring aids and to become aware of occurring controversies, as it encourages students to develop meaningful discussion threads and contributes to improve their learning.

On the other hand, Chu et al. (2017) examine the effectiveness of using wikis based on the theoretical framework of Wiki-supported project-based learning to compare the perceptions and actions among HE students that use Wikis in their assignments. Although students mostly hold positive attitudes toward the use of wikis, key differences appear for motivation and knowledge management aspects, and the level of participation and core actions on the wikis vary among students, likely due to the variations in prior learning experiences, technical backgrounds, and the relationship between learning goals and collaborative learning.

Otherwise, the Framework for Analyzing Student Comments in Wikis of Group Writing to Inform Learning Assessment is designed by Hu, Cheong, and Chu (2018) to measure three coding categories (e.g., student interaction, meaning construction, and thinking development) in the writing process edited on Wiki page comments, where relations among coding categories are discovered and interpreted by means of statistics and data mining. Lastly, Alghasab, Hardman, and Handley (2019) apply the discourse analysis framework to examine teacher and student online interaction (e.g., discussion posts and text edits) during wiki-mediated collaborative English writing. As a finding, two roles adopted by teachers are revealed: (1) directive, students mainly interact with teachers rather than with peers and write by themselves individually, and (2) dialogic, students interact and collaborate to produce texts.

3.3.4   YouTube-Based Education

YouTube is the most popular digital media platform, where worldwide users upload and access videos of a broad diversity of topics, which are daily reproduced during one billion of hours. YouTube as a public SNS is a channel for scholars interested in editing and distributing educational content to be accessed by their learners, as the next account of five works explains, which represent the 5% of the sample.

1. Orús et al. (2016) report a project based on learner-generated videos uploaded to YouTube, where students assess learning outcomes and satisfaction. The conclusions reveal that although participation did not directly increase subjective learning or satisfaction, it had a direct bias on academic performance and the perceived acquisition of cross-curricular competencies.
2. Hong et al. (2016) apply cognitive-affective theory of learning with media to enable students to control the pace of learning through YouTube by repeating playback, fast forwarding, or rewinding functions. Hence, they explore the relationship between cognitive and affective factors in learning with SNSs. Outcomes reveal Internet cognitive failure is negatively correlated to self-efficacy of learning and interest learning with social media, but self-efficacy and learning interest are positively correlated to learning satisfaction with social media.
3. Choi and Behm-Morawitz (2017) apply the uses and gratifications theory and social cognitive theory to propose a Predicted Model of YouTube Video Production to associate video exposure to attractiveness and thereby foster motivation and efficacy. The aim is to examine YouTube beauty content gurus as digital literacy educators and how they can represent Science, Technology, Engineering, and Mathematics (STEM) beauty videos to share information and boost skills related to video production. The study reveals beauty gurus used adding text and social media links and included positive educational messages in their videos. An online trial showed that beauty guru's videos motivate viewers to produce a video, which is mediated through source attractiveness.
4. Shoufan (2019) considers the cognitive theory of multimedia learning to investigate how far educational videos on YouTube support cognitive features (e.g., signaling, personalization, pre-training, etc.) and how far these traits support students' learning to understand the semantics of "likes" and "dislikes" of YouTube's educational videos. Then a sample of 105 videos is examined with respect to their cognitive attributes, giving as a result that voice, coherence, spatial contiguity, and redundancy are the most significant characteristics.

5. Saurabh and Gautam (2019) design an approach that relies on YouTube analytics data, time-series analysis, and entropy-based decision tree classifier to reveal the trends, seasonality, and temporal patterns for educational videos on YouTube. The idea is to reveal aspects such as the relationship between video uploading activity, the features that are most important for the popularity of videos, and the channel's age and its popularity. The analysis reports that video rank and number of views follow the Zipf law distribution for educational videos, as well as holding a strong correlation between the geographical location of viewers and the location of industry the channel caters to.

3.3.5   Education Based on Diverse SNS

Several SNSs and tools have been also ideal platforms for supporting educational affairs of scholars, learners, and communities. In the sample of related works, 12% report the labor developed through the use of six instances such as LinkedIn, a SNS oriented to business and professional aims; Google+, a SNS that is actually operating for corporative users; Google Docs, which offers word processing and spreadsheet whose files are accessed by several users; Blog, a binnacle whose posts evolve as personal diary that is object of readers' reply; WhatsApp, a multimedia service that facilitates asynchronous and synchronous communication; and Pinterest, a tool to author and manage thematic digital boards. A brief of their utilization is stated as follows:

Concerning *LinkedIn*, Benson and Filippaios (2015) identify the level of awareness and degree of application of professional usage of Facebook and LinkedIn among 645 HE students and graduates in the UK. They find that work experience and age play a key role in their use of OSN for professional purposes, knowledge, and career management as younger students are more SN savvy to identify business opportunities, while older graduates are less confident. In addition Lim and Richardson (2016) analyze diverse data (e.g., demographic items, frequency and intensity of SNS usage, and perceptions for EN) to interpret the students' SNS experiences (e.g., mainly in Facebook and LinkedIn), measure the impact on their social presence in online courses, and identify perceptions of using SNSs for education. Outcomes reveal that although students confess daily, frequent, and active usage of SNSs and manifest positive perceptions of using SNSs for education, there is no key correlation for students' perceived social presence in online learning.

In regard to *Google+*, Liao, Huang, Chen, and Huang (2015) follow the collective efficacy theory to adapt the technology acceptance model with the aim of examining the influential factors in students' use of such a SNS to learn in ubiquitous-learning settings as well as to evaluate their learning attitudes and usage effects. The result shows that collective efficacy and personal innovativeness affect learner attitudes through perceived playfulness, usefulness, and ease of use, in addition to satisfaction, self-perceived usage effects, and continued use intention. Moreover, Schwarz and Caduri (2016) identify pedagogies used by teachers in secondary schools while interacting with their students through OSNs, as well as the kind of community that emerges from the teaching OSN practices. Thus, they undertake narrative inquiry with several teachers and analyze logs of interactions with their students tracked in Facebook and Google+, giving as a result, respectively, social learning, autonomy, and active engagement among students are the most common teaching guides and the appearance of a learning community of inquiry.

Concerning *Google Docs*, Ishtaiwa and Aburezeq (2015) examine its effect on improving four kinds of graduate student collaboration (e.g., student with student, instructor, content, and interface) and the factors that constrain student collaboration. The results provide a finding for each kind of collaboration as respectively follows: benefit from the experience of other students, improvement of the interaction with instructor, collaboration enhancement, and the ease and simplicity of using the tool. Lack of teamwork and technological skills and preference for other messaging options limit the use of Google Docs for collaboration duties. What is more, Liu and Lan (2016) study the differences in English learners' motivation, perceptions, and vocabulary gain on using the Google Docs in collaborative and individual modes. The outcomes show *collaborators* achieved a better vocabulary gain than the *individuals*, likely because the former was more encouraged to learn and considered more attractive the learning experience than the latter.

As for *blogging*, Lee and Bonk (2016) estimate measures of learners' perceived emotional closeness with others to examine the SN of the students' relationships and their online interactions (i.e., the quantity of replies that a learner posts to and receives from others' postings) in a course using weblogs for writing and sharing reflective journals. As the course progresses, the SN patterns and values assessed by peer relationships change at the end of the period in contrast to that at the beginning due to the impact of blogging. Kuznetcova, Glassman, and Lin (2019) investigate how online tools, as Second Life, in conjunction with blogging and lecturing facilitate the formation of a distributed learning setting. Student interactions recorded on the Blogger platform are analyzed by SNA. The results show the network was more connected and distributed than the one that lacks the online tool.

In relation to *WhatsApp*, Asterhan and Bouton (2017) analyze teenagers' knowledge sharing practices (i.e., upload and download of content between peers) in SNSs, mostly Facebook and WhatsApp, through two surveys composed of six types of questions (e.g., concerning: Whether students share? Where, when, and why do students share? What do they share with each other in SN groups? Who are the sharers?). The findings reveal that knowledge shares is a common and widespread custom that entails some type of knowledge, including copying solved homework tasks and other assignments, where future reciprocation is an argument for its practice.

With respect to the *Pinterest* tool, Song, Williams, Pruitt, and Schallert (2017) examine how it supports graduate students to explore language use in everyday life for a class assignment by means of remixing, contextualizing, and interconnecting multimodal texts while creating affinity spaces with class members and the internet public. As a result, the social nature of the participation in the Pinterest board emerges, and some kinds of language use are stimulated on the Pinterest board too, as well as the Pinterest activity as digital curation produces a participatory culture that fosters students' collaboration and informal learning.

As for *online forums*, Chao, Lai, Liu, and Lin (2018) build a proactive discussion forum, which contains question categories and question menus, to proactively request help for problem solving through the delivery of messages to student's friends as well as explore how such a proactive mode influences students' participation and their perceptions. As a consequence, experimental group shows more frequent participation in problem solving and more positive attitudes toward help seeking.

3.3.6 Studies of How Several SNSs Support Education

As a preliminary finding of the present review, the most common practice is the study, where researchers focus on specific issues concerning the use of SNSs that a numerous community of users respond to a given survey. Therefore, in this review, 11 works, 12% of the sample, report the outcomes of the following inquests.

During 2015 a couple of works are worthy of being mentioned; the first is a model composed of artificial immune systems, ontologies, and a recommender system, which is designed by Stantchev, Prieto-González, and Tamm (2015) to infer knowledge and interest from individuals through the exploration of aggregated data tracked from their interaction in crowdsourcing and collaborative learning environments, such as SNSs. The aim is to assess the user's knowledge level in diverse topics and suggest specific education guide concerning his/her studies to get a given job. Though the incipient results, authors expect to apply such a model to enhance the recommendations made to users of those learning settings.

Besides, Alwagait, Shahzad, and Alim (2015) analyze the impact of SNSs on HE students' academic performance in Saudi Arabia, as well as which SNSs are the most popular in the community. The outcome reveals there is no linear relationship between grade point average score and social media usage in a week, but time management negatively biases learners' studies. Inclusive Facebook, Twitter, Instagram, and LinkedIn are recognized as the most popular SNSs.

Furthermore, in 2016, several studies were published to identify how SNSs influence academic users' educational practices as the following account reveals:

- Sobaih, Moustafa, Ghandforoush, and Khan (2016) report the use of SNSs for sharing social media for academic aims in Egypt and as a teaching and learning tool, where Facebook and WhatsApp are the most commonly used platforms by faculty, who make a minimal use of SNSs due to diverse perils, barriers, and concerns such as privacy, entertainment, socialization, loss of control, monitoring of student activities, ethical issues, and awareness.
- Dermentzi, Papagiannidis, Toro, and Yannopoulou (2016) consider the decomposed theory of planned behavior and the uses and gratifications theory to analyze the perceptions and willingness that academic staff confess for using SNSs.

Results reveal that academics consider SNSs suitable for networking and promoting a professional profile, while online technologies are useful for making new acquaintances in their field and seeking academic information.

- Manca and Ranieri (2016a, 2016b) gather empirical evidence concerning HE scholars' personal, teaching, and professional use of social media fostered by diverse SNSs, as well as the relationship between kinds of usage, frequency, and learning settings. The survey unveils Facebook as the most daily SNS used for personal and professional aims, whereas blogs and wikis are daily preferred for teaching duty. Moreover, frequency of personal use is associated with the one of professional usage more than with the frequency of teaching use, while the experience with e-learning is positively related with social media utilization.

Lately, three representative studies focus on performance, SNS adoption, and study visibility as their respective summary reports then:

- Liu, Kirschner, and Karpinski (2017) offer evidence about the negative relationship between SNS use and academic performance in a meta-analysis that compiles 28 studies that gather the opinion of $N = 101,441$ HS and HE students, whose results reveal a high negative relationship between SNS use and academic performance, which is stronger in females and college students.
- Akçayır (2017) promotes the SNS adoption for teaching through a study of how 658 faculty members use SNSs for educational aims, where opinions, motivations, and constraints to use them in courses are considered, given as a result that nearly 50% report the use of SNSs for education because they offer a means for fast and effective communication, while privacy represents the main issue.
- Waycott, Thompson, Sheard, and Clerehan (2017) take into account the community of practice framework and the notion of a Virtual Panopticon to examine how students strive to make assessable work visible online to others. Results unveil that students feel they are part of a cohesive learning community, although they wonder how visible their profile and own work should be.

A short time ago, Cao, Masood, Luqman, and Ali (2018) apply the stress-strain-outcome model and the dual-system theory to examine how excessive use of SNSs induces cognitive emotional preoccupation as a stressor and turns into strain in the way of techno-exhaustion, life and privacy invasion, and further influencing the academic performance of students. A result of the survey applied to 505 HE students confirmed such hypothesis, where negative consequences diminish the academic performance of addictive SNS users. Molinillo, Anaya-Sánchez, Aguilar-Illescas, and Vallespín-Arán (2018) study the relationships between perceived benefits and attitude toward collaborative learning through a case study, where HE groups of students conceive solutions, interact, argue, and negotiate to solve business problems, whose proposals have been developed through the use of Moodle and some collaboration tools (e.g., wikis, forums, WordPress, Blogger, Facebook, Twitter). As a result, perceived benefits, flow experience, and active learning are found as relevant factors that affect attitude toward collaborative learning.

3.3.7 Education Based on Commercial SNSs

Concerning the use of commercial general-purpose SNSs that are being used for educational affairs, *Ning* represents an option that HE institutions such as University of California at Riverside and Kent State University have exploited. In this SNS users play the online tutor role, schools create a web site, and training courses are offered. All of this after a trial period, users choose a basic, performance, or ultimate account to preserve their own OSN on the Ning platform.

As an example of the Ning use for EN, the current review describes just one work, 1% of the sample, published by Rutten, Ros, Kuijpers, and Kreijns (2016), who design an affordance-based framework to identify the properties of the relationship between the usability features of Facebook, Ning, and two Virtual Learning Environments and the traits of how they stimulate the teen students' practice of online career skills (e.g., introducing oneself, sharing experiences, interacting, and connecting). The outcomes assert that the two SNSs apparently offer a stronger setting for practicing online career skills than those given by the couple of environments, although students do not show more online career behavior in these sites.

3.3.8 Education Based on Open SNSs

With regard to the use of open general-purpose SNSs, which are being considered for educational services, *Elgg* is the most demanded. It is an open source social networking engine in PHP and MySQL that is available for users interested in downloading the software to deploy their own SNS using their own computer and communication resources, or hosting them. A growing community of developers and users offers support for novices who aim to develop their OSN. As an example of such experiences in EN, in this section three works are summarized, including one related to a tool to represent the 4% of the sample:

- Ergün and Usluel (2016) apply SNA to assess the communication structure concerning time and the involvement of HE students and instructors in an OSN supported by Elgg, whose forum provides the raw data to be analyzed to estimate density and centrality measures. Results show that the lowest density occurs during the first week, while the highest density happens when the instructor participates. As for centrality, students found in the center and those on the surroundings differ on the basis of time and the instructor's involvement.
- de-Marcos, Garcia-Lopez, and Garcia-Cabot (2016) evaluate four learning settings for stimulating students' learning performance during a benchmark study, where students learn a computer course through educational games (i.e., delivery of a narrative storyline), gamification (i.e., focuses on competition through trophies, badges, challenges, leaderboard), usage of OSNs based on *Elgg* as SNS (i.e., promotes cooperation by means of blogging, questions, answers, liking, built-in twitter, dashboard), and social gamification (i.e., combines both

OSN and gamification). Results reveal social gamification reached better results in terms of immediacy and for diverse assessments.

- de-Marcos, García-López, García-Cabot, Medina-Merodio et al. (2016) apply SNA to predict academic performance of undergraduate students who take a gamified course delivered through Elgg, analyzing the structure of the network graph that resembles a small world and the influence that student's position has on learning achievement, where individual centrality measures are estimated and used as predictors of students' results.

Ultimately, as an open tool available for supporting OSNs, the iPad social networking application *Story & Painting House* has been also used by Liu, Chen, and Tai (2017) who study how elementary teenagers collaborated to author English multimedia stories in a networked creation community. They attempt to interpret the impact of the SN activities on students' engagement during the networked creation community, particularly engagement and collaboration modes. The achieved results are examined through SNA, which uncovers that students' knowledge level exerts on their structural positions in the SN, whereas students of lower proficiency are more active in the SN while their knowledge level is not an influence of their popularity. Besides, students' structural positions in the SN significantly exert their flow perception and motivation in the networked creation activity; while students that appear at the central position in the out-degree centrality perceive higher level perceptual engagement, those who stand at the central positions in the in-degree centrality perceive lower level perceptual engagement.

3.3.9   Education Based on Public Education-Oriented SNSs

As a novel trend, education-oriented SNS emerges to provide a broad set of functionalities that are suitable for educational affairs in contrast to general-purpose SNSs which were not thought for such a specific domain. Up to date, *Edmodo* represents a public education-oriented SNS that offers specialized support for teachers, students, and parents, who respectively share class content, collaborate to develop assignments, and supplement home learning activity. As a sample, two related works, equivalent to the 2% of the sample, which use Edmodo are introduced:

1. Won, Evans, Carey, and Schnittka (2015) analyze how middle school youths leverage a SNS as a part of the iterative design process for STEM problem solving, as well as whether Edmodo fosters students' collaborative design and boosts facilitator interaction to assist in progression of the design process. Thus, they deploy Studio STEM content to expose a problem involving animal species that students should collaboratively solve through a design process. As a result, 50% of youths report they are willing to use EN and collaborate through the discussion of ideas, comparing methods, and posting questions. In addition, the posts coded by facilitators for casual facilitation, encouragement, and assessment motivate learners to make questions and discuss ideas.

2. Ellahi (2018) wonders if the use of a SNS enhances learners' interest and increases their satisfaction and perceived learning performance. Hence, he develops a teaching case composed of three courses delivered by *Edmodo*, which boost the mediating role of students' interest and moderating role of instructor's support. Outcomes show that SNS usage is a key predictor of learning satisfaction and learning performance, while student's interest exerts a positive influence, even though the instructor's support effect is inconsistent, but not useless.

3.3.10 Education Based on Private Education-Oriented SNSs

Diverse proper education-oriented SNSs and tools have been built to facilitate the deployment of OSN to spread educational services as an alternative for classic e-learning systems. Such investment aims at meeting specific educational needs as the following five related works, which correspond to 5% of the sample, report:

1. Lee and Kim (2016) analyze how the OSN support given by *KakaoTalk* improves students' skills for writing English sentences during in and out-of-class tasks in a multimodal learning setting. Such a private SNS leads learners to edit their sentences based on suitable content and a to-verb grammar-driven that fosters students' recalling, writing, motivation, and collaboration.
2. Scott, Sorokti, and Merrell (2016) deploy *The Hive*, as a SNS of the Enterprise 2.0 system that integrates into a platform some tools and functions similar to the public versions of Web 2.0. The aim is to provide direct communication, private learning, collaboration spaces, social learning, knowledge spaces, and public program information sharing spaces. Such a system facilitates students' learning outside classroom, supports dialogue between students, and reinforces the program learning community. Outcomes show that most of students participated in learning activities within these public and non-mandatory spaces beyond the classroom, as well as revealing the practice of self-regulated learning and high levels of cognitive and learning presence in their posts.
3. Mahnane (2017) designs the *New Educational Social Network* to recommend learning activities to students. Such a SNS assembles clusters of students according to their learning styles, habits, and interests shown during frequent episodes by means of the Apriori algorithm, in addition to dynamically grouping users based on collaborative filtering. Author asserts the recommendations reach better precisions than other data mining algorithms used in classic SNS.
4. Karataev and Zadorozhny (2017) tailor a social learning framework to further extensible personalization, course adaptation, and student participation in a learning process according to student needs related to collective learning experiences. Such a framework follows concepts of crowdsourcing, OSN technology, and adaptive systems to boost users for authoring content as mini-lessons, learning such lessons according to adaptive learning pathways, and interacting with peers. Authors deploy the framework as a self-adaptive learning education-oriented SNS and evaluate its advice through classroom studies. Outcomes aim

at using collective learning experiences in adaptive social learning and reveal that students form clusters and have their own best learning pathway.

5. Hatzipanagos and John (2017) brief students' view of the King's Social Harmonisation Project SNS that boosts sense of community and serves as a space that models digital professionalism. The survey reveals that the private SNS is less favored than Facebook and boosts interactions between peers, as well as the claims of how communications are scrutinized by staff and the lack of system notifications of activity through instant messaging, including its diffusion.

3.4 Analysis of the Educational Networking Field

Once the profile of the 92 works that compose the sample has been stated, this review also considers both quantitative and qualitative analysis to highlight the relevance that corresponds to the categories of the TEN, as well as the features that depict the PDENA. Thus, in this section, a series of annual frequencies is shown and accompanied by their respective interpretation. In addition, a discussion of the EN arena is given according to its strengths, weaknesses, opportunities, and threats; posteriorly the response to the problem and research questions are revealed.

3.4.1 Statistical Outcomes

The statistics represent annual frequencies of the values that instantiate the traits (e.g., category, feature, instance value) that depict the 92 works gathered to tailor the EN landscape. Such estimations are shown through a series of tables, whose content is ordered according to the sequence of apparition observed by the items that compose the TEN and PDENA, in the way they are edited and stated as follows.

As a first instance, the impact of the diverse *kinds of SNSs* defined in the TEN is manifested by means of the account estimated for the ten cases of SNSs described in Sect. 3.3, whose annual frequency is edited at the upper level of Table 3.4 in the same order of the aforementioned exposition. As can be seen, *Facebook* represents 41% of the SNSs used to deploy EN, while an incipient exploitation of *education-oriented* only weights 8%. This means that nearly 92% of the EN field relies on *general-purpose* SNS according to the following distribution: (1) *public*: *popular* 63%, *diverse* 12%, and *several* 12%; (2) *commercial* 1%; and (3) *open* 4%.

Concerning the *kinds of works* category of the TEN, the annual frequencies are provided at the bottom of Table 3.4, where the distinct categories appear in ascending sequence. As a result, the *tools* category is the topic less reported with just 2% of the sample, while the *study* is the most published category as it represents 76%, in contrast with *experiment*, *review*, *logistic*, and *approach* that, respectively, contribute with 3%, 3%, 7%, and 9%.

Table 3.4 Annual frequencies of 92 related works for the SNSs and kinds of works categories that compose the taxonomy for educational networking

Category	2015	2016	2017	2018	2019	Sum
Social network sites	**21**	**25**	**22**	**11**	**13**	**92**
Public SNSs: Facebook	12	8	7	4	7	38
Public SNSs: Twitter	2	1	2	2	2	9
Public SNSs: Wikis-Wikipedia	1	0	3	1	1	6
Public SNSs: YouTube	0	2	1	0	2	5
Public SNSs: diverse	3	4	2	1	1	11
Public SNSs: several	2	4	3	2	0	11
Commercial SNSs	0	1	0	0	0	1
Open SNSs	0	3	1	0	0	4
Educational SNSs: public	1	0	0	1	0	2
Educational SNSs: proprietary	0	2	3	0	0	5
Kinds of works	**21**	**25**	**22**	**11**	**13**	**92**
Tool	0	0	2	0	0	2
Experiment	0	2	1	0	0	3
Review	0	0	2	1	0	3
Logistic	1	0	2	3	0	6
Approach	1	2	4	0	1	8
Study	19	21	11	7	12	70

Regarding the features that compose the PDENA, the series of Tables 3.5, 3.6, and 3.7 provide in ascendant order the annual frequencies for their instances that reach the higher account. Inclusive with the aim of considering cases lesser relevant, "groups of instances" are defined to cluster homogenous instances whose frequency is small. Such groups are defined between quotes in the body of the table and at the bottom as a note where the specific instances that compose a given *group* are identified. Having said that, the succession of Tables 3.5, 3.6, and 3.7 is introduced, interpreted, and edited according to the order in which they appear in the PDENA as follows:

- *School*: *HE* is the most significant instance with 68%, followed by "several" (e.g., *HS*, *K–12*, *Middle*, etc.) with 20% and letting 12% for *Graduate*.
- *Domain*: *Education*, *idioms*, and *computer sciences* are the most considered specific instances of the sample with 7%, 10%, and 13%, respectively, whereas with 33%, the predominant group is *diverse* that considers any educational topic not included in the other specific instances or "group instances."
- *Learning*: Many instances compose the sample, where the majority are gathered in the group "several" (e.g., *learning styles*, *sharing content*, etc.) to reach 45%, while the most applied instances *OSN-based learning & teaching*, *social learning*, and *social constructivism* respectively weight 13%, 15%, and 16%.
- *Function*: In addition to "several" (e.g., *problem-solving*, *social media*, etc.), *use*, *collaboration*, and *communication* instances get the 37%, 20%, 16%, and 14% of the sample, respectively.

Table 3.5 Annual frequencies of four features that compose the Pattern to Depict Educational Networking Applications, where each "group of instances" appears between quotes and the instances it gathers are defined at the bottom

Category	2015	2016	2017	2018	2019	Sum
School	**21**	**25**	**22**	**11**	**13**	**92**
Graduate	5	3	1	1	1	11
"Several"	2	5	8	1	2	18
HE	14	17	13	9	10	63
Domain	**21**	**25**	**22**	**11**	**13**	**92**
"Health & mind"	0	2	1	0	2	5
Education	1	2	1	2	0	6
"Human expression"	5	1	1	1	1	9
Idiom	0	2	5	1	1	9
"Human affairs"	4	2	1	3	0	10
"Science & technology"	3	3	4	0	1	11
Computer sciences	3	4	2	0	3	12
Diverse	5	9	7	4	5	30
Learning	**21**	**25**	**22**	**11**	**13**	**92**
Active learning	0	2	0	1	0	3
Argumentative learning	2	0	1	0	0	3
Community of inquiry	0	2	1	1	0	4
OSN-based learning & teaching	0	2	7	2	1	12
Social learning	5	5	0	1	3	14
Social constructivism	7	4	3	0	1	15
"Several"	7	10	10	6	8	41
Function	**21**	**25**	**22**	**11**	**13**	**92**
Interaction	1	2	0	1	1	5
Community of learning practice	0	2	3	1	1	7
Communication	3	2	4	2	2	13
Collaboration	8	1	2	1	3	15
Use	4	8	3	2	1	18
"Several"	5	10	10	4	5	34

School: "Several": diverse, elementary, HS, K–12, middle, secondary, tertiary, etc.
Domain: "Health & mind": learning, physiotherapy, psychology, skills; "Human expression": architecture, culture, communication, music, social media; "Idiom": English, language; "Human affairs": business, management, marketing, social sciences, tourism; "Science & technology": chemistry, engineering, math, paleontology, sciences, STEM, technology, TIC, etc.
Learning: "Several": learning styles, sharing content, social cognitive theory, teaching–learning strategies, uses and gratifications theory, etc.
Function: "Several": instructor support, problem solving, social gamification, social media, etc.

- *Effect*: Progressively, *knowledge construction, acceptance, discussion, engagement, performance,* and "Several" (e.g., *attitudes, social presence,* etc.) illustrate the 9%, 9%, 9%, 11%, 15%, and 42% of the consequences produced on users of OSNs, where *perceptions* own the remaining 5% of the sample.

Table 3.6 Annual frequencies of three features that compose the Pattern to Depict Educational Networking Applications, where each "group of instances" appears between quotes and the instances it gathers are defined at the bottom

Category	2015	2016	2017	2018	2019	Sum
Effect	**21**	**25**	**22**	**11**	**13**	**92**
Perceptions	2	2	0	0	1	5
Knowledge construction	3	2	2	0	1	8
Acceptance	1	2	3	0	2	8
Discussion	3	1	2	2	0	8
Engagement	2	1	2	2	3	10
Performance	5	4	2	2	1	14
"Several"	5	13	11	5	5	39
Setting	**21**	**25**	**22**	**11**	**13**	**92**
Collaborative learning	3	0	1	0	0	4
e-Learning	0	2	2	0	1	5
Blended learning	3	7	0	0	0	10
"Several"	3	4	2	1	2	12
OSN	4	1	7	5	1	18
Classroom	8	11	10	5	9	43
Country	**21**	**25**	**22**	**11**	**13**	**92**
Australia	0	1	1	0	2	4
International (various countries)	0	1	2	0	1	4
Hong Kong	0	0	2	1	1	4
Malaysia	2	0	1	0	1	4
Israel	2	1	2	0	0	5
Spain	0	3	1	1	0	5
Germany	3	1	2	0	0	6
Turkey	2	1	1	1	1	6
Taiwan	1	2	1	2	1	7
USA	5	4	3	1	2	15
"Several"	6	11	6	5	4	32

Effect: "Several": attitudes, distraction, self-efficacy, skills development, social presence, teacher learning, writing, etc.
Setting: "Several": conference, forum, informal learning, multimedia learning, etc.
Country: "Several": Algeria, Egypt, Greece, India, Italy, Korea, Pakistan, Serbia, Thailand, UK, etc.

- *Setting*: *Classroom* portrays the setting where users interact with SNSs as 47% of the sample reports, followed by OSN, "several" (e.g., *conference, forum*, etc.), and blended learning that respectively mean 20%, 13%, and 11%.
- *Country*: "Several" nations and the *USA* cover with half of the sample (e.g., 35% and 16%, respectively), while *Asian* (e.g., Taiwan, Turkey, Israel, Malaysia, Hong Kong) and *Europe* (e.g., Germany and Spain) countries contribute with 28% and 12%, respectively.
- *Sample*: The 92 related works assigned to the six ranges of the sample size reveal a $\mu = 14.83$ and $\sigma = 3.01$, where the interval most used is [51, 100] with 20% and the interlude less considered is [1001, 101,847] whose percentage is 11.

Table 3.7 Annual frequencies of two features that compose the Pattern to Depict Educational Networking Applications, where each "group of instances" appears between quotes and the instances it gathers are defined at the bottom

Category	2015	2016	2017	2018	2019	Sum
Sample	**21**	**25**	**22**	**11**	**13**	**92**
Not defined	1	0	0	1	1	3
[1, 50]	5	2	4	2	1	14
[51, 100]	3	4	5	1	5	18
[101, 200]	2	11	1	1	2	17
[201, 400]	7	5	3	1	1	17
[401, 1000]	2	0	5	4	2	13
[1001, 101,847]	1	3	4	1	1	10
Subjects	**21**	**25**	**22**	**11**	**13**	**92**
"Several"	1	1	0	0	1	3
Students & faculty	1	0	1	1	1	4
Teachers	2	0	2	1	1	6
Faculty	1	5	1	1	2	10
Students	16	19	18	8	8	69

Subjects: "Several": Professional, subscribers, users, etc.

- *Subjects*: Samples composed only of *students* represent 75%, while samples where only *faculty* and *teachers* participate correspond to 17%.

3.4.2 Discussion of the Educational Networking Arena

With the aim of discussing the results generated by the present review, an examination of the state that EN reveals is outlined in this section. Hence, the reported labor and estimated statistics are considered to ground the analysis according to the following perspectives: strengths, weaknesses, opportunities, and threats.

3.4.2.1 Strengths

Related to the *strengths* of the EN field, these are mainly due to the fact that ICTs, Web 2.0, and TETL represent the ideal bench for the development and expansion of OSN, where education is an open target of research, development, and application, as the following account manifests:

- Most of the world population pertains to millennials and Gen Z generations, who have a tendency by nature to use, assimilate, and harness emergent technologies, such as OSNs, which easily incorporate in their daily life, and therefore, they conform the majority of users that compose the EN community.
- OSNs foster human's natural ways of living, where socialization is the reflex of a gregarious sense that permits people to survive, grow, and conquer; hence,

education as the formal way of personal development constitutes a propitious practice worthy to deploy through EN settings to profit such a regular behavior.

- The apparition and spread of diverse technologies, settings, and tools that surround OSNs, where most of them are for free, facilitate the incursion of new users, the intensive usage of savvy addicts, and the transformation of their everyday practices such as communication, collaboration, teaching, and learning.
- Diverse SNSs that are rent-free offer technological capacities that simultaneously connect millions of users, spread social media, synchronously and asynchronously broadcast communication, and facilitate the dual role of content publication, access, and update, which together enable EN to enhance classic e-learning.
- Curriculum design's criteria, toil, and scope evolve as a result of considering EN social functionalities that inspire the introduction of novel forms for defining learning objectives, content authoring, lecture sequencing, assessment, help seeking, feedback, and support.
- EN facilitates the incorporation, reinforcement, and innovation of learning theories, teaching styles, and pedagogical paradigms that privilege students' cooperative, collaborative, and social labor, as well as interaction with academic staff.
- EN boosts students to achieve meaningful learning outcomes by socially and reflectively working with both peers and instructors within and outside of classroom encouraging a strong sense of belonging and social capital.
- Through communitarian habits, EN foments students' attitudes, feelings, motivation, engagement, and resilience that contribute to improve their learning strategies, behavior, and accomplishments.

3.4.2.2   Weaknesses

As for the weaknesses, EN mostly deals with diverse subjective and factual-inspired fears that warn, argue, and constrain the consideration to deploy formal education through OSN, where some of the issues to face are the following:

- The lack of users' knowledge for utilizing OSNs, as well as the functionalities offered by each SNS, which are needed to efficiently exploit social exposure, communication, and collaboration tasks that are demanded for EN.
- Adults' general assumption that OSNs are only useful for personal and private affairs, but not suitable for professional and formal duties such as education.
- The belief that OSNs are a source of distraction, cheating, gossip, bullying, and fraudulent engagement that broadcast fake news and unreliable information that result inappropriate for serious affairs such as formal education.
- The shortage of an ethical code that regulates the usage of OSNs, as well as the kind of social media that is admissible to distribute among users, which attempts honorable practices, such as those for academic teaching and learning.
- Management, property, rights, and possession of social media are transferred to the owner of the used public SNS, something that could damage the exploitation and monetization of the investment made for educational purposes.

- Apprehensiveness of potential intromission of strangers that hack the SNSs, steal private information, supplant user identity, observe behaviors to discover patterns that manipulate people, spread untrue social media on behalf of a given user, and illegally access personal ICT resources as a result of users interaction with OSNs, fears that are not desirable in educational settings.
- Access and appropriation of a personal or organizational OSN by intruders, who deceive, discredit, hook, and manipulate SN users with their contacts and followers that damage their identity and prestige, issue that attempts EN use.
- Distortion of the former message and social media shared by OSN users, when such a content is replicated, interpreted, and discussed by others, as it could produce misconceptions that affect the purpose of the sender, issue that cannot happen in educational content, instructor assignments, and learners' work.
- Incorporation of formal education through OSNs demands a huge effort and hard procedure to respectively convince academic authorities and officially register the course, which represent a "long and winding road" that condemns EN labor for research and experimental purposes.
- Quite a few authentic SNSs designed for educational affairs, which are public or open to the academic community, as well as private ones; even worst most of the EN labor relies on popular public general-purpose SNSs, specifically Facebook which has not been exempt from security and legal issues.
- An assumption that the delivery of education through OSNs could deduct formalism, seriousness, and prestige that characterizes formal academic practice.
- Parents' skepticism and distrust of benefits that OSNs offer to users produce they are unwilling to admit EN practice as a reliable option to educate children.

3.4.2.3 Opportunities

Concerning the *opportunities* that appear before the EN domain, there are a plenty of diverse targets and lines awaiting to be taken into account for inspiring new research because emergent technologies foster the irruption of OSN in the daily human life, particularly in educational duties as the next account reveals:

- The irruption of new technologies, specifically ICT and TETL, in the ordinary human life propels OSNs to transform basic social practices, where education cannot remain indifferent and resist to evolve for including such innovations.
- Wireless communication, smartphones, and OSNs recreate a setting that permanently facilitates users to spontaneously generate virtual communities; therefore, such a modern behavior should be harnessed for educational affairs.
- Before the spread and evolution of online education, SNSs are a new wave that complements, and in some cases replaces, traditional systems, where social view guides the design, development, and assessment of learners' activities.
- Real-time, online, offline, synchronous, and asynchronous communication modes that OSNs easily offer to users open the deployment of diverse learning, pedagogic, teaching, content authoring, and assessment approaches.

- Due to being quite a few education-oriented SNSs and reports of their application for EN affairs, a need to cope such a deficit claims educational software developers to build suitable solutions that satisfy a potential growing demand.
- The fact that most of the EN publications correspond to studies provides empirical evidence of how OSNs impact the students' apprenticeship, as well as attitudes, engagement, and behavior, which ground the investment in EN.
- As most of the EN practice is devoted to the HE tier, a need for considering other academic levels, such as K–12, emerges to engage younger and adult learners in novel learning practices, whose OSN technology is familiar.
- The variety of domain knowledge that is delivery through EN reveals that such a modality is worthy to be considered for contributing to formal education.
- As a diversity of learning paradigms is being taken into account for leading students' apprenticeship during their interaction with EN, a call for conceiving new theories that describe, explain, and predict how learning happens is made.
- Because only basic functionalities are implemented in SNSs, a broad variety of approaches (e.g., intelligence, student-centered, adaptation, etc.) tackled by other e-learning systems (e.g., intelligent tutoring systems, educational data mining, learning analytics, etc.) await to be included as part of EN to enhance its impact.
- Resulting from primary and collateral effects that EN produces on users' inner and outer world, new trends of study for education, pedagogy, sociology, psychology, cognition, and social communication disciplines are open.
- Classroom is the most frequent setting where EN complements academic labor; hence, its practice inspires novel styles of teaching and learning habits.
- Although practically there are OSN users all over the world, the application of EN through the nations is limited (i.e., according to the review's sample, 16% of the countries); therefore, a wide potential market expects to be satisfied.

3.4.2.4 Threats

With relation to the *threats* to be considered for the development of the EN labor, they correspond to the growing and extending use of ICTs that risks their usage because of delinquency and dishonesty among other reasons, which demand the consideration of several measures to appropriately prevent and react incidents, such as the following:

- Cyber risks (e.g., attacks, viruses, hijack, malware, frauds, etc.) claim that the usage of OSNs should be backed by a permanent and advance security infrastructure to warrant privacy and secure access, especially for educational affairs to preserve students' identity, academic practices, and school digital resources.
- Lack of international, national, and institutional laws and norms that regulate, supervise, and sanction OSN practices and infringe the investment in EN projects.

- Educational regulations, guidelines, ethics, facilities, and customs are oriented to classroom settings, and even though they have been slowly adapted for e-learning, they have not been updated yet to include EN labor; thus, such a practice is prevented from receiving formal accreditation and acknowledgment.
- Scarce consciousness and literacy of society, professional market, and family about how OSNs are able to contribute to spread formal education discredits serious initiatives to extend the usage of educational-based social networking.
- Incertitude of security, protection, and privacy that assure the reliable control of EN facilities averts academic officers' disposition to invest in EN projects.
- Although EN at classroom is the most usual practice, its usage still requests academics to be convinced and trained to guide students during such activities.
- Academic dishonesty habits (e.g., in the sense of giving, receiving, and using solved solutions, completed homework assignments, and answers to test items) at classrooms can be easily increased through the use of EN; hence, technical and ethical measures should be established to lower them.
- As OSNs facilitate users' exposure to diverse sorts of social media and interact with familiar and unknown people, suddenly they could distract, spend time for leisure, and engage in informal activities that affect their concentration, reduce performance, and risk learning goals; thus, diverse arrangements are needed.
- Since EN relies on social learning and social constructivism paradigms, giving as a result social capital, knowledge construction, and learning derivatives exchange among students, such outcomes demand suitable curriculum design, smart teaching guide, and best learning practices to clearly identify individual's apprenticeship achievements and merits to rightly assess each pupil.
- Actually, OSNs are an agent of disruption and innovation that is transforming the life of world society; thus, this trend offers an opportunity window that must be harnessed to spread EN to contribute to improve academic practice.

3.4.3 Responses to Research Questions

Once the landscape of EN has been stated through a summary of 92 relevant publications, a series of frequencies estimated for the features that compose the PDENA that characterize the works of the sample, and a discussion of the field from four points, the responses to the research problem and questions are provided as follows.

As regards to the problem of "How are academic provosts, officers, and staff being encouraged to pragmatically consider the development and use of SNSs to enhance and extend current academic offer?," the response is: this review offers a tale of how OSNs evolve since the emergence of pioneer works up to date, where empirical evidence is gathered from recent labor to identify targets of application, theoretical baseline, scope of the outcomes, and academic achievements, as well as pros and cons of the investment in EN projects. This fresh reference is worthy to be considered by school officers for decision-making about the convenience to surf a cool wave that spreads novel forms of education delivery based on ICTs.

As for the responses for the research questions made in the Introduction, both inquiries and answers are edited in the incoming relation:

1. *How EN has evolved?* The background of the EN domain, traced in Sect. 3.2.2, reveals its roots, progress, and results in order to tailor the baseline that backs recent educational-based social networking.
2. *What are the key lines that distinguish EN labor?* The TEN, proposed in Sect. 3.2.4, distinguishes two categories, where one corresponds to the kind of SNS that furthers educational practice, while another represents the type of research that is fulfilled to examine the impact of OSNs on involved scholars.
3. *What are the relevant traits that characterize EN research?* The PDENA, introduced in Sect. 3.2.5, embraces a series of features that depict the sample of related works to highlight diverse perspectives of the EN duty.
4. *What about the recent labor carried out in EN?* Inspired in the TEN, a profile of 92 key articles is pointed out, in Sect. 3.3, to sketch a landscape of fresh evidence of how OSNs contribute to improve academic aims.
5. *What are the main affairs considered by recent EN works and how they have been approached?* A series of annual frequencies with their respective interpretation, edited in Sect. 3.4.1, unveils the common subjects of study (e.g., oriented academic tier, domain, effect), as well as the factors (e.g., learning paradigm, functionalities) taken into account to deal with, including supplementary aspects of the work (e.g., SNS, setting, country, sample size, subjects).
6. *What to expect concerning the situation and development of EN?* The discussion of the stated landscape of the EN arena is addressed from four views, edited in Sect. 3.4.2, that picture advantages, disadvantages, chances, and risks to explain the current situation and visualize cases to be considered.

3.5 Conclusions

In this review, a timeline has been traced to know, understand, and analyze how OSNs emerged to disrupt the world society' customs and affairs, transforming and innovating human practices of communication, interaction, and engagement in diverse sorts of relations and commitments, such as education. In this context, EN represents a fresh surge that enhances the richness, scope, and impact of distance and online teaching and learning, where social view is the main policy that leads the development, application, and study of educational-based social networking.

The present work shapes a context where EN is situated as a result of the concurrence of ICTs, Web 2.0, TETL, SNs, OSNs, and SNSs, which together back the appearance of the twenty-first century education. With this in mind, a landscape of

EN was proposed as a solution to solve the research problem of contributing to spread the field based on the answers given to six research inquiries, which side by side outline the share of this chapter that correspond to: the tale of EN record, the TEN that organizes the labor, the PDENA that distinguishes each research, a briefing of recent EN duty, the distinction of the most common attended affairs and the forms they are being approached based on annual frequencies of the PDENA features with their analysis, and a discussion of the status and progress of EN from four outlooks.

This labor was not exempt of shortages, especially the source considered for gathering related works because it only takes into account research published in indexed journals by the Journal Citation Reports®. Hence, other kinds of media are advisable to include such as books, edited books, papers published in proceedings, PhD thesis, technical reports, and chapters, which altogether will contribute to tailor a broader and more significant EN landscape. Another constraint corresponds to the type of work surveyed in this review, because only scientific research was considered, missing other contributions related to field studies and industry leaders' opinion. Deeper analysis is needed to describe and infer correlations among the features' instances that compose the PDENA, as well as patterns that cluster homogenous works and graphs that facilitate the identification of relationships among entities.

First of all, this review of how EN recently has advanced should include the two aforementioned missing perspectives, where one corresponds to the inclusion of market studies that quantitatively report EN usage, effects, communities of users, geographical development, and how the human life is transforming. The other reveals the opinion of owners and developers of SNSs, as well as their vision to extend the disruption of society, and particularly education, is needed to visualize the tendencies that are in progress of transforming human society. As for the analysis of results, pattern discovery to identify cohorts of common features that distinguish the related works is needed, as well as analytics to interpret the raw repository of PDENA records to infer labor evolution and tendency. What is more, SNA is welcome to find out patterns of connections among the features that characterize the works, all of this in order to analyze the structure of the network.

Acknowledgments This research was partially funded by grants CONACYT SNI 36453, IPN-SIP 2018-1407, IPN-SIP 2019-5563, IPN-SIP/DI/DOPI/EDI-154/18, and IPN-SIBE-2019-2020. The author gives testimony of the anointing, strength, and wisdom given by his Father, Brother Jesus, and Helper, as part of the research projects of World Outreach Light to the Nations Ministries (WOLNM). Finally, a special mention and gratitude is given to Lawrence Whitehill, a British English expert reviewer, who tunes the manuscript.

Appendix

Table 3.8 Description of the sample of EN works based on the Pattern to Depict Educational Networking Applications

Author	SNS	Work	School	Domain	Learning	Function	Effect	Setting	Country	Sample	Subjects
Güler (2015)	Facebook	Study	HE	Social media	Social constructivism	Design studio	Performance	Classroom	Turkey	75	Students
Ainin et al. (2015)	Facebook	Study	HE	Diverse	Social learning	Use	Performance	OSN	Malaysia	1165	Students
Hamid et al. (2015)	Facebook	Study	HE	Computer science	Social constructivism	Collaboration	Perceptions	Classroom	Malaysia	46	Students
Imlawi et al. (2015)	Facebook	Study	HE	Computer science	Social learning	Communication	Engagement	OSN	USA	266	Students
Michikyan et al. (2015)	Facebook	Study	HE	Social media	Social learning	Use	Performance	Classroom	USA	261	Students
Asterhan and Hever (2015)	Facebook	Study	HE	Social science	Learning by dialogue	Communication	Discussion	Forum	Israel	60	Students
Greenhow et al. (2015)	Facebook	Study	Diverse	Sciences	Argumentative learning	Collaboration	Knowledge construction	Collaborative learning	USA	346	Students
Puhl et al. (2015)	Facebook	Study	Graduate	Social science	Group awareness	Collaboration	Knowledge construction	Collaborative learning	Germany	63	Teachers
Sarapin and Morris (2015)	Facebook	Study	HE	Diverse	Uses and gratifications theory	Communication	Perceptions	Classroom	USA	308	Faculty
Tsovaltzi et al. (2015)	Facebook	Study	HE	Education	Argumentative learning	Collaboration	Knowledge construction	Collaborative learning	Germany	40	Students
Sendurur et al. (2015)	Facebook	Study	Graduate	Computer science	Social learning	Use	Threats	Classroom	Turkey	412	Teachers
Miron and Ravid (2015)	Facebook	Study	Graduate	Diverse	Social learning	Teaching	Acceptance	Blended	Israel	398	Students

(continued)

Table 3.8 (continued)

Author	SNS	Work	School	Domain	Learning	Function	Effect	Setting	Country	Sample	Subjects
Akcaoglu and Bowman (2016)	Facebook	Study	HE	Diverse	Affective learning	Instructor support	Perceptions	Classroom	USA	87	Students
Gupta and Irwin (2016)	Facebook	Study	HE	Diverse	Multiple resource theory	Lecture	Distraction	Classroom	Australia	150	Students
Sharma et al. (2016)	Facebook	Approach	HE	Diverse	Social constructivism	Use	Prediction	Classroom	Oman	215	Students
Kaewkitipong et al. (2016)	Facebook	Study	HE	Computer science	Active learning	Interaction	Acceptance	Classroom	Thailand	169	Students
Lambić (2016)	Facebook	Study	HE	Education	Social learning	Use	Performance	Computer-assisted learning	Serbia	139	Students
Manasijević et al. (2016)	Facebook	Study	HE	Engineering	Social learning	Use	Perceptions	Classroom	Serbia	226	Students
Čičević et al. (2016)	Facebook	Study	HE	Engineering	Social learning	Use	Behavior	Classroom	Serbia	200	Students
Argyris and Xu (2016)	Facebook	Study	HE	Diverse	Social cognitive theory	Career development	Self-efficacy	Classroom	USA	260	Students
Topîrceanu (2017)	Facebook	Approach	HE	Computer science	Academic dishonesty	Assessment	Cheating	Classroom	Romania	586	Students
Rap and Blonder (2017)	Facebook	Study	HS	Chemistry	Community of inquiry	Community of learning practice	Attitudes	OSN	Israel	707	Students
Lantz-Andersson et al. (2017)	Facebook	Study	Secondary	Language	Flipped classroom	Community of learning practice	Discussion	Classroom	Sweden	13,000	Teachers
Tsovaltzi et al. (2017)	Facebook	Study	HE	Education	Argumentative learning	Collaboration	Knowledge construction	Collaborative learning	Germany	128	Teachers

Marzo et al. (2017)	Facebook	Study	HE	Social media	Adaptive learning	Personalization	Motivation	OSN	Spain	74	Students
Naqshbandi et al. (2017)	Facebook	Study	HE	Diverse	Big 5 model of personality	Use	Performance	Classroom	Malaysia	1165	Students
Dhir et al. (2017)	Facebook	Study	HS	Diverse	Uses and gratifications theory	Use	Acceptance	Classroom	India	942	Students
Datu et al. (2018)	Facebook	Study	HE	Diverse	4-factor model academic engagement	Use	Engagement	Classroom	Philippines	700	Students
Lin (2018)	Facebook	Study	HE	Education	Peer feedback	Assessment	Attitudes	OSN	Taiwan	32	Students
Keles (2018)	Facebook	Logistic	Graduate	Education	Community of inquiry	Community of learning practice	Discussion	OSN	Turkey	92	Teachers
Atroszko et al. (2018)	Facebook	Logistic	HE	Diverse	Social learning	Communication	Addiction	OSN	Poland	1157	Students
Chen et al. (2019)	Facebook	Study	HE	Diverse	Social learning	Community of learning practice	Acceptance	OSN	Taiwan	764	Students
Thai et al. (2019)	Facebook	Study	HE	Psychology	Social learning	Collaboration	Engagement	e-Learning	Australia	67	Students
Toker and Baturay (2019)	Facebook	Study	HE	Diverse	Social learning	Collaboration	Performance	Classroom	Turkey	120	Students
Saini and Abraham (2019)	Facebook	Study	Graduate	Math	Social constructivism	Collaboration	Engagement	Classroom	India	66	Teachers
Awidi et al. (2019)	Facebook	Study	HE	Architecture	Model improving student learning	Design-based learning	Engagement	Classroom	Australia	108	Students

(continued)

Table 3.8 (continued)

Author	SNS	Work	School	Domain	Learning	Function	Effect	Setting	Country	Sample	Subjects
Moorthy et al. (2019)	Facebook	Study	HE	Diverse	Teaching-learning strategies	Teaching	Perceptions	Classroom	Malaysia	298	Students
Feng et al. (2019)	Facebook	Study	HE	Diverse	Distraction-conflict theory	Use	Distraction	Classroom	Hong Kong	92	Students
Aramo-Immonen et al. (2015)	Twitter	Study	Graduate	Social media	Social constructivism	Collaboration	Discussion	Conference	Finland	225	Professional
Menkhoff et al. (2015)	Twitter	Study	HE	Management	Pedagogical tweeting	Interaction	Engagement	Blended	Singapore	41	Students
Rehm and Notten (2016)	Twitter	Study	Diverse	Diverse	Social capital	Communication	Teacher learning	Classroom	Germany	4196	Faculty
Tang and Hew (2017)	Twitter	Review	Diverse	Diverse	OSN-based learning & teaching	Communication	Acceptance	Classroom	Diverse	8502	Students
Lackovic et al. (2017)	Twitter	Study	HE	Physiotherapy	OSN-based learning & teaching	Social media	Engagement	Classroom	UK	43	Students
Nagle (2018)	Twitter	Review	HE	Diverse	OSN-based teaching	Communication	Teacher learning	Classroom	Canada	Undefined	Faculty
Denker et al. (2018)	Twitter	Study	HE	Communication	OSN-based learning	Social media	Engagement	Classroom	USA	483	Students
Carpenter et al. (2019)	Twitter	Study	K–12	Diverse	OSN-based teaching	Communication	Identity development	Classroom	USA	33,184	Faculty
Prestridge (2019)	Twitter	Study	HE	Computer science	Self-generating online professional learning	Personal learning networks	Knowledge construction	Classroom	Diverse	15	Faculty

Brailas et al. (2015)	Wikipedia	Logistic	HE	Culture	Learning as adaptation	Social media	Acculturation	Classroom	Greece	14	Students-Faculty
Chu et al. (2017)	Wiki, PBworks	Tool	Secondary	Social science	Social constructivism	Evaluation	Writing	Classroom	Hong Kong	219	Students
Heimbuch and Bodemer (2017)	Wiki	Experiment	HE	Paleontology	OSN-based learning	Communication	Knowledge construction	Classroom	Germany	81	Students
Chu et al. (2017)	Wiki	Logistic	HE	English	Project-based learning	Project development	Discussion	Classroom	Hong Kong	71	Students
Hu et al. (2018)	Wiki	Logistic	Secondary	Management	Wiki-based learning	Interaction	Writing	Classroom	Hong Kong	245	Students-Faculty
Alghasab et al. (2019)	Wiki	Study	HS	English	Socio-cultural theory	Interaction	Writing	Classroom	Kuwait	53	Students-Faculty
Orús et al. (2016)	YouTube	Study	HE	Marketing	Active learning	Video generation	Performance	Iterative learning	Spain	125	Students
Hong et al. (2016)	YouTube	Study	Diverse	Music	Cognitive-affective theory of learning with media	Self-directed learning	Self-efficacy	OSN	Taiwan	117	Users
Choi and Behm-Morawitz (2017)	YouTube	Logistic	HE	STEM	Social cognitive theory	Video generation	Skills development	OSN	USA	374	Students-Faculty
Shoufan (2019)	YouTube	Study	HE	Computer science	Cognitive theory of multimedia learning	Video evaluation	Cognitive value	Massive open online course	UAE	428	Students
Saurabh and Gautam (2019)	YouTube	Approach	HE	Computer science	Teaching-learning strategies	Video access	Acceptance	Iterative learning	India	Undefined	Subscribers

(continued)

Table 3.8 (continued)

Author	SNS	Work	School	Domain	Learning	Function	Effect	Setting	Country	Sample	Subjects
Benson and Filippaios (2015)	Facebook & LinkedIn	Study	HE	Business	Social constructivism	Collaboration	Awareness	OSN	UK	600	Students
Lim and Richardson (2016)	Facebook & LinkedIn	Study	Graduate	Technology	OSN-based learning	Use	Social presence	e-Learning	USA	82	Students
Liao et al. (2015)	Google+	Study	HE	Culture	Social constructivism	Collaboration	Performance	u-Learning	Taiwan	321	Students
Schwarz and Caduri (2016)	Facebook & Google+	Study	Secondary	Diverse	Community of inquiry	Community of learning practice	Discussion	Classroom	Israel	5	Students
Ishtaiwa and Aburezeq (2015)	Google Docs	Study	Graduate	TIC	Social constructivism	Content authoring	Teacher learning	Blended	UAE	178	Students
Liu and Lan (2016)	Google Docs	Study	Tertiary	English	Social constructivism	Collaboration	Knowledge construction	Classroom	Taiwan	65	Students
Lee and Bonk (2016)	Blog	Study	Graduate	Computer science	Reflection	Interaction	Writing	Blended	Korea	22	Students
Kuznetcova et al. (2019)	Blog	Study	HE	Psychology	Social capital	Communication	Distributed learning	Classroom	USA	56	Students
Asterhan and Bouton (2017)	Facebook & WhatsApp	Study	Secondary	Diverse	OSN-based learning	Sharing	Knowledge sharing	OSN	Israel	721	Students
Song et al. (2017)	Pinterest	Study	Graduate	Language	Social constructivism	Collaboration	Language use	Informal learning	USA	12	Students
Chao et al. (2018)	SN forum	Study	HE	English	Help seeking	Problem-solving	Discussion	Forum	Taiwan	37	Students
Stantchev et al. (2015)	SNSs several	Approach	HE	Diverse	Undefined	Use	Recommendation	OSN	Germany	Undefined	Students

Study	Tool	Type	Level	Field	Theory	Collaboration	Performance	Setting	Country	N	Population
Alwagait et al. (2015)	SNSs several	Study	HE	Diverse	Social constructivism	Collaboration	Performance	Classroom	Saudi Arabia	108	Students
Sobaih et al. (2016)	SNSs several	Study	HE	Tourism	Social learning	Communication	Acceptance	Classroom	Egypt	190	Faculty
Dermentzi et al. (2016)	SNSs several	Study	HE	Diverse	OSN-based teaching	Use	Engagement	e-Learning	Diverse	370	Faculty
Manca and Ranieri (2016a)	SNSs several	Study	HE	Diverse	Sharing content	Use	Academic practices	Blended	Italy	6139	Faculty
Manca and Ranieri (2016b)	SNSs several	Study	HE	Diverse	Sharing content	Use	Academic practices	Blended	Italy	6139	Faculty
Liu et al. (2017)	SNSs several	Review	Diverse	Diverse	OSN-based learning	Use	Performance	OSN	Diverse	101,847	Students
Akçayır (2017)	SNSs several	Study	HE	Diverse	OSN-based teaching	Communication	SNS adoption for teaching	Classroom	Turkey	658	Faculty
Waycott et al. (2017)	SNSs several	Study	HE	Language	Social theory of learning	Community of learning practice	Student visibility	e-Learning	Australia	20	Students
Cao et al. (2018)	SNSs several	Study	HE	Diverse	Cognitive emotional preoccupation	Use	Performance	OSN	China	505	Students
Molinillo et al. (2018)	SNSs several	Study	HE	Business	Active learning	Collaboration	Attitudes	Classroom	Spain	486	Students
Rutten et al. (2016)	Facebook & Ning	Study	Prevocational	Skills	Externalizing	Expression	Skills development	Virtual learning	The Netherlands	103	Students
Ergün and Usluel (2016)	Elgg	Study	HE	Education	Social constructivism	Instructor support	Knowledge construction	Blended	Turkey	114	Students
de-Marcos et al. (2016)	Elgg	Experiment	HE	Computer science	Social learning	Social gamification	Performance	Blended	Spain	167	Students

(continued)

Table 3.8 (continued)

Author	SNS	Work	School	Domain	Learning	Function	Effect	Setting	Country	Sample	Subjects
de-Marcos et al. (2016)	Elgg	Experiment	HE	Computer science	Learning styles	Social gamification	Performance	Blended	Spain	379	Students
Liu et al. (2017)	Story & Painting House	Tool	Elementary	English	Social constructivism	Community creation	Engagement	Multimedia	Taiwan	26	Students
Won et al. (2015)	Edmodo	Study	Middle	STEM	Design-based learning	Problem-solving	Discussion	Classroom	USA	44	Students
Ellahi (2018)	Edmodo	Study	HE	Business	Networked learning	Instructor support	Performance	OSN	Pakistan	150	Students
Lee and Kim (2016)	KakaoTalk	Study	HE	English	Social constructivism	Writing	Achievement	Multimodal learning	Korea	62	Students
Scott et al. (2016)	The Hive	Approach	Graduate	Learning	Community of inquiry	Community of learning practice	Social presence	Blended	USA	164	Students
Mahnane (2017)	EO-P	Approach	HE	Math	Learning styles	Recommending	Personalization	OSN	Algeria	80	Students
Karataev and Zadorozhny (2017)	Self-adaptive learning	Approach	HE	Computer science	Learning by teaching	Adaptive learning	Collective learning	e-Learning	USA	260	Students
Hatzipanagos and John (2017)	KINSHIP	Study	HE	Diverse	OSN as a support for learning	Communication	Acceptance	OSN	UK	67	Students

References

Ainin, S., Naqshbandi, M. M., Moghavvemi, S., & Jaafar, N. I. (2015). Facebook usage, socialization and academic performance. *Computers & Education, 83*, 64–73.

Akcaoglu, M., & Bowman, N. D. (2016). Using instructor-led Facebook groups to enhance students' perceptions of course content. *Computers in Human Behavior, 65*, 582–590.

Akçayır, G. (2017). Why do faculty members use or not use social networking sites for education? *Computers in Human Behavior, 71*, 378–385.

Al-Ali, S. (2014). Embracing the selfie craze: exploring the possible use of Instagram as a language mlearning tool. *Issues and Trends in Educational Technology, 2*(2), 1–16.

Alghasab, M., Hardman, J., & Handley, Z. (2019). Teacher-student interaction on wikis: Fostering collaborative learning and writing. *Learning, Culture and Social Interaction, 21*, 10–20.

Allan, M. (2004). A peek into the life of online learning discussion forums: Implications for web-based distance learning. *The International Review of Research in Open and Distributed Learning, 5*(2), 1–18.

Alwagait, E., Shahzad, B., & Alim, S. (2015). Impact of social media usage on students academic performance in Saudi Arabia. *Computers in Human Behavior, 51*, 1092–1097.

Aramo-Immonen, H., Jussila, J., & Huhtamäki, J. (2015). Exploring co-learning behavior of conference participants with visual network analysis of Twitter data. *Computers in Human Behavior, 51*, 1154–1162.

Argyris, Y. E. A., & Xu, J. D. (2016). Enhancing self-efficacy for career development in Facebook. *Computers in Human Behavior, 55*, 921–931.

Asterhan, C. S., & Hever, R. (2015). Learning from reading argumentive group discussions in Facebook: Rhetoric style matters (again). *Computers in Human Behavior, 53*, 570–576.

Asterhan, S. C. S., & Bouton, E. (2017). Teenage peer-to-peer knowledge sharing through social network sites in secondary schools. *Computers & Education, 110*, 16–34.

Atroszko, P. A., Balcerowska, J. M., Bereznowski, P., Biernatowska, A., Pallesen, S., & Andreassen, C. S. (2018). Facebook addiction among Polish undergraduate students: Validity of measurement and relationship with personality and well-being. *Computers in Human Behavior, 85*, 329–338.

Awidi, I. T., Paynter, M., & Vujosevic, T. (2019). Facebook group in the learning design of a higher education course: An analysis of factors influencing positive learning experience for students. *Computers & Education, 129*, 106–121.

Benson, V., & Filippaios, F. (2015). Collaborative competencies in professional social networking: Are students short changed by curriculum in business education? *Computers in Human Behavior, 51*, 1331–1339.

Blair, R., & Serafini, T. M. (2014). Integration of education: Using social media networks to engage students. *Systemics, Cybernetics, and Informatics, 6*(12), 1–4.

Boyd, D. M., & Ellison, N. B. (2008). Social network sites: Definition, history, and scholarship. *Journal of Computer-Mediated Communication, 13*, 201–230.

Brady, K. P., Holcomb, L. B., & Smith, B. V. (2010). The use of alternative social networking sites in higher educational settings: A case study of the e-learning benefits of Ning in education. *Journal of Interactive Online Learning, 9*(2), 151–170.

Brailas, A., Koskinas, K., Dafermos, M., & Alexias, G. (2015). Wikipedia in education: Acculturation and learning in virtual communities. *Learning, Culture and Social Interaction, 7*, 59–70.

Cain, J. (2008). Online social networking issues within academia and pharmacy education. *American Journal of Pharmaceutical Education, 72*(1), 10.

Cao, X., Masood, A., Luqman, A., & Ali, A. (2018). Excessive use of mobile social networking sites and poor academic performance: Antecedents and consequences from stressor-strain-outcome perspective. *Computers in Human Behavior, 85*, 163–174.

Carpenter, J. P., Kimmons, R., Short, C. R., Clements, K., & Staples, M. E. (2019). Teacher identity and crossing the professional-personal divide on twitter. *Teaching and Teacher Education, 81*, 1–12.

Chambers, J. A., & Poore, R. V. (1975). Computer networks in higher education: socio-economic-political factors. *Communications of the ACM, 18*(4), 193–199.

Chao, P. Y., Lai, K. R., Liu, C. C., & Lin, H. M. (2018). Strengthening social networks in online discussion forums to facilitate help seeking for solving problems. *Journal of Educational Technology & Society, 21*(4), 39–50.

Charlton, T., Marshall, L., & Devlin, M. (2008). Evaluating the extent to which sociability and social presence affects learning performance. *ACM SIGCSE Bulletin, 40*(3), 342–342.

Chen, P. (2011). From CMS to SNS: Educational networking for urban teachers. *Journal of Urban Learning, Teaching, and Research, 7*, 50–61.

Chen, S. Y., Kuo, H. Y., & Hsieh, T. C. (2019). New literacy practice in a facebook group: The case of a residential learning community. *Computers & Education, 134*, 119–131.

Choi, G. Y., & Behm-Morawitz, E. (2017). Giving a new makeover to STEAM: Establishing YouTube beauty gurus as digital literacy educators through messages and effects on viewers. *Computers in Human Behavior, 73*, 80–91.

Chu, S. K., Capio, C. M., van Aalst, J. C., & Cheng, E. W. (2017). Evaluating the use of a social media tool for collaborative group writing of secondary school students in Hong Kong. *Computers & Education, 110*, 170–180.

Chu, S. K. W., Zhang, Y., Chen, K., Chan, C. K., Lee, C. W. Y., Zou, E., et al. (2017). The effectiveness of wikis for project-based learning in different disciplines in higher education. *The Internet and Higher Education, 33*, 49–60.

Čičević, S., Samčović, A., & Nešić, M. (2016). Exploring college students' generational differences in facebook usage. *Computers in Human Behavior, 56*, 83–92.

Craig, E. M. (2007). Changing paradigms: managed learning environments and Web 2.0. *Campus-Wide Information Systems, 24*(3), 152–161.

Datu, J. A. D., Yang, W., Valdez, J. P. M., & Chu, S. K. W. (2018). Is facebook involvement associated with academic engagement among Filipino university students? A cross-sectional study. *Computers & Education, 125*, 246–253.

Davies, S., Mullan, J., & Feldman, P. (2017). *Rebooting learning for the digital age: What next for technology-enhanced higher education*? Technical Report. Oxford: Higher Education Policy Institute.

Dede, C. (2010). Comparing frameworks for 21st century skills. In J. A. Bellanca & R. Brandt (Eds.), *21st century skills: Rethinking how students learn* (pp. 51–76). Bloomington, IN: Solution Tree Press.

de-Marcos, L., Garcia-Lopez, E., & Garcia-Cabot, A. (2016). On the effectiveness of game-like and social approaches in learning: Comparing educational gaming, gamification & social networking. *Computers & Education, 95*, 99–113.

de-Marcos, L., García-López, E., García-Cabot, A., Medina-Merodio, J. A., Domínguez, A., Martínez-Herráiz, J. J., et al. (2016). Social network analysis of a gamified e-learning course: Small-world phenomenon and network metrics as predictors of academic performance. *Computers in Human Behavior, 60*, 312–321.

Denker, K. J., Manning, J., Heuett, K. B., & Summers, M. E. (2018). Twitter in the classroom: Modeling online communication attitudes and student motivations to connect. *Computers in Human Behavior, 79*, 1–8.

Dermentzi, E., Papagiannidis, S., Toro, C. O., & Yannopoulou, N. (2016). Academic engagement: Differences between intention to adopt social networking sites and other online technologies. *Computers in Human Behavior, 61*, 321–332.

Dhir, A., Khalil, A., Lonka, K., & Tsai, C. C. (2017). Do educational affordances and gratifications drive intensive Facebook use among adolescents? *Computers in Human Behavior, 68*, 40–50.

Drexler, W., Baralt, A., & Dawson, K. (2008). The teach Web 2.0 Consortium: A tool to promote educational social networking and Web 2.0 use among educators. *Educational Media International, 45*(4), 271–283.

Ellahi, A. (2018). Social networking sites as formal learning environments in business social networking sites as formal learning environments in business. *Educational Technology & Society, 21*(4), 64–75.

Ergün, E., & Usluel, Y. K. (2016). An analysis of density and degree-centrality according to the social networking structure formed in an online learning environment. *Journal of Educational Technology & Society, 19*(4), 34–46.

Eurich-Fulcer, R., & Schofield, J. W. (1995). Wide-area networking in K-12 education: Issues shaping implementation and use. *Computers & Education, 24*(3), 211–220.

Ezumah, B. A. (2013). College students' use of social media: Site preferences, uses and gratifications theory revisited. *International Journal of Business and Social Science, 4*(5), 27–34.

Feng, S., Wong, Y. K., Wong, L. Y., & Hossain, L. (2019). The internet and facebook usage on academic distraction of college students. *Computers & Education, 134*, 41–49.

Ferlander, S. (2007). The importance of different forms of social capital for health. *Acta Sociologica, 50*(2), 115–128.

Fonte, V., & Davis, B. (1979). Networking for life-long learning in graduate education. *Alternative Higher Education, 4*(1), 24–31.

Goldfarb, A., Pregibon, N., Shrem, J., & Zyko, E. (2011). *Informational brief on social networking in education*. Technical Report. New York Comprehensive Center.

Goodyear, P., & Retalis, S. (Eds.). (2010). *Technology-enhanced learning Design patterns and pattern languages*. Rotterdam: Sense Publishers.

Greenhow, C., Gibbins, T., & Menzer, M. M. (2015). Re-thinking scientific literacy out-of-school: Arguing science issues in a niche Facebook application. *Computers in Human Behavior, 53*, 593–604.

Grosseck, G., & Holotescu, C. (2008). Can we use Twitter for educational activities. In *4th International Scientific Conference, eLearning and Software for Education*, Bucharest, Romania.

Güler, K. (2015). Social media-based learning in the design studio: A comparative study. *Computers & Education, 87*, 192–203.

Gupta, N., & Irwin, J. D. (2016). In-class distractions: The role of Facebook and the primary learning task. *Computers in Human Behavior, 55*, 1165–1178.

Guraya, S. Y. (2016). The usage of social networking sites by medical students for educational purposes: A meta-analysis and systematic review. *North American Journal of Medical Sciences, 8*(7), 268.

Hamid, S., Waycott, J., Kurnia, S., & Chang, S. (2015). Understanding students' perceptions of the benefits of online social networking use for teaching and learning. *The Internet and Higher Education, 26*, 1–9.

Hargadon, S., (2010) *Educational networking: The important role Web 2.0 will play in education*. Technical Report. Elluminate.

Harrison, R., & Thomas, M. (2009). Identity in online communities: Social networking sites and language learning. *International Journal of Emerging Technologies and Society, 7*(2), 109–124.

Hatzipanagos, S., & John, B. A. (2017). Do institutional social networks work? Fostering a sense of community and enhancing learning. *Technology, Knowledge and Learning, 22*(2), 151–159.

Heaney, C. A., & Israel, B. A. (2008). Social networks and social support. Health behavior and health education. In K. Glanz, B. K. Rimer, & K. Vswanath (Eds.), *Health behavior and health education: Theory, research, and practice* (Vol. 4, pp. 189–210). San Francisco: Jossey-Bass.

Heimbuch, S., & Bodemer, D. (2017). Controversy awareness on evidence-led discussions as guidance for students in wiki-based learning. *The Internet and Higher Education, 33*, 1–14.

Holcomb, L. B., Brady, K. P., Smith, B. V., & Bethany, V. (2010). The emergence of "educational networking": Can non-commercial, education-based social networking sites really address the privacy and safety concerns of educators. *Journal of Online Learning and Teaching, 6*(2), 475–481.

Hong, J. C., Hwang, M. Y., Szeto, E., Tsai, C. R., Kuo, Y. C., & Hsu, W. Y. (2016). Internet cognitive failure relevant to self-efficacy, learning interest, and satisfaction with social media learning. *Computers in Human Behavior, 55*, 214–222.

Hu, X., Cheong, C. W. L., & Chu, S. K. W. (2018). Developing a multidimensional framework for analyzing student comments in wikis. *Journal of Educational Technology & Society, 21*(4), 26–38.

Imlawi, J., Gregg, D., & Karimi, J. (2015). Student engagement in course-based social networks: The impact of instructor credibility and use of communication. *Computers & Education, 88*, 84–96.

Ishtaiwa, F. F., & Aburezeq, I. M. (2015). The impact of Google Docs on student collaboration: a UAE case study. *Learning, Culture and Social Interaction, 7*, 85–96.

Issa, T., Isaias, P., & Kommers, P. (Eds.). (2015). *Social networking and education: Global perspectives* (Lecture Notes in Social Networks). Cham: Springer.

Jerald, C. D. (2009). *Defining a 21st century education*. Center for Public Education, 16.

Johnston, T. C. (2002). Teaching with a weblog: How to post student work online. In *Allied academies international conference: Academy of educational leadership* (Vol. 7, No. 1, p. 33). Jordan Whitney Enterprises, Inc.

Kaewkitipong, L., Chen, C. C., & Ractham, P. (2016). Using social media to enrich information systems field trip experiences: Students' satisfaction and continuance intentions. *Computers in Human Behavior, 63*, 256–263.

Karataev, E., & Zadorozhny, V. (2017). Adaptive social learning based on crowdsourcing. *IEEE Transactions on Learning Technologies, 10*(2), 128–139.

Kaufman, P. B. (2007). Video, education and open content: Notes toward a new research and action agenda. *First Monday, 12*(4).

Keles, E. (2018). Use of Facebook for the Community Services Practices course: Community of inquiry as a theoretical framework. *Computers & Education, 116*, 203–224.

Kuznetcova, I., Glassman, M., & Lin, T.-J. (2019). Multi-user virtual environments as a pathway to distributed social networks in the classroom. *Computers & Education, 130*, 26–39.

Lackovic, N., Kerry, R., Lowe, R., & Lowe, T. (2017). Being knowledge, power and profession subordinates: Students' perceptions of Twitter for learning. *Internet and Higher Education, 33*, 41–48.

Lambić, D. (2016). Correlation between Facebook use for educational purposes and academic performance of students. *Computers in Human Behavior, 61*, 313–320.

Lange, J., Casey, R., Dunqua, B., Dunnet, C., Knezek, G., Flanigan, J., et al. (1984). ATS-1: Social service satellite networks in the Pacific. In *Proceedings of the sixth Pacific telecommunications conference* (pp. 109–118). Honolulu, HI: Pacific Telecommunications Council.

Lantz-Andersson, A., Peterson, L., Hillman, T., Lundin, M., & Rensfeldt, A. B. (2017). Sharing repertoires in a teacher professional Facebook group. *Learning, Culture and Social Interaction, 15*, 44–55.

Lee, J., & Bonk, C. J. (2016). Social network analysis of peer relationships and online interactions in a blended class using blogs. *The Internet and Higher Education, 28*, 35–44.

Lee, K. S., & Kim, B. G. (2016). Cross space: The exploration of SNS-based writing activities in a multimodal learning environment. *Journal of Educational Technology & Society, 19*(2), 57–76.

Liao, Y. W., Huang, Y. M., Chen, H. C., & Huang, S. H. (2015). Exploring the antecedents of collaborative learning performance over social networking sites in a ubiquitous learning context. *Computers in Human Behavior, 43*, 313–323.

Lim, J., & Richardson, J. C. (2016). Exploring the effects of students' social networking experience on social presence and perceptions of using SNSs for educational purposes. *The Internet and Higher Education, 29*, 31–39.

Lin, G. Y. (2018). Anonymous versus identified peer assessment via a Facebook-based learning application: Effects on quality of peer feedback, perceived learning, perceived fairness, and attitude toward the system. *Computers & Education, 116*, 81–92.

Lin, N. (2001). *Social capital*. Cambridge, UK: Cambridge University Press.

List, J. S., & Bryant, B. (2009). Integrating interactive online content at an early College High School: An exploration of Moodle, Ning and Twitter. *Meridian A Middle School Computer Technologies Journal, 12*(1), 1–4.

Liu, C. C., Chen, Y. C., & Tai, S. J. D. (2017). A social network analysis on elementary student engagement in the networked creation community. *Computers & Education, 115*, 114–125.

Liu, D., Kirschner, P. A., & Karpinski, A. C. (2017). A meta-analysis of the relationship of academic performance and social network site use among adolescents and young adults. *Computers in Human Behavior, 77*, 148–157.

Liu, S. H. J., & Lan, Y. J. (2016). Social constructivist approach to web-based EFL learning: Collaboration, motivation, and perception on the use of Google docs. *Educational Technology & Society, 19*(1), 171–186.

Lőrincz, L., Koltai, J., Győr, A. F., & Takács, K. (2019). Collapse of an online social network: Burning social capital to create it? *Social Networks, 57*, 43–53.

Mahnane, L. (2017). Recommending learning activities in social network using data mining algorithms. *Journal of Educational Technology & Society, 20*(4), 11–23.

Manasijević, D., Živković, D., Arsić, S., & Milošević, I. (2016). Exploring students' purposes of usage and educational usage of Facebook. *Computers in Human Behavior, 60*, 441–450.

Manca, S., & Ranieri, M. (2016a). "Yes for sharing, no for teaching!": Social Media in academic practices. *Internet and Higher Education, 29*, 63–74.

Manca, S., & Ranieri, M. (2016b). Facebook and the others. Potentials and obstacles of social media for teaching in higher education. *Computers & Education, 95*, 216–230.

Marin, A., & Wellman, B. (2011). Social network analysis: An introduction. In J. Scott & P. J. Carrington (Eds.), *The SAGE handbook of social network analysis* (pp. 331–339). Los Angeles: SAGE Publications.

Marzo, A., Ardaiz, O., Sanz de Acedo, M. T., & Sanz de Acedo, M. L. (2017). Personalizing sample databases with Facebook information to increase intrinsic motivation. *IEEE Transactions on Education, 60*(1), 16–21.

McAnge, T. R., Jr. (1990). *A survey of educational computer networks*. Technical Report. Virginia Polytechnic Institute.

Menkhoff, T., Chay, Y. W., Bengtsson, M. L., Woodard, C. J., & Gan, B. (2015). Incorporating microblogging ("tweeting") in Higher Education: Lessons learnt in a knowledge management course. *Computers in Human Behavior, 51*, 1295–1302.

Michikyan, M., Subrahmanyam, K., & Dennis, J. (2015). Facebook use and academic performance among college students: A mixed-methods study with a multi-ethnic sample. *Computers in Human Behavior, 45*, 265–272.

Miron, E., & Ravid, G. (2015). Facebook groups as an academic teaching aid: Case study and recommendations for educators. *Journal of Educational Technology & Society, 18*(4), 371–384.

Mitchell, E., & Watstein, B. S. (2007). The places where students and scholars work, collaborate, share and plan: Endless possibilities for us! *Reference Services Review, 35*(4), 521–524.

Molinillo, S., Anaya-Sánchez, R., Aguilar-Illescas, R., & Vallespín-Arán, M. (2018). Social media-based collaborative learning: Exploring antecedents of attitude. *Internet and Higher Education, 38*(1), 18–27.

Moorthy, K., T'ing, L. C., Wei, K. M., Mei, P. T. Z., Yee, C. Y., Wern, K. L. J., et al. (2019). Is facebook useful for learning? A study in private universities in Malaysia. *Computers & Education, 130*, 94–104.

Nagle, J. (2018). Twitter, cyber-violence, and the need for a critical social media literacy in teacher education: A review of the literature. *Teaching and Teacher Education, 76*, 86–94.

Naqshbandi, M. M., Ainin, S., Jaafar, N. I., & Shuib, N. L. M. (2017). To Facebook or to Face Book? An investigation of how academic performance of different personalities is affected through the intervention of Facebook usage. *Computers in Human Behavior, 75*, 167–176.

Nee, C. (2014). The effect of educational networking on students' performance in biology. In *Technology, colleges and community conference* (pp. 73–97).

O'reilly, T. (2007). What is Web 2.0: Design patterns and business models for the next generation of software. *Communications & Strategies, 65*, 17–37.

Orús, C., Barlés, M. J., Belanche, D., Casaló, L., Fraj, E., & Gurrea, R. (2016). The effects of learner-generated videos for YouTube on learning outcomes and satisfaction. *Computers & Education, 95*, 254–269.

Park, Y., Heo, G. M., & Lee, R. (2008). Cyworld is my world: Korean adult experiences in an online community for learning. *International Journal of Web Based Communities, 4*(1), 33–51.

Pellegrino, J. W., & Hilton, M. L. (2013). *Education for life and work: Developing transferable knowledge and skills in the 21st century*. Washington, DC: The National Academies Press.

Prestridge, S. (2019). Categorising teachers' use of social media for their professional learning: A self-generating professional learning paradigm. *Computers & Education, 129*, 143–158.

Puhl, T., Tsovaltzi, D., & Weinberger, A. (2015). Blending Facebook discussions into seminars for practicing argumentation. *Computers in Human Behavior, 53,* 605–616.

Rap, S., & Blonder, R. (2017). Thou shall not try to speak in the Facebook language: Students' perspectives regarding using Facebook for chemistry learning. *Computers & Education, 114,* 69–78.

Rehm, M., & Notten, A. (2016). Twitter as an informal learning space for teachers!? The role of social capital in Twitter conversations among teachers. *Teaching and Teacher Education, 60,* 215–223.

Riel, M. (1990). Computer-mediated communication: A tool for reconnecting kids with society. *Interactive Learning Environments, 1*(4), 255–263.

Rielgelsberger, J., & Sasse, M. A. (2000). Trust me, I'm a. com. *Intermedia, 28*(4), 23–27.

Rutten, M., Ros, A., Kuijpers, M., & Kreijns, K. (2016). Usefulness of social network sites for adolescents' development of online career skills. *Journal of Educational Technology & Society, 19*(4), 140–150.

Saini, C., & Abraham, J. (2019). Implementing Facebook-based instructional approach in pre-service teacher education: An empirical investigation. *Computers & Education, 128,* 243–255.

Sarapin, S. H., & Morris, P. L. (2015). Faculty and Facebook friending: Instructor–student online social communication from the professor's perspective. *The Internet and Higher Education, 27,* 14–23.

Saurabh, S., & Gautam, S. (2019). Modelling and statistical analysis of YouTube's educational videos: A channel Owner's perspective. *Computers & Education, 128,* 145–158.

Schwarz, B., & Caduri, G. (2016). Novelties in the use of social networks by leading teachers in their classes. *Computers & Education, 102,* 35–51.

Scott, K. S., Sorokti, S. H., & Merrell, J. D. (2016). Learning "beyond the classroom" within an enterprise social network system. *Internet and Higher Education, 29,* 75–90.

Sendurur, P., Sendurur, E., & Yilmaz, R. (2015). Examination of the social network sites usage patterns of pre-service teachers. *Computers in Human Behavior, 51,* 188–194.

Sharma, S. K., Joshi, A., & Sharma, H. (2016). A multi-analytical approach to predict the Facebook usage in higher education. *Computers in Human Behavior, 55,* 340–353.

Sharples, M. (2006). How can we address the conflicts between personal informal learning and traditional classroom education. In M. Sharples (Ed.), *Big issues in mobile learning.* Report of a Workshop by the KNEMLI, 21–24.

Shoufan, A. (2019). Estimating the cognitive value of YouTube's educational videos: A learning analytics approach. *Computers in Human Behavior, 92,* 450–458.

Siddiq, F., Gochyyev, P., & Wilson, M. (2017). Learning in digital networks–ICT literacy: A novel assessment of students' 21st century skills. *Computers & Education, 109,* 11–37.

Sobaih, A. E. E., Moustafa, M. A., Ghandforoush, P., & Khan, M. (2016). To use or not to use? Social media in higher education in developing countries. *Computers in Human Behavior, 58,* 296–305.

Song, K., Williams, K., Pruitt, A. A., & Schallert, D. (2017). Students as pinners: A multimodal analysis of a course activity involving curation on a social networking site. *Internet and Higher Education, 33,* 33–40.

Stantchev, V., Prieto-González, L., & Tamm, G. (2015). Cloud computing service for knowledge assessment and studies recommendation in crowdsourcing and collaborative learning environments based on social network analysis. *Computers in Human Behavior, 51,* 762–770.

Tang, Y., & Hew, K. F. (2017). Using Twitter for education: Beneficial or simply a waste of time? *Computers & Education, 106,* 97–118.

Thai, M., Sheeran, N., & Cummings, D. J. (2019). We're all in this together: The impact of Facebook groups on social connectedness and other outcomes in higher education. *The Internet and Higher Education, 40,* 44–49.

Toker, S., & Baturay, M. H. (2019). What foresees college students' tendency to use facebook for diverse educational purposes? *International Journal of Educational Technology in Higher Education, 16*(1), 1–20.

Topîrceanu, A. (2017). Breaking up friendships in exams: A case study for minimizing student cheating in higher education using social network analysis. *Computers & Education, 115,* 171–187.

Tsovaltzi, D., Judele, R., Puhl, T., & Weinberger, A. (2015). Scripts, individual preparation and group awareness support in the service of learning in Facebook: How does CSCL compare to social networking sites? *Computers in Human Behavior, 53,* 577–592.

Tsovaltzi, D., Judele, R., Puhl, T., & Weinberger, A. (2017). Leveraging social networking sites for knowledge co-construction: Positive effects of argumentation structure, but premature knowledge consolidation after individual preparation. *Learning and Instruction, 52,* 161–179.

Voogt, J., & Roblin, N. P. (2012). A comparative analysis of international frameworks for 21st century competences: Implications for national curriculum policies. *Journal of Curriculum Studies, 44*(3), 299–321.

Warf, B., Vincent, P., & Purcell, D. (1999). International collaborative learning on the World Wide Web. *Journal of Geography, 98*(3), 141–148.

Waycott, J., Thompson, C., Sheard, J., & Clerehan, R. (2017). A virtual panopticon in the community of practice: Students' experiences of being visible on social media. *The Internet and Higher Education, 35,* 12–20.

Wellman, B., Salaff, J., Dimitrova, D., Garton, L., Gulia, M., & Haythornthwaite, C. (1996). Computer networks as social networks: Collaborative work, telework, and virtual community. *Annual Review of Sociology, 22*(1), 213–238.

Won, S. G., Evans, M. A., Carey, C., & Schnittka, C. G. (2015). Youth appropriation of social media for collaborative and facilitated design-based learning. *Computers in Human Behavior, 50,* 385–391.

Yang, H. L., & Tang, J. H. (2003). Effects of social network on students' performance: A web-based forum study in Taiwan. *Journal of Asynchronous Learning Networks, 7*(3), 93–107.

Part II
Conceptual

Chapter 4
The Platform Adoption Model (PAM): A Theoretical Framework to Address Barriers to Educational Networking

Dawn B. Branley-Bell

Abstract Social networking platforms are widely adopted as a social tool in everyday life. Research has identified that networking platforms may also bring many benefits when used in an educational environment. For instance, educational networking may lead to enhanced communication skills; increased teamwork and collaboration; greater comprehension of alternative viewpoints; improved creativity, productivity, and work efficiency; and increased learning speed. However, uptake of educational networking has been slower than expected, and there is an absence of theoretical models or frameworks to help understand and address this. The aim of this chapter is threefold: firstly, to provide a comprehensive overview of the current body of knowledge, identifying and collating barriers toward educational networking adoption and increased usage; secondly, to identify key theories of behavior change relevant to educational networking uptake; and thirdly, to build a comprehensive model and framework for overcoming the identified barriers and improving the contribution of educational networking platforms. This chapter achieves this by drawing upon four of the leading behavioral theories from psychology, computer science, and behavioral economics (theory of planned behavior, technology acceptance model, information system success model, and protection motivation theory) to build the platform adoption model. The model provides a theory-driven basis for future research that will be beneficial within academia and the wider audience, e.g., developers, educators, and learning establishments.

Keywords Theoretical model, Blended learning, Platform adoption · Platform adoption model · Behavior change

D. B. Branley-Bell (✉)
Northumbria University, Newcastle, UK
e-mail: dawn.branley-bell@northumbria.ac.uk

© Springer Nature Switzerland AG 2020
A. Peña-Ayala (ed.), *Educational Networking*, Lecture Notes in Social Networks, https://doi.org/10.1007/978-3-030-29973-6_4

Abbreviations

ENP	Educational networking platform
ISS	Information system success model
PAM	Protection acceptance model
PMT	Protection motivation theory
SEM	Structural equation modelling
SNP	Social networking platform
TAM	Technology acceptance model
TPB	Theory of planned behavior

4.1 Introduction

Social networking platforms (SNPs) are widely used in the general population as an everyday tool to facilitate social connections with peers, family, colleagues, and/or other likeminded individuals (Branley, 2015). The term educational networking refers to the use of networking platforms within an educational environment – whether this is the use of traditional SNPs within this context (e.g., Facebook), or the use of platforms designed specifically for this purpose (i.e., dedicated educational networking platforms). Educational networking represents a shift from the traditional "one-way" classroom model of educators as the sole source of information to a more social constructivist approach, allowing students and educators to work together to share and create content (Caron & Brennaman, 2014; Chen, 2011). Despite many potential benefits – such as increased teamwork, improved productivity, and increased learning speed – uptake of ENPs has been slower than expected. As educational networking is a relatively new area, existing research can be sparse, and there is an absence of theoretical frameworks to aid in the understanding of adoption and usage.

The aim of this chapter is firstly to review the current literature in order to identify the potential benefits of educational networking and existing barriers to ENP adoption; secondly, to identify key multidisciplinary behavioral theories from across psychology, computer science, and behavioral economics; thirdly, to build upon the existing theories to create a more comprehensive theory-driven model of platform adoption; and finally, to provide a detailed framework for future research aiming to increase ENP adoption and usage. In addition to providing a basis for further academic research to apply, test, and/or refine the model, this work also has value across the wider audience including educators, learning establishments, and platform developers. For instance, advanced understanding of the benefits of ENPs can help encourage educational establishments to successfully introduce this technology, while greater awareness of potential misuse or obstacles allows pro-active mechanisms and training to be implemented to ensure that adoption goes smoothly. Awareness of barriers to uptake can also help developers create more suitable and more appealing ENPs and/or refine existing platform features (such as security and privacy settings).

This chapter starts with a section summarizing *current perspectives on educational networking* including identified benefits of educational networking for students and educators and potential barriers to adoption. This section also discusses the factors that educators should take into account when creating learning materials designed for ENP delivery. The next section discusses the development of *a framework for the future* by incorporating elements from existing behavioral models to develop a more comprehensive model applicable to ENP adoption. The final section summarizes the *implications* of this work including steps for future research.

4.2   Current Perspectives on Educational Networking

Prior to considering the development of behavior change models to increase educational networking, it is important to understand the potential positives and negatives of ENP usage. Therefore, this section provides a detailed overview of the existing research, highlighting benefits of ENP usage for students and educators and identifying potential barriers to usage (including concerns around security, productivity, and ethical issues).

4.2.1   Benefits for Students

Online social networking is widely adopted in everyday life (Lim & Richardson, 2016). It is estimated that over 3.2 billion people actively use SNPs such as Facebook and Twitter – representing approximately 42% of the global population (HootSuite, 2019; We Are Social, 2018). More recently, research has identified that social networking may also be beneficial within educational contexts. It has been suggested that educational networking is particularly relevant to our society where "learning on demand" is a way of life and individuals regularly turn to technology to seek and/ or share information (Dabbagh & Kitsantas, 2012). Educational networking can provide many benefits including allowing students to learn at their own pace and providing the flexibility to work education around their home life and other commitments (Chen, 2011; Peacock & DePlacido, 2018).

Students are already voluntarily adopting a degree of educational usage using traditional SNPs (Masic, Sivic, & Pandza, 2012; Yuen & Yuen, 2008). Sadowski, Pediaditis, and Townsend (2017) found university students were using SNPs as a tool for fostering peer connectedness with fellow students. Sadowski et al. suggest that, with further research, specifically designed ENPs could be "harnessed to maximize both student connectedness and academic learning processes and outcomes" (p. 88). Mazman and Usluel (2010) investigated educational usage of popular social networking site, Facebook. They found that 50% of the variance in educational usage was explained by initial, non-educational Facebook adoption. They describe this as showing "that people first use an innovation in their everyday lives for different purposes while considering its usefulness, ease of use or social influence. Then, these purposes initiate and shape the educational usage of that innovation with its compatibility, potential and interaction with the surrounding educational context." This suggests that ENPs, platforms specifically designed for educational networking, may be positively adopted by users already familiar with SNPs (Welsher et al., 2017).

Some ENP options currently available include Ning for education (NING, 2019), Edmodo (Edmodo, 2019), and Twiducate (livelingua, 2019). There have even been some limited examples of educational establishments creating their own networking platforms (Carr, 2013). There are many promising reasons for introducing ENPs. For instance, Ken Nee (2014) conducted a study investigating students' use of Edmodo as a supplement to their face-to-face teaching. They found that students who used Edmodo performed significantly better in Biology than those who did not use the platform and received only traditional face-to-face teaching. They identified six key factors attributing to this improvement: (1) educational networking fostering self-paced learning, (2) reduced boredom due to more "hands-on" learning, (3) improved complex conceptual understanding, (4) increased interest and motivation, (5) exposure to additional information, and (6) enhanced communication and interactivity.

Other studies have linked educational networking to enhanced communication skills and critical thinking (Habibi et al., 2018; Hamid, Waycott, Kurnia, & Chang, 2015; Mazman & Usluel, 2010; Thaiposri & Wannapiroon, 2015); increased teamwork and collaboration (Ajjan & Hartshorne, 2008; Caron & Brennaman, 2014; de-Marcos, Garcia-Lopez, & Garcia-Cabot, 2016; Habibi et al., 2018; Hamid et al., 2015; Sadowski et al., 2017); greater comprehension of alternative viewpoints (Brady, Holcomb, & Smith, 2010; Salmerón, Macedo-Rouet, & Rouet, 2016); and improved engagement, creativity, productivity, and learning efficiency (de-Marcos et al., 2016; Habibi et al., 2018; Hamid et al., 2015; Lim & Richardson, 2016). George and Dellasega (2011) found that SNPs can enhance creativity, bolster real time communication outside the classroom, increase students links with experts, and foster opportunities for collaboration, suggesting that these benefits will also exist for dedicated ENPs (Ajjan & Hartshorne, 2008). Educational networking can also increase student interest in educational topics and improve teaching effectiveness (Ken Nee, 2014). Furthermore, the ability to easily incorporate visual elements, such as simulation and animation, can aid student understanding of difficult or abstract concepts (Barak, Ashkar, & Dori, 2011). For example, Welsher et al. (2017) found that simulation-based skill learning delivered via an ENP was beneficial to geographically distributed medical trainees.

Self-regulated learning may be improved by the use of ENPs (Habibi et al., 2018) by providing pupils with the ability and opportunity to take ownership of their learning, for example, by creating their own topics/forums based upon their specific needs and interests (Brady et al., 2010; de-Marcos et al., 2016; Hamid et al., 2015). ENPs make it easier for collective knowledge to be easily stored, organized, and searched – providing valuable resources for the future. This provides a wealth of instantly accessible information that students can turn to if they need clarification about particular subjects (Al-Mukhaini, Al-Qayoudhi, & Al-Badi, 2014). This can also be useful for highlighting particularly relevant or helpful contributions to guide pupils toward quality content and/or provide examples of good work (de-Marcos et al., 2016). In addition, ENPs may provide greater opportunities for discussion between educator and student(s) when required, for example, allowing and encouraging students to ask questions as they arise – or helping those who may not have felt comfortable doing so in a group classroom environment.

In addition to direct influences upon learning, the mere act of personal interaction can lead to social benefits such as increased feelings of social inclusion and connectedness (Hung & Yuen, 2010) – factors that have been linked to greater well-being (Sadowski et al., 2017). Connectedness could be especially important for home learning and geographically distributed students (Brady, Holcomb, & Smith, 2002; Welsher et al., 2017), introverted students (Hamid et al., 2015), or students with disabilities which make face-to-face interaction difficult. Educational networking may also have benefits beyond a singular environment or discipline and may pave the way to wider and/or improved networking and collaboration (e.g., between academic departments, educational establishments and other sectors). Safargaliev and Vinogradov (2015) propose that educational networking tools could be particularly beneficial within an innovation-led economy by creating an active and more accessible link between industry and academia. This could lead to graduates more suited to industry jobs; something like this is particularly valuable given that there are concerns that many students lack relevant work experience when leaving education (Fogel & Nehmad, 2009).

There are many psychological theories that support the benefits of educational networking. For example, social learning theory (Bandura, 1977) suggests that we learn from observing others – suggesting that online interaction with other students and/or interactive online content (such as interactive, instructional videos for distance learning students) could be beneficial. Likewise, social development theory (Vygotsky, 1978) suggests that social interaction actually precedes learning and cognitive development. Interacting and networking with others initially enables students to complete tasks with the help of others. As learning continues, interacting with others allows the user to start to demonstrate that they can now complete tasks independently. This also fits within social constructivism (Piaget, 1953), i.e., that culture and environment are key elements of learning. Students who become a part of a wider community of students – such as the community potentially provided by educational networking – are more likely to learn more effectively. Potential benefits, such as self-efficacy, have been shown by Zimmerman (2000) to be a highly effective predictor of student motivation and learning. Therefore, there is a sound theoretical underpinning in the introduction and use of such platforms.

4.2.2   Benefits for Educators

There are also many benefits of ENPs for educators, such as fostering more active, student-centered learning (Minocha, 2009) and increasing interaction between educators and students via a "dual-channel delivery of information" (Ken Nee, 2014). Educators are always seeking to improve upon student engagement, and there is a wealth of research showing that interactive sessions are significantly more beneficial for student learning than traditional lectures with little or no interaction between educator and student. A study by Knight and Wood (2005) compared two styles of teaching for the same syllabus over two semesters: the traditional lecture format and a more interactive format with student participation, active problem solving, and frequent in-class assessment of student understanding. They found that students in the interactive classes had significantly higher levels of learning and better conceptual understanding. Knight and Wood were able to replicate these findings over 2 academic years, and other studies have found similar results supporting interactive learning (Abdel Meguid & Collins, 2017; Huxham, 2005).

The benefits of introducing ENPs need not be restricted to within the individual institution – the platforms can also provide a method through which to disseminate or showcase students' work to the wider population via a secure, moderated channel. There is also the opportunity to collaborate with students from other educational establishments who may be working on similar topics (Caron & Brennaman, 2014).

ENPs also have the potential to have a positive impact upon educator workload and allow them to focus their time more effectively (Ken Nee, 2014). For example, a proportion of student feedback may be able to be delivered automatically and/or instantaneously. Positive methods for reducing teacher workload are critical given increased levels of stress and burnout among educators, particularly early career teachers (Hultell, Melin, & Gustavsson, 2013; Väisänen, Pietarinen, Pyhältö, Toom, & Soini, 2018).

4.2.3   Barriers to Adoption

Despite the identified benefits, adoption of educational networking is much lower – and slower – than would be expected (Brady et al., 2010; Sobaih, Moustafa, Ghandforoush, & Khan, 2016). A meta-analysis by Guraya (2016) found that 20% of medical students reported using social networking for educational purposes (i.e., for sharing academic or educational information), but actual figures for reported educational use ranged from 1.7% to 54% across the studies included in their meta-analysis. Research has demonstrated that the majority of students report positive perceptions of educational networking (Holcomb, Brady, & Smith, 2010; Lim & Richardson, 2016; Neier & Zayer, 2015), suggesting that adoption of the technology itself is not the issue and barriers to adoption lie elsewhere.

Low uptake may be partially due to a lack of awareness around ENPs (Welsher et al., 2017) particularly in relation to ways in which dedicated educational platforms can address concerns that have originated from mainstream SNPs and the potential advantages they can offer in addition to traditional classroom-based learning. The following sections explore potential concerns around the use of educational networking in more detail, including concerns around security and privacy, malicious usage, impact on academic performance, isolation, decreased interaction, and ethical issues.

4.2.3.1   Security and Privacy

Some of the main concerns around general social networking use relate to security and privacy (Holcomb et al., 2010). Many SNPs are easily accessible by third parties outside of the users' immediate connections. This can lead to concerns around the viewing of unsuitable content posted by others (Branley & Covey, 2017b) or the security and privacy of the user's own data (Hoofnagle, King, Li, & Turow, 2010; Small, Kasianovitz, Blanford, & Celaya, 2012). For example, recent coverage in the mass media around the Cambridge Analytica scandal may have led to increased concerns, or at least increased salience of potential risks associated with SNPs (Information Age, 2018). The scandal occurred when Cambridge Analytica, a British political consulting firm, breached Facebook's terms of service when acquiring and using personal data of approximately 87 million Facebook users. As a consequence of this event, trust in the platform dropped and concerns increased around third-party applications using SNPs to harvesting personal information (Kanter, 2018).

There are also concerns that online content may not be suitable for all users, whether this is of an adult, illegal, extremist, and/or violent nature or related to potentially "triggering" content that may have an impact upon a subgroup of vulnerable users (Branley & Covey, 2017a, 2017b; Branley & Covey, 2018; Gerstenfeld, Grant, & Chiang, 2003). In relation to educational networking specifically, Hung and Yuen (2010) found that some educators are concerned about the threat of spam and phishing attacks. Therefore, Hung and Yuen suggest the use of dedicated, private ENPs rather than mainstream SNPs. Although user trust may be difficult to earn, providing a purpose-built ENP that is secure (e.g., encrypted, restricted access, and moderated) is a step in the right direction.

4.2.3.2   Malicious Usage

There has been wide coverage of bullying through SNPs – often referred to as cyberbullying – in both the academic literature and the mass media (e.g., Chisholm, 2006; Law, Shapka, Hymel, Olson, & Waterhouse, 2011; Smith et al., 2008; Wright, 2018). This may further contribute to negative perceptions of educational networking. In order to address this, it is necessary to ensure that some form of content moderation is implemented and that any inappropriate content is addressed quickly and efficiently. Again the use of specific ENPs can help to address this via restricted access to students, educators, and validated users only. However, it is still important that reporting and moderation procedures are put into place. Any internal cyberbullying incidents (e.g., between students) must be addressed appropriately, just as they should be in the offline environment. A perhaps unexpected benefit of ENPs is that the use of online platforms may actually make it easier to detect (and/or report) incidents of bullying that may otherwise go unnoticed.

To further address incidents of bullying, educators should decide whether they wish to allow any anonymous use of the platform, e.g., whether students can post an anonymous question or comment. Anonymity may potentially increase the likelihood that a student will ask a question (e.g., due to lowered potential for embarrassment or lowered apprehension about how their comment or question may reflect upon them as a student). On the other hand, anonymity has been associated with increased anti-social and/or bullying behavior (Mishna, Saini, & Solomon, 2009; Smith et al., 2008), and therefore, ensuring that all users are identifiable may help to limit negative behavior. It is the responsibility of the establishment to identify whether they feel that anonymous contributions are necessary on the ENP and if so, how to moderate this accordingly (or perhaps restrict anonymous usage to certain tasks).

4.2.3.3 Impact Upon Academic Performance

There are also some concerns around the impact of social networking usage on academic performance. For example, Paul, Baker, and Cochran (2012) found a significant negative relationship between time spent by students on SNPs and their academic performance. Kirschner and Karpinski (2010) also found that greater Facebook usage was related to decreased academic performance. This is a controversial area and other studies have not found any evidence of a link between usage and decreased academic performance (e.g., Pasek, More, & Hargittai, 2009). Also, it is important to note that both Paul et al. and Kirschner and Karpinski were looking at the effect of *non-education specific* SNPs where the emphasis is not primarily upon learning, and therefore, the platforms may have been more likely to pose an external distraction from the students' studies. Previous research, such as that by Hung and Yuen (2010), suggests that dedicated ENPs such as Ning provide more focused learning environments.

4.2.3.4 Isolation and Decreased Interaction

The adoption of internet and technology-based interventions may raise concerns around a reduction in face-to-face interaction and whether this may lead to increasingly isolated students. Previous research shows that social interaction is an important factor in successful learning. Lave and Wenger (1991) coined the conceptual term "community of practice" to refer to frameworks of social interaction that individuals belong to. The concept is that individuals generally are involved in several communities of practice, including at home, school, work, and within other social contexts. Research into communities of practice within the educational environment has shown that "engagement in social practice is the fundamental principle by which we learn" (Wenger, 1998). Therefore, it is not unreasonable that people may have negative or pessimistic opinions of educational networking if it is perceived to be a tool to replace social interaction. However, rather than reducing social interaction, ENPs should be designed to *supplement* traditional social interaction (Hung & Yuen, 2010) and further strengthen overall interaction between students and educators. Social interaction can be increased by allowing group participation and discussion even when away from the classroom (Habibi et al., 2018). Hung and Yuen (2010) found that the students in their study had an overwhelmingly positive response to the use of educational networking to supplement their face-to-face teaching. They contributed this largely due to the information sharing and interaction features of the ENP. This suggests that implemented correctly, ENPs should not lead to isolation or decreased social interaction. Rather, ENPs should be used to increase social connectedness. The students in Hung and Yuen's study reported that use of an ENP led to them feeling more connected to their educators and fellow students and also made them more comfortable interacting with them in the offline, classroom environment. Therefore, the social benefits of educational networking may extend beyond the online environment by having a positive, strengthening influence upon face-to-face interaction. This increased sense of social connectedness could be particularly beneficial for distance-learning students who have limited opportunities for face-to-face interaction. As described by Rovai (2002a, 2002b): "If online learners feel a sense of community, it is possible that this emotional connectedness may provide the support needed for them not only to complete successfully a class or a program but also to learn more" (p. 321).

4.2.3.5   Impact Upon the "Digital Divide"

The UK and the USA have steep social gradients in educational performance. Educational inequality can have an effect across the entire economic spectrum – even children from affluent areas perform better in societies with less educational inequality (Wilkinson & Pickett, 2009). It is vital that the introduction of ENPs does not emphasize any inequalities in learning and educational outcome, for example, by providing an advantage to students who have greater access to technology and/ or are more computer literate than others. Although Hung and Yuen (2010) found that the majority of students had an overwhelmingly positive experience with educational networking, they did find that a small subset of users struggled with some of the technological aspects. This was especially true for users who were unfamiliar with SNPs. Many past studies have identified inequalities in access to technology, known as the "digital divide" (Livingstone & Helsper, 2007; Norris, 2001; van Dijk,

2006). While inequalities in access to the internet have decreased considerably in some populations, current research suggests that there is now a "second level digital divide" (Hargittai, 2002). Rather than relating purely to access to the technology, this second level divide relates to *how* people are using the internet, for example, the breadth of the activities they do online, their digital skills, and the type of opportunities or tools they are able to access online. Previous research shows that individuals from less-advantaged households have fewer digital skills (Gui & Argentin, 2011; Hargittai, 2002). Even in those populations where it may appear to be the "norm" to access social networking platforms, research suggests that there are still likely to be inequalities in the type of usage (Micheli, 2016).

The adoption of ENPs may affect the digital divide, however in which direction is currently unknown. For example, although it is possible that those with higher digital literacy skills may find ENPs easier to use and therefore benefit more than those with lower digital literacy, it is also possible that the wider adoption of ENPs within schools and learning establishments may actually decrease the digital divide between students. What is clear is that it is critical that equitable access to learning content (and associated opportunities) is available to all students (Smart Education Networks by Design, 2014). Future research should aim to investigate the impact of educational networking adoption on learning inequalities and how to address any identified issues.

4.2.3.6 Work-Life Balance

There are also ethical concerns around work-life balance and the introduction of ENPs, i.e., if learners or educators feeling that their work-life balance is compromised by this technology. Although more contact time between educators and students could be beneficial, it is important that students and staff are still advised on how to effectively manage and respect a healthy work-life balance (Habibi et al., 2018). Using specific ENPs, rather than SNPs, may help to provide a clearer distinction between educational networking and personal social networking (Caron & Brennaman, 2014; Hung & Yuen, 2010; Ken Nee, 2014; Selwyn, 2009; Tsai, Lin, & Tsai, 2001). However, as students often struggle with effectively managing their work and private time (Peacock & DePlacido, 2018), future research should aim to investigate whether the 24-hour accessibility of ENPs could have the potential to be problematic for some students, particularly those feeling under academic pressure. Academic staff are also under high workloads – which in turn has the potential to affect their job satisfaction and well-being (Houston, Meyer, & Paewai, 2006). High workload has also been identified as a barrier to the adoption of technology (Gregory & Lodge, 2015). Healthy work-life balance must be fostered within all work sectors, and this is something that must be considered throughout the design, adoption, and ongoing usage of educational networking technologies. If a problem with work-life balance is identified, it is possible that some limits on platform accessibility (either in relation to time or location) could be implemented.

4.2.4 Blended Learning: Successfully Developing ENP Learning Materials

The effectiveness of an ENP in relation to improving student outcomes is likely to be significantly affected by the quality of the learning content/materials that it provides (Welsher et al., 2017). Online platforms should not be regarded as a quick, sub-par alternative to traditional teaching. Online content needs to reflect the same diligent level of attention to detail and quality as traditional, offline teaching materials. One of the main problematic areas when creating online content is that educators attempt to directly adapt their traditional face-to-face teaching materials (Peacock & DePlacido, 2018). However, materials need to be created specifically for the online environment in order to ensure that students benefit from the unique interactive elements offered by online platforms. For example, simply uploading static PDF files of lecture slides completely undermines the meaning of the term "networking." Online content should be interactive, dynamic, and engaging. Due to the unique environment of online networking platforms and their relatively new introduction to education, there is a need to adequately support educators when first adopting this technology (Peacock & DePlacido, 2018). ENPs also introduce a wide range of digital curation tools and platforms that could be overwhelming for educators to navigate (Mott, 2010). Learning establishments should ensure that relevant and effective training programs are put into place. A recent report stresses the importance of "substantial training and support of teachers and technical staff" and the need for educational establishments to have a clear strategy and plan for the introduction of ENPs (Smart Education Networks by Design, 2014). The authors of the report suggest that this training should begin at least 6 months prior to platform adoption.

Previous research has investigated the best methods of integrating traditional face-to-face teaching and online learning. As identified by Caron and Brennaman (2014), the adoption of ENPs reflects a change in learning paradigms. Chen (2011) describes learning in this context as demand driven, social, embedded in rich social contexts, reflective, and involving the internalization from social to the individual. An experiential design framework by Pine and Gilmore (1999) proposes that experiences encompass four realms: (1) esthetic, whether the design is inviting, interesting, and comfortable; (2) escapist, whether the design focuses on immersive activities; (3) educational, whether the design encourages active learning and exploration; and (4) entertainment, whether the design allows for fun and enjoyment to help sustain attention and motivation. Using this framework, Chen suggests that platforms should aim to include aspects from across all four realms. Chen highlights that the primary purpose of ENPs is to engage rather than entertain. Therefore, caution should be utilized in relation to the escapist and entertainment aspects of ENPs in order to ensure that learning remains the primary focus and that users are not led into over-involvement or counterproductive outcomes. For example, students could become overwhelmed or over-involved by the amount of activities available and/or the social elements of the platform.

4.3   A Framework for the Future: Incorporating Theoretical Models to Overcome Barriers to ENP Adoption

Existing research within this field tends to be lacking in theoretical underpinning, with no established model or framework available to specifically address uptake of ENPs by educators, learning establishments, and students. In order to address this gap in the current literature, this section draws upon four of the leading behavioral theories from psychology and other disciplines: *theory of planned behavior* (TPB: Ajzen, 1991), *technology acceptance model* (TAM: Davis, Bagozzi, & Warshaw, 1989), *information system success model* (ISS: DeLone & McLean, 1992, 2003), and *protection motivation theory* (PMT: Rogers, 1975). Each of these theories is introduced and discussed in relation to the factors which may influence educational networking. Key factors from across the models have been integrated to build a new model – the *platform adoption model* (PAM). This section ends by explaining how the PAM can be used to guide future research and provide a framework for the future.

4.3.1   Theory of Planned Behavior

The theory of planned behavior (TPB; Ajzen, 1991) originates from the field of psychology and is one of the most widely applied models of human behavior. The model asserts that behavior is a direct function of behavioral intention, and behavioral intention is a function of *attitude*, *subjective norms*, and *perceived behavioral control*. The model is shown in Fig. 4.1.

Applying the model to ENPs suggests that adoption will be influenced by users' general positive or negative *attitude* toward the platform, i.e., whether they perceive the ENP to be a good or bad thing; their perceived social pressure to participate in ENP usage (*subjective norm*); and their perceived behavioral control, i.e., how confident they feel in their ability to successfully use the ENP. These three factors will feed into their *intention* to use the platform and subsequently their future usage.

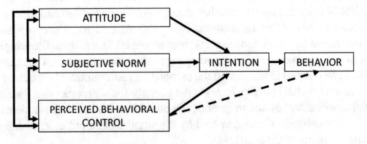

Fig. 4.1 The theory of planned behavior model

Despite its widespread use within behavioral research, there is limited research applying the TPB specifically to educational networking. However, Hartshorne and Ajjan (2009) and Mazman and Usluel (2010) have shown the TPB factors to be significant predictors of Facebook use for educational reasons. This demonstrates that these factors are likely to be relevant to education-specific platforms. One of the potential limitations of solely applying the TPB model to educational networking relates to the lack of factors specifically concerning the technological elements of the platform (e.g., functionality, design, and interface). While some of these issues may potentially be captured by factors such as *perceived behavioral control*, theoretical models more specific to technology could be integrated to build a more comprehensive model.

4.3.2   Technology Acceptance Model and Information System Success Model

Two technology-specific models are the technology acceptance model (TAM; Davis et al., 1989) and the information system success model (ISS; DeLone & McLean, 1992, 2003). The TAM proposes that *perceived usefulness* and *perceived ease of use* are two of the main factors predicting technology acceptance (Fig. 4.2). Similar to the TPB, these factors have an influence on behavior via *intention to use*.

Perceived usefulness and *perceived ease of use* have both been found to be positively related to Facebook adoption (Mazman & Usluel, 2010) and intention to use Web 2.0 technologies including social networking (Ajjan & Hartshorne, 2008). More specific to educational networking, *perceived usefulness* has been shown to be significant predictor of using traditional SNPs such as Facebook *for educational purposes* (Mazman & Usluel, 2010). Therefore, ENP platforms are likely to be influenced by similar factors.

Similar to TAM, the ISS taps into factors around the usability of the platform (Fig. 4.3). The ISS presents three dependent variables: *intention to use*, *effective use*, and *user satisfaction*, as well as *net benefits* derived from use of the system. These factors are preceded by *information quality*, *system quality*, and *service quality*.

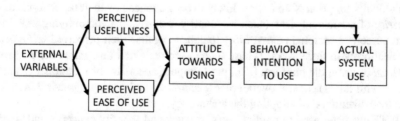

Fig. 4.2 The technology acceptance model

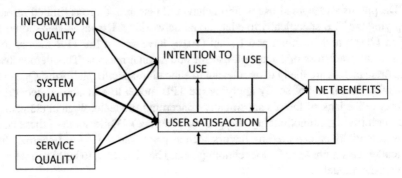

Fig. 4.3 Information system success model

One of the main limitations of the TAM and ISS, in relation to ENP uptake, is that they do not account for interpersonal factors which have been shown to predict platform usage and behavior, such as social norms (Banks et al., 2010). Due to social interaction being at the heart of social and educational networking, it is vital that any theoretical model applied to these platforms incorporates *both* technological and social factors.

4.3.3 Protection Motivation Theory

Many of the potential concerns around ENPs are based upon perceptions of risk (e.g., concerns over privacy and security and the risk of cyberbullying). A comprehensive model of ENP adoption should incorporate elements relating to risk and decision-making. It is suggested that individuals often make decisions by weighing up the costs and benefits of a behavior, sometimes referred to as the cost-benefit analysis or risk-benefit tradeoff (Branley, 2015; Branley & Covey, 2018). This is something that is often overlooked when applying theory-driven models to technology adoption (despite risk-benefit tradeoffs being linked to online behavior, e.g., Branley & Covey, 2018). One of the main risk perception models is protection motivation theory (PMT; Rogers, 1975; Fig. 4.4). Research has started to apply PMT to areas such as cybersecurity (e.g., Blythe, Coventry, & Little, 2015), and it is possible that PMT could be used to explain adoption of ENPs. PMT suggests that an individual's adoption of an ENP is likely to be influenced by their perceptions of the *severity* of associated risks (such as security, privacy, or cyberbullying risks) and their perceived personal *vulnerability* (balanced against their perceived *self-efficacy* for dealing with these risks). These factors are weighted up against the perceived *rewards* of using the platform (e.g., peer support, access to information, communication with tutors). Users' overall *threat appraisal* and *coping appraisal* then feed into their likelihood of adopting the technology.

PMT has been used to predict users' privacy and security concerns and sharing of personal information on SNPs (Marett, McNab, & Harris, 2011; Mohamed &

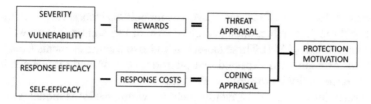

Fig. 4.4 Protection motivation theory

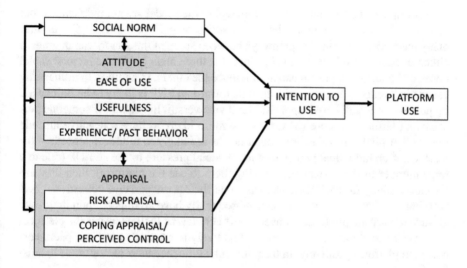

Fig. 4.5 The platform adoption model

Ahmad, 2012). *Self-efficacy* has also been shown to predict intention to adopt Web 2.0 technologies by faculty members (Ajjan & Hartshorne, 2008). The PMT factors are likely to influence users' appraisal – and subsequent usage – of ENPs. Therefore, these factors should be incorporated into models of adoption.

4.3.4   Proposed Framework and the Platform Adoption Model

By incorporating key elements from each of the discussed models, it is possible to create a more comprehensive model of ENP adoption. This allows gaps in the previous models to be addressed, e.g., the TPB is a good overall model for predicting human behavior; however, it does not include technology specific factors (such as those from the TAM and ISS). The technology focused models do not incorporate threat and risk appraisal (covered within PMT). The proposed model incorporates all of these elements and is termed the *platform adoption model* (PAM: Fig. 4.5).

PAM identifies the role of *social norm*, *attitude*, and *risk/coping appraisal* in influencing educational networking adoption and usage. Therefore capturing the key elements from TPB, PMT and TAM. The model also incorporates technological influence through its effect upon perceived ease of use and usefulness which feed into users' *attitude* (TAM). These factors should also capture elements from the ISS model; for example, it is expected that platform, content, and service quality will feed into users' attitudes toward the platform.

A description of each of the factors included within the PAM model is provided in Table 4.1, including how each factor is likely to influence usage (via intention to use).

It is important to note that the factors within the model are not independent but instead influence one another through a series of pathways (Fig. 4.5). As with many theoretical models, the pathways between some of the factors can operate in either, or both, directions. In this instance, the three main predictive factors: social norm, attitude, and appraisal can all influence the others, in addition to influencing usage via intention. For example, attitude toward an ENP is likely to be influenced by positive or negative risk appraisal (and vice versa). In general, if someone perceives a platform as more risky, they are more likely to have a negative attitude toward that platform. Likewise, social norms are likely to influence attitude – for instance, if an individual perceives a lot of social pressure to use an ENP (due to a large number of their peers encouraging them to use the platform), then they are also more likely to reach the conclusion that their peers perceive educational networking as a "good thing." This can subsequently have an impact upon their own attitude toward the platform. Conversely, if very few of their peers use the platform (i.e., weak social norm), then they are less likely to conclude that their peers perceive the platform positively. In turn, this may influence their own attitude against the platform.

Table 4.1 Platform adoption model factors

Factor	Potential influence upon platform uptake/usage
Social norm	Direct positive or negative effect upon intention to use through social pressure. Can also potentially affect intention via *attitude* (e.g., "all my peers are using it so it must be good") and *appraisal* (e.g., "all my friends are using it so it can't be that risky or difficult")
Attitude	Direct positive or negative effect upon intention to use and/or indirect influence through *social norm* or *appraisal*:
(i) Ease of use	If an ENP is perceived as easy to use, users are more likely to intend to use the platform
(ii) Usefulness	If an ENP is perceived as useful, intention, intention to use is likely to be increased
(iii) Experience/ past behavior	Past experience with SN or ENPs or collaborative learning experiences may have an influence on a users' attitude toward the use of educational networking (positively or negatively)
Appraisal	Direct positive or negative effect upon intention to use and/or indirect influence through *social norm* or *appraisal*:
(i) Risk appraisal	If ENP usage is perceived as risky (e.g., security risks, cyberbullying), intention to use is likely to be reduced
(ii) Coping appraisal/perceived control	If users feel they have control over their ENP use (e.g., they feel confident adjusting privacy settings), intention to use is likely to be increased

Of course, as this is a preliminary version of the model, it is suggested that future research seeks to test the model and causal pathways and to build upon (and refine) it accordingly. Using this new model, it is possible to identify a guidance framework for targeting uptake of educational networking. The following sections provide clear examples of how each of the PAM factors could potentially be influenced to increase educational networking uptake for both students and educators.

4.3.4.1   Influencing Social Norms Around ENPs

One potential method to help strengthen perceived social norms is to demonstrate positive perceptions from other students, for example, by asking a class to offer examples of when and/or how they have found the platform useful (or if they are not yet using ENP, to think about ways in which they could imagine such a platform being helpful). This could help to foster positive social norms to help encourage ENP adoption. Similarly, to help bolster positive social norms among educators, it could be beneficial to present users with information on how other educators and educational establishments have successfully used ENPs. In keeping with the social norms approach to behavior change, it is likely that this type of intervention will be most beneficial when the shared experiences come from respected/credible sources and/or from individuals and/or environments that the educator/establishment regards as similar to themselves (Elgaaied-Gambier, Monnot, & Reniou, 2018).

4.3.4.2   Influencing Attitude to ENPs

As proposed in the model (Fig. 4.5), attitude toward ENP adoption can be influenced by three factors: ease of use, perceived usefulness, and experience/past behavior. There are multiple ways in which each of these factors could potentially be influenced to encourage student and educator uptake – these are detailed in Table 4.2.

4.3.4.3   Influencing Risk and Coping Appraisal of ENPs

To lower students' *risk appraisal* and encourage ENP adoption, it may be beneficial to ask what they perceive the risks to be and explain how these risks are addressed by the platform (or associated policies). Overall appraisal can also be improved by increasing *coping appraisal* by identifying steps toward limiting risks and helping students feel more confident in their ENP use. For instance, a student concerned about security and privacy may feel more confident in using ENPs if they are taught how to ensure they have a secure, encrypted connection to the platform.

In a similar manner, educators' risk appraisal can potentially be improved by identifying the risks that they are most concerned about (these may be risks that affect themselves and/or their students) and explain how these risks are addressed and/or identify steps toward limiting these risks. If educators are taught how to use

Table 4.2 Influencing student and educator attitude to educational networking using the platform adoption model

Ease of use	
Student	(i) Design interfaces that are intuitive and where possible incorporate familiar features, e.g., similar to SNPs (Welsher et al., 2017) (ii) Provide training before adoption and/or following platform changes (iii) Adopt user-centric design processes to actively involve students throughout the design and development of ENPs
Educator	(i) Ensure that the educator interface is user-friendly (ii) Provide educator training prior to adoption. This will encourage educators to adopt in the first instance and maintain a positive attitude toward the ENP, while providing them with the ability to address student concerns. Ensure training is regularly updated (iii) Adopt user-centric design processes to involve educators throughout the design and development of ENPs
Usefulness	
Student	(i) Ensure benefits are clearly explained to students prior to introducing the platform (ii) Reinforce usefulness by regularly highlighting learning progress. Platform developers and educators may wish to research the most effective methods of producing this feedback – online visual "progress markers," face-to-face feedback, etc.
Educator	(i) Prior to introducing the ENP, ensure that educators are informed of the benefits of the platform as a *complementary tool* to existing teaching (ii) Provide ongoing training on how to design content specifically for online platforms (iii) Demonstrate how ENPs *differ* from traditional SNPs and how this can increase their effectiveness (e.g., learning focused)
Experience/past behavior	
Student	Discuss any concerns with students prior to introducing ENPs and demonstrate a clear way to address these concerns
Educator	As some educators may hold a negative or ambivalent view of networking platforms, it is important to explain and demonstrate how ENPs differ from traditional SNPs and how this can address concerns from previous experiences (e.g., privacy or bullying)

the platform in a confident manner and how to limit any perceived risk, their coping appraisal and perceived control should be increased. For educators, this includes detailed training on what to do if something does not go to plan when their students are using the ENP (e.g., how to moderate content, how to react to cyberbullying or unauthorized access).

4.4   Summary and Implications

This chapter has addressed an important gap in the existing knowledge around educational networking by:

1. Reviewing the existing literature to identify potential benefits of educational networking, above and beyond traditional teaching methods

2. Identifying current student and educator attitudes toward ENP usage and identifying potential barriers to adoption, including the identification of ethical issues that must be addressed prior to introducing ENPs to the learning environment
3. Introducing four leading theoretical models and illustrating how each model contains factors relevant to ENP adoption
4. Collating key information from across all four models to produce a preliminary, comprehensive model specific to platform uptake: the *platform adoption model* (PAM)
5. Utilizing the PAM to produce a preliminary theory-driven framework for increasing student and educator uptake of ENPs and providing the basis for future research

Educational networking has the potential for many positive outcomes for users – from improved learning to social benefits and improved well-being. By harnessing these positives and addressing potential concerns, educational networking can continue to go from strength to strength. Understanding the benefits of ENPs can help educational establishments to appreciate and successfully introduce and manage this technology. Similarly, awareness of potential misuse and concerns allows protective mechanisms to be implemented from the start. Insight into other barriers to uptake can help to identify training and reassurance that may be needed by staff and students. The proposed model and framework provides a solid basis for future research (including testing and refinement of the model) and has the potential to be beneficial within academia and the wider audience, such as developers, educators, and learning establishments. Developers can also use the model and framework to help guide the design of future platforms.

The author suggests that future studies should aim to test the model across different samples using a confirmatory technique such as structural equation modelling (SEM). Future research may also wish to investigate the role of individual differences – for example, individual learning styles and differences in motivation, as it is important to identify if ENPs may provide an advantage or disadvantage to subgroups of users. Furthermore, future studies should investigate student and educator work-life balance and what factors may mediate whether ENP usage has a positive – rather than detrimental – effect upon workload.

Although the proposed framework aims to increase effective use of ENPs, it is important to ensure that this use supplements – rather than replaces – more traditional interaction (Hung & Yuen, 2010). Replacing face-to-face communication can be detrimental to well-being, for example, as a consequence of not nurturing the strong sense of social presence that is found in direct interaction (Brady et al., 2010). Recent research by Sadowski et al. (2017) highlights the need to maintain a balance between digital and face-to-face learning and connectedness. It is also vital to include physical assessment for "hands-on" skills, to complement online simulation (Welsher et al., 2017). Future research should aim to identifying the most productive ratio of face-to-face learning and ENP usage, across a range of contexts (this ratio is not likely to be one, universal "golden" number for all teaching environments, educators, and students). It is also important to respect student (and educator) adjustment; social interaction cannot be forced and users may take time to adjust and familiarize with their new learning environment – just as they would in the offline world (Brady et al., 2010). Some users may judge ENPs negatively if they feel forced or coerced into their use (Neier & Zayer, 2015; Sadowski et al., 2017); therefore, the introduction of new ENPs must be done in a phased and sympathetic manner, with adequate familiarization and training.

One last word of caution is to remember that educational methods of any description are only as good as the content that they deliver. It is worth highlighting to educators that "using social networking to supplement face-to-face courses can become time intensive" (Hung & Yuen, 2010, p. 713). Therefore, educators should ensure that they can dedicate the necessary time to correctly designing and structuring the learning materials prior to adopting this technology in their classroom. Poor teaching materials and/or inadequate teaching techniques will not be overcome by even the best, most well-designed, user-friendly ENP. Previous research shows that students express frustration when educators appear to have little presence or interest (Peacock & DePlacido, 2018). The key is to strengthen all aspects of education, using educational networking as a *complementary* addition to an already high-quality teaching program. ENPs must not to be regarded as a substitute or replacement for good quality, attentive teaching, but rather as another step in the quest toward teaching excellence and equality in learning.

References

Abdel Meguid, E., & Collins, M. (2017). Students' perceptions of lecturing approaches: Traditional versus interactive teaching. *Advances in Medical Education and Practice, 8*, 229–241. https://doi.org/10.2147/AMEP.S131851

Ajjan, H., & Hartshorne, R. (2008). Investigating faculty decisions to adopt Web 2.0 technologies: Theory and empirical tests. *The Internet and Higher Education, 11*(2), 71–80. https://doi.org/10.1016/J.IHEDUC.2008.05.002

Ajzen, I. (1991). The theory of planned behavior. *Organizational Behavior and Human Decision Processes, 50*(2), 179–211. https://doi.org/10.1016/0749-5978(91)90020-T

Al-Mukhaini, E. M., Al-Qayoudhi, W. S., & Al-Badi, A. H. (2014). Adoption of social networking in education: A study of the use of social networks by higher education students in Oman. *Journal of International Education Research (JIER), 10*(2), 143. https://doi.org/10.19030/jier.v10i2.8516

Bandura, A. (1977). Social learning theory. In *Social learning theory* (pp. 1–46). New York: General Learning Press.

Banks, M. S., Onita, C. G., Meservy, T. O., Banks, M., Shane, & Onita, C. G. (2010). Risky behavior in online social media: Protection motivation and social influence. In *AMCIS 2010 Proceedings*. Retrieved from http://aisel.aisnet.org/amcis2010/372

Blythe, J. M., Coventry, L., & Little, L. (2015). *Unpacking security policy compliance: The motivators and barriers of employees' security behaviors*. Retrieved from https://www.usenix.org/conference/soups2015/proceedings/presentation/blythe

Brady, K. P., Holcomb, L. B., & Smith, B. V. (2002). The use of alternative social networking sites in higher educational settings: A case study of the E-learning benefits of Ning in education. *Journal of Interactive Online Learning, 9*(2), 151–170. Retrieved from https://www.learntechlib.org/p/109412/

Brady, K. P., Holcomb, L. B., & Smith, B. V. (2010). The use of alternative social networking sites in higher educational settings: A case study of the E-learning benefits of Ning in education. *Journal of Interactive Online Learning, 9*(2), 151–170. Retrieved from www.ncolr.org/jiol

Branley, D. (2015). *Risky behaviour: Psychological mechanisms underpinning social media users' engagement.* Durham University.

Branley, D. B., & Covey, J. (2017a). Is exposure to online content depicting risky behavior related to viewers' own risky behavior offline? *Computers in Human Behavior, 75*, 283–287. https://doi.org/10.1016/j.chb.2017.05.023

Branley, D. B., & Covey, J. (2017b). Pro-ana versus pro-recovery: A content analytic comparison of social media users' communication about eating disorders on Twitter and Tumblr. *Frontiers in Psychology, 8*, 1356. https://doi.org/10.3389/fpsyg.2017.01356

Branley, D. B., & Covey, J. (2018). Risky behavior via social media: The role of reasoned and social reactive pathways. *Computers in Human Behavior, 78*, 183–191. https://doi.org/10.1016/j.chb.2017.09.036

Caron, J., & Brennaman, K. (2014). Co-evolution of emerging technology, learning theory, and instructional design. *Theories of Educational Technology, 9*, 185. https://doi.org/10.1177/0973258614528614

Carr, D. (2013, July 31). Universities create their own social networks for students – InformationWeek. *InformationWeek.* Retrieved from https://www.informationweek.com/universities-create-their-own-social-networks-for-students/d/d-id/1110997

Chen, P. (2011). *From CMS to SNS: Educational networking for urban teachers.* Retrieved from https://files.eric.ed.gov/fulltext/EJ952059.pdf

Chisholm, J. F. (2006). Cyberspace violence against girls and adolescent females. *Annals of the New York Academy of Sciences, 1087*, 74–89. https://doi.org/10.1196/annals.1385.022

Dabbagh, N., & Kitsantas, A. (2012). Personal Learning Environments, social media, and self-regulated learning: A natural formula for connecting formal and informal learning. *The Internet and Higher Education, 15*, 3–8. https://doi.org/10.1016/j.iheduc.2011.06.002

Davis, F. D., Bagozzi, R. P., & Warshaw, P. R. (1989). User acceptance of computer technology: A comparison of two theoretical models. *Management Science, 35*, 982. https://doi.org/10.1287/mnsc.35.8.982

DeLone, W. H., & McLean, E. R. (1992). Information systems success: The quest for the dependent variable. *Information Systems Research, 3*, 60. https://doi.org/10.1287/isre.3.1.60

DeLone, W. H., & McLean, E. R. (2003). The DeLone and McLean model of information systems success: A ten-year update. *Journal of Management Information Systems, 19*, 9. https://doi.org/10.1080/07421222.2003.11045748

de-Marcos, L., Garcia-Lopez, E., & Garcia-Cabot, A. (2016). On the effectiveness of game-like and social approaches in learning: Comparing educational gaming, gamification & social networking. *Computers & Education, 95*, 99–113. https://doi.org/10.1016/J.COMPEDU.2015.12.008

Edmodo. (2019). *Connect with students and parents in your paperless classroom | Edmodo.* Retrieved May 6, 2019, from https://www.edmodo.com/

Elgaaied-Gambier, L., Monnot, E., & Reniou, F. (2018). Using descriptive norm appeals effectively to promote green behavior. *Journal of Business Research, 82*, 179. Retrieved from https://hal.archives-ouvertes.fr/hal-01630909/file/Using-descriptive-norm-appeals-effectively-to-promote-green-behavior.pdf

Fogel, J., & Nehmad, E. (2009). Internet social network communities: Risk taking, trust, and privacy concerns. *Computers in Human Behavior, 25*(1), 153–160. https://doi.org/10.1016/j.chb.2008.08.006

George, D. R., & Dellasega, C. (2011). Use of social media in graduate-level medical humanities education: Two pilot studies from Penn State College of Medicine. *Medical Teacher, 33*(8), e429–e434. https://doi.org/10.3109/0142159X.2011.586749

Gerstenfeld, P. B., Grant, D. R., & Chiang, C.-P. (2003). Hate online: A content analysis of extremist internet sites. *Analyses of Social Issues and Public Policy, 3*(1), 29–44. https://doi.org/10.1111/j.1530-2415.2003.00013.x

Gregory, M. S.-J., & Lodge, J. M. (2015). Academic workload: The silent barrier to the implementation of technology-enhanced learning strategies in higher education. *Distance Education, 36*(2), 210–230. https://doi.org/10.1080/01587919.2015.1055056

Gui, M., & Argentin, G. (2011). Digital skills of internet natives: Different forms of digital literacy in a random sample of northern Italian high school students. *New Media & Society, 13*(6), 963–980. https://doi.org/10.1177/1461444810389751

Guraya, S. (2016). The usage of social networking sites by medical students for educational purposes: A meta-analysis and systematic review. *North American Journal of Medical Sciences, 8*(7), 268. https://doi.org/10.4103/1947-2714.187131

Habibi, A., Mukinin, A., Riyanto, Y., Prasohjo, L. D., Sulistiyo, U., Sofwan, M., et al. (2018). Building an online community: Student teachers' perceptions on the advantages of using Social Networking Services in a teacher education program. *Turkish Online Journal of Distance Education, 19*(1), 46–61. Retrieved from https://eric.ed.gov/?id=EJ1165898

Hamid, S., Waycott, J., Kurnia, S., & Chang, S. (2015). Understanding students' perceptions of the benefits of online social networking use for teaching and learning. *The Internet and Higher Education, 26*, 1–9. https://doi.org/10.1016/J.IHEDUC.2015.02.004

Hargittai, E. (2002). Second-level digital divide: Differences in people's online skills. *First Monday, 7*(4). https://doi.org/10.5210/fm.v7i4.942

Hartshorne, R., & Ajjan, H. (2009). Examining student decisions to adopt Web 2.0 technologies: Theory and empirical tests. *Journal of Computing in Higher Education, 21*(3), 183–198. https://doi.org/10.1007/s12528-009-9023-6

Holcomb, L. B., Brady, K. P., & Smith, B. V. (2010). The emergence of educational networking: can non-commercial, education-based social networking sites really address the privacy and safety concerns of educators? *MERLOT Journal of Online Learning and Teaching, 6*(2), 475–481.

Hoofnagle, C. J., King, J., Li, S., & Turow, J. (2010). How different are young adults from older adults when it comes to information privacy attitudes and policies? *Social Science Research Network, 4*(19), 10. https://doi.org/10.2139/ssrn.1589864

HootSuite. (2019). *Digital in 2019 – global report*. Retrieved from https://hootsuite.com/pages/digital-in-2019

Houston, D., Meyer, L. H., & Paewai, S. (2006). Academic staff workloads and job satisfaction: Expectations and values in academe. *Journal of Higher Education Policy and Management, 28*(1), 17–30. https://doi.org/10.1080/13600800500283734

Hultell, D., Melin, B., & Gustavsson, J. P. (2013). Getting personal with teacher burnout: A longitudinal study on the development of burnout using a person-based approach. *Teaching and Teacher Education, 32*, 75–86. https://doi.org/10.1016/J.TATE.2013.01.007

Hung, H.-T., & Yuen, S. C.-Y. (2010). Educational use of social networking technology in higher education. *Teaching in Higher Education, 15*(6), 703–714. https://doi.org/10.1080/13562517.2010.507307

Huxham, M. (2005). Learning in lectures. *Active Learning in Higher Education, 6*(1), 17–31. https://doi.org/10.1177/1469787405049943

Information Age. (2018). Online privacy concerns in a post-Cambridge Analytica scandal era. *Information Age*. Retrieved from https://www.information-age.com/online-privacy-cambridge-analytica-123477564/

Kanter, J. (2018, April 17). Facebook trust levels collapse after Cambridge Analytica data scandal. *Business Insider*. Retrieved from https://www.businessinsider.com/facebook-trust-collapses-after-cambridge-analytica-data-scandal-2018-4?r=UK&IR=T

Ken Nee, C. (2014). The effect of educational networking on students' performance in biology. *International Journal on Integrating Technology in Education, 3*(1), 21–41. https://doi.org/10.5121/ijite.2014.3102

Kirschner, P. A., & Karpinski, A. C. (2010). Facebook® and academic performance. *Computers in Human Behavior, 26*(6), 1237–1245. https://doi.org/10.1016/J.CHB.2010.03.024

Knight, J. K., & Wood, W. B. (2005). Teaching more by lecturing less. *Cell Biology Education, 4*(4), 298–310. https://doi.org/10.1187/05-06-0082

Lave, J., & Wenger, E. (1991). Chapter 1: Legitimate peripheral participataion. In *Situated learning: Legitimate peripheral participation*. Cambridge/New York: Cambridge University Press. https://doi.org/10.1017/CBO9780511803932

Law, D. M., Shapka, J. D., Hymel, S., Olson, B. F., & Waterhouse, T. (2011). The changing face of bullying: An empirical comparison between traditional and internet bullying and victimization. *Computers in Human Behavior, 28*(1), 232–226. https://doi.org/10.1016/j.chb.2011.09.004

Lim, J., & Richardson, J. C. (2016). Exploring the effects of students' social networking experience on social presence and perceptions of using SNSs for educational purposes. *The Internet and Higher Education, 29*, 31–39. https://doi.org/10.1016/J.IHEDUC.2015.12.001

livelingua. (2019). *Twiducate – social networking & media for schools :: Education 2.0*. Retrieved May 6, 2019, from https://www.livelingua.com/twiducate/

Livingstone, S., & Helsper, E. (2007). Gradations in digital inclusion: Children, young people and the digital divide. *New Media & Society, 9*(4), 671–696. https://doi.org/10.1177/1461444807080335

Marett, K., McNab, A. L., & Harris, R. B. (2011). Social networking websites and posting personal information: An evaluation of protection motivation theory. *AIS Transactions on Human-Computer Interaction, 3*(3), 170. Retrieved from https://aisel.aisnet.org/thci/vol3/iss3/2

Masic, I., Sivic, S., & Pandza, H. (2012). Social networks in medical education in Bosnia and Herzegovina. *Materia Socio Medica, 24*(3), 162. https://doi.org/10.5455/msm.2012.24.162-164

Mazman, S. G., & Usluel, Y. K. (2010). Modeling educational usage of Facebook. *Computers & Education, 55*(2), 444–453. https://doi.org/10.1016/J.COMPEDU.2010.02.008

Micheli, M. (2016). Social networking sites and low-income teenagers: Between opportunity and inequality. *Information, Communication & Society, 19*(5), 565–581. https://doi.org/10.1080/1369118X.2016.1139614

Minocha, S. (2009). Role of social software tools in education: A literature review. *Education + Training, 51*(5/6), 353–369. https://doi.org/10.1108/00400910910987174

Mishna, F., Saini, M., & Solomon, S. (2009). Ongoing and online: Children and youth's perceptions of cyber bullying☆. *Children and Youth Services Review, 31*(12), 1222–1228. https://doi.org/10.1016/j.childyouth.2009.05.004

Mohamed, N., & Ahmad, I. H. (2012). Information privacy concerns, antecedents and privacy measure use in social networking sites: Evidence from Malaysia. *Computers in Human Behavior, 28*(6), 2366–2375. https://doi.org/10.1016/J.CHB.2012.07.008

Mott, J. (2010). Envisioning the post-LMS Era: The Open Learning Network. *EDUCAUSE Quarterly*.

Neier, S., & Zayer, L. T. (2015). Students' perceptions and experiences of social media in higher education. *Journal of Marketing Education, 37*(3), 133–143. https://doi.org/10.1177/0273475315583748

NING. (2019). *Create an education website*. Retrieved May 6, 2019, from https://www.ning.com/create-educational-website/

Norris, P. (2001). *Digital divide: Civic engagement, information poverty, and the Internet worldwide*. Cambridge: Cambridge University Press.

Pasek, J., More, E., & Hargittai, E. (2009). Facebook and academic performance: Reconciling a media sensation with data. *First Monday, 14*(5). https://doi.org/10.5210/fm.v14i5.2498

Paul, J. A., Baker, H. M., & Cochran, J. D. (2012). Effect of online social networking on student academic performance. *Computers in Human Behavior, 28*(6), 2117–2127. https://doi.org/10.1016/J.CHB.2012.06.016

Peacock, S., & DePlacido, C. (2018). Supporting staff transitions into online learning: A networking approach. *Journal of Perspectives in Applied Academic Practice, 6*(2), 67–75. https://doi.org/10.14297/jpaap.v6i2.336

Piaget, J. (1953). The origins of intelligence in children. *Journal of Consulting Psychology, 17*, 467. https://doi.org/10.1037/h0051916

Pine, B. J., & Gilmore, J. H. (1999). *The experience economy: Work is theatre & every business a stage*. Boston: Harvard Business School Press.

Rogers, R. W. (1975). A protection motivation theory of fear appeals and attitude change. *The Journal of Psychology, 91*(1), 93–114. https://doi.org/10.1080/00223980.1975.9915803

Rovai, A. P. (2002a). Building sense of community at a distance. *International Review of Research in Open and Distance Learning, 3*(1), 1–16. Retrieved August 29, 2019 from https://www.learntechlib.org/p/49515/

Rovai, A. P. (2002b). Sense of community, perceived cognitive learning, and persistence in asynchronous learning networks. *Internet and Higher Education, 5*, 319. https://doi.org/10.1016/S1096-7516(02)00130-6

Sadowski, C., Pediaditis, M., & Townsend, R. (2017). University students' perceptions of social networking sites (SNSs) in their educational experiences at a regional Australian university. *Australasian Journal of Educational Technology, 33*(5), 77–90. Retrieved from https://www.learntechlib.org/p/181380/

Safargaliev, E. R., & Vinogradov, V. L. (2015). Educational networking as key factor of specialist training in universities. *International Education Studies, 8*(10), 200. Retrieved from https://www.questia.com/library/journal/1P3-3840402951/educational-networking-as-key-factor-of-specialist

Salmerón, L., Macedo-Rouet, M., & Rouet, J.-F. (2016). Multiple viewpoints increase students' attention to source features in social question and answer forum messages. *Journal of the Association for Information Science and Technology, 67*(10), 2404–2419. https://doi.org/10.1002/asi.23585

Selwyn, N. (2009). Faceworking: Exploring students' education-related use of Facebook. *Learning, Media and Technology, 34*(2), 157–174. https://doi.org/10.1080/17439880902923622

Small, H., Kasianovitz, K., Blanford, R., & Celaya, I. (2012). What your tweets tell us about you: Identity, ownership and privacy of Twitter data. *International Journal of Digital Curation, 7*(1), 174–197. https://doi.org/10.2218/ijdc.v7i1.224

Smart Education Networks by Design. (2014). Washington, DC. Retrieved from www.cosn.org

Smith, P. K., Mahdavi, J., Carvalho, M., Fisher, S., Russell, S., & Tippett, N. (2008). Cyberbullying: Its nature and impact in secondary school pupils. *Journal of Child Psychology and Psychiatry, and Allied Disciplines, 49*(4), 376–385. https://doi.org/10.1111/j.1469-7610.2007.01846.x

Sobaih, A. E. E., Moustafa, M. A., Ghandforoush, P., & Khan, M. (2016). To use or not to use? Social media in higher education in developing countries. *Computers in Human Behavior, 58*, 296–305. https://doi.org/10.1016/J.CHB.2016.01.002

Thaiposri, P., & Wannapiroon, P. (2015). Enhancing students' critical thinking skills through teaching and learning by inquiry-based learning activities using social network and cloud computing☆. *Procedia – Social and Behavioral Sciences, 174*, 2137–2144. https://doi.org/10.1016/j.sbspro.2015.02.013

Tsai, C.-C., Lin, S. S., & Tsai, M.-J. (2001). Developing an Internet Attitude Scale for high school students. *Computers & Education, 37*(1), 41–51. https://doi.org/10.1016/S0360-1315(01)00033-1

Väisänen, S., Pietarinen, J., Pyhältö, K., Toom, A., & Soini, T. (2018). Student teachers' proactive strategies for avoiding study-related burnout during teacher education. *European Journal of Teacher Education, 41*(3), 301–317. https://doi.org/10.1080/02619768.2018.1448777

van Dijk, J. A. G. M. (2006). Digital divide research, achievements and shortcomings. *Poetics, 34*(4–5), 221–235. https://doi.org/10.1016/J.POETIC.2006.05.004

Vygotsky, L. S. (1978). *Mind in society: The development of higher psychological processes*. London: Harvard University Press.

We Are Social. (2018). *Digital in 2018*. Retrieved from https://digitalreport.wearesocial.com/download

Welsher, A., Rojas, D., Khan, Z., VanderBeek, L., Kapralos, B., & Grierson, L. E. M. (2017). The application of observational practice and educational networking in simulation-based and distributed medical education contexts. *Simulation in Healthcare: The Journal of the Society for Simulation in Healthcare, 13*(1), 3–10. https://doi.org/10.1097/SIH.0000000000000268

Wenger, E. (1998). *Communities of practice: Learning as a social system*. The Systems Thinker. Retrieved from https://thesystemsthinker.com/communities-of-practice-learning-as-a-social-system/

Wilkinson, R., & Pickett, K. (2009). The spirit level new edition: Why equality is better for everyone. *Penguin Sociology*.

Wright, M. F. (2018). Cyberbullying: Prevalence, characteristics, and consequences. In Z. Yan (Ed.), *Analyzing human behavior in cyberspace* (p. 333). Hershey, PA: IGI Global.

Yuen, S. C., & Yuen, P. (2008). Social networks in education. In *E-learn: World conference on e-learning in corporate, government, healthcare, and higher education* (Vol. 2008, pp. 1408–1412). Association for the Advancement of Computing in Education. Retrieved from https://www.learntechlib.org/noaccess/29829/

Zimmerman, B. J. (2000). Self-efficacy: An essential motive to learn. *Contemporary Educational Psychology, 25*, 82. https://doi.org/10.1006/ceps.1999.1016

Part III
Projects

Chapter 5
Groups and Networks: Teachers' Educational Networking at B@UNAM

Francisco Cervantes-Pérez, Guadalupe Vadillo, and Jackeline Bucio

Abstract The purpose of this chapter is to present an integrated proposal that can provide the setting for a virtual academic community of practice among teachers of a Mexican online high school involved in specific projects. The design process of an initial social site called *Academic-match* is described, as well as its evaluation. Given the challenges related to the tool's networking potential that were met, a second proposal is provided. It integrates three distinct but interrelated elements: teacher recruitment, teacher selection, and collaboration among them. Recruitment of teachers is done through a call for participation in each of the specific academic projects involved through existing academic social networks. The selection process, according to the established criteria for each project, is done through *Academic-match*, in its improved version that involves new functionalities. Finally, collaborative work among the selected teachers is achieved through a virtual academic community of practice. This final proposal takes advantage of the qualities of each tool and allows for specific tailoring depending on the project at hand and the initial number of potential participants. This proposal lets teachers feel free to engage in tasks, yet it provides a structure for organizing and a space for collaboration.

Keywords Educational networking · Social network · Online social network · Social network sites · Social learning

F. Cervantes-Pérez (✉)
Universidad Nacional Autónoma de México (UNAM), CUAED and ENES-Morelia, Mexico City, Mexico
e-mail: francisco_cervantes@cuaed.unam.mx

G. Vadillo · J. Bucio
UNAM-CUAED-B@UNAM, Mexico City, Mexico

© Springer Nature Switzerland AG 2020
A. Peña-Ayala (ed.), *Educational Networking*, Lecture Notes in Social Networks, https://doi.org/10.1007/978-3-030-29973-6_5

161

Abbreviations

ANUIES	*Asociación Nacional de Instituciones de Educación Superior*
ASN	*Academic Social Networks*
B@UNAM	*Bachillerato a Distancia de la UNAM*
ESN	Educational Social Networks
GNLE	Globally Networked Learning Environment
ICT	Information and Communication Technologies
iOS	iPhone Operating System
NPS	Networked Participatory Scholarship
SN	Social Networks
SNS	Social Networking Sites
UNAM	*Universidad Nacional Autónoma de México*
vACoP	Virtual Academic Community of Practice

5.1 Introduction: Justification, Purpose, and Relevance

In one hand, school faculty represents a talented group that frequently is not fully acknowledged when academic projects are planned, designed, and implemented, and, on the other hand, teachers often feel they might contribute more to their schools and that their lack of involvement may hinder the success of activities and resources they feel are imposed upon them. Both situations used to occur in B@UNAM, UNAM's online high school program, but that changed during B@UNAM's curriculum renovation.

This chapter aims to present the architecture of an academic social network that can provide the setting for a virtual academic community of practice (vACoP) among B@UNAM teachers. A series of characteristics and requirements were established so it could be a space where teams of teachers could be organized for specific educational purposes by an academic leader coordinating a new project. This leader may be a designated teacher to solve a specific problem, to examine other possibilities in some areas of opportunity, or to innovate teaching and/or learning practices. Also, the academic staff of the program is interested in nurturing the collaboration among teachers, so their voices could be heard in institutional projects. The main concern was for teachers to feel at ease, interact freely, and enjoy a collaborative experience to develop academic solutions. Besides, the idea was to transform groups of teachers into a network, naturally proactive and engaged, as Dron and Anderson (2014, 2015) define.

The chapter is organized as follows: first, a review of the literature about communities of practice and academic social networks is presented; second, highlights of the distinguishing characteristics of B@UNAM are described, as well as the challenges faced while renovating the program and trying to transform group activity into network interactions, and, finally, the first attempt to solve the need for teacher involvement in the analysis and improvement of syllabi; third, we present the evaluation of this initial experience and the proposal of the architecture of an academic social network to meet the challenges faced during the pilot phase; and fourth, an evaluation of our proposal is discussed taking into account other theories and models found in the literature. Finally, in the conclusions section, contributions of the proposal are presented, as well as its road map and potential further developments.

5.2 Background and Related Works

This section reviews concepts, important findings, and challenges related to educational social networks and academic communities of practice. The emphasis is placed on the use of these resources in teacher training and practices.

5.2.1 Educational Social Networks

One of the first definitions of social network sites was proposed by Boyd and Ellison (2007, p. 1083): "web-based services that allow individuals to (1) construct a public or semi-public profile within a bounded system, (2) articulate a list of other users with whom they share a connection, and (3) view and traverse their list of connections and those made by others within the system." Educational networking allows users in this area to interact with other professionals, share information, and create an active virtual community (Insani, Suherdi, & Gustine, 2018) through the use of internal and external messaging, chats, internal email, and some tools that enable the user to build groups. All these possibilities look appealing, but interaction features do not necessarily mean collaboration, as James (2014) has studied regarding the use of ICT collaborative technologies for scholarly work in Australia.

Social networks (SN) and educational social networks (ESN) may be part of a continuum, as Firpo, Zhang, Olfman, Sirisaengtaksin, and Roberts (2019) suggest. They establish that there is a fine line between academic and social networking; thus, professional relationships can evolve into personal ones, and the other way around.

Their functions for teachers, according to Kelly and Antonio (2016), are as follows:

- Providers of feedback
- Modelers of practice
- Supporters of reflection
- Agents of relationships
- Agents of socialization
- Advocates of the practical

In their study, they found that teachers shared resources and responded to one another with pragmatic advice, while using a public network, but failed to do much of reflection on practice, feedback, or modeling (probably because it was a public environment).

According to Martín, Hernán-Losada, and Haya (2016), ESN have two elements:

- The social component that enables socialization (presentations of individuals, groups, or schools and universities, as well as communication tools)
- The educational component that supports facilities to teach and learn (educational resources, portfolios, assessment tools, project proposals, etc.)

As its use increases, the importance of exchanging information for academic purposes as LinkedIn does (Porcel, Ching-López, Lefranc, Loia, & Herrera-Viedma, 2018) is even greater. This type of networking constitutes an emerging field that observes and analyzes collaboration among people involved in different aspects of the educational spectrum. Academic social networking can be considered a form of digital professional activity or it can be seen as networked participatory scholarship (NPS) as Veletsianos and Kimmons (2012) have proposed. Its use can be widespread throughout a school or system or restricted to a specific group, for example, IGGY, a social educational network specifically developed for gifted students (Charalampidi & Hammond, 2016).

Klimova and Poulova (2015) put forward an SN classification, and they identify five types of social networks:

- Profile-based social networks, such as Facebook and LinkedIn, where social interaction is the most important outcome.
- Content-based social networks, like YouTube or Instagram, where social interaction is not as important as sharing content.
- Microblog social networks, such as Twitter, where one can publish a short message and add a video.
- Multiuser virtual networks, for example, World of Warcraft, where participants communicate with others through avatars.
- Finally, we have the white label or private label social network, where a social network can be built based on an offered platform to create a mini-community.

This initial classification portrays the multiplicity of aspects and combinations in the mix for a taxonomy that shows the diversity of social media we can find in cyberspace. There are even other particularities that can be added: stability allows to identify ephemeral social media (messages that self-destroy in only a few seconds) or networks that are a part of the service of an email account (Google Plus), the open counterparts of commercial open networks (e.g., Twitter vs. Mastodon), or those based on the user's geolocation (Tinder).

Social networks have also been used to promote parent engagement in student learning. For example, Willis and Exley (2018) researched its use in an Australian school and found that social media both enabled and impeded their engagement. It was fostered, among other things, by making visible their child's learning and by making it possible to share information and ideas between school and home. However, the lack of knowledge of how to participate in social media on behalf of the parents could have decreased their participation.

Until now, the main concern of the use of ESN has been students' interactions and communications, as well as how they build learning communities and how this information helps teachers to a better assessment of learners' participation in the learning activities they design (Rabbany, Elatia, Takaffoli, & Zaïane, 2014). For instance, ESN is useful in a variety of ways: they provide an academic space different from the institutional web educational platform since they encompass a sense of freedom and may lead to a higher degree of message dispersion.

The classification of social learning arrangements by Dron and Anderson (2014, 2015) identifies the following types: dyads, sets, groups, networks, or collectives. Dyads are the simplest form of learning: only two people participate, and they tend to be very effective, yet the experience may be very expensive. Sets are collections of people who share certain attributes (e.g., their interest in a certain subject). Groups generally have a hierarchical structure and membership is explicit. Networks are constituted in terms of connections with others. They are not designed, they appear as a result of connections with others, and they are based on trust and social capital. They offer more freedom to their participants than groups. Collectives have to do with many-to-one communication and act as a single agent that may affect many people. Collectives can be formed unintentionally, but technologies allow for crowds to gain more agencies via the manipulation of algorithms that enable the aggregation of crowd behaviors. These influential authors assert that sets, groups, and nets may overlap and contribute to a collective. Siemens (2019) adds systems to this frame: "systems matter more than networks. Networks don't exist in a vacuum. They exist and are shaped by the environments in which they exist. Networks are ephemeral. Systems exist to preserve (n/p)."

5.2.2 Academic Communities of Practice (CoP)

Wenger (2010) defines communities of practice as social learning systems. A CoP is a simple social system, while a complex social system comprises interrelated CoP. This concept is rooted in the assumption that learning happens in the relationship established between the individual and the world. Trust and Horrocks (2019) assert that members can be at the same time learners and teachers and that "[s]ince teaching is a multifaceted profession, educators might benefit from large CoPs that consist of members with diverse professional knowledge and expertise. Additionally, since participation in a CoP is fluid and dynamic, CoPs must change and grow as membership changes" (p. 109).

The model by Nistor et al. (2014) shows that expertise or quality of interventions in a virtual CoP determines the number of interventions and the expert status (or centrality).

5.2.3 Evidence of the Positive Impact of Academic Networking and Academic CoP

There is evidence that academic networking has positive effects in teachers' practices (e.g., Van Waes, De Maeyer, Moolenaar, Van Petegem, & Van den Bossche, 2018; Smyrnova-Trybulska, Morze, and Kuzminska (2019). The development of school networks began in the 1970s. During the last three decades of the twentieth century, there was greater autonomy of individual schools, as well as increasing demand for public accountability. These environments fostered the development of educational networks that serve a variety of functions: a political function (like-minded educators come together and serve as lobby groups), an information function (through rapid exchange of information among participants), a psychological function (innovators who tend to be isolated in their schools can group with others), and a skills function (in order to learn skills not taught through traditional professional development programs, from other participants) (Sliwka, 2003). When peer exchange is involved, these networks generally lack hierarchy and require structure and some form of organizational leadership, so the networking process begins, principles and guidelines are defined, recruitment of members is achieved, and communication infrastructure is created.

The usefulness of ESN for learning has been established (e.g., Nazir & Brouwer, 2019). They identified the cognitive, social, and teacher presence of the Community of Inquiry framework while working with Facebook, as well as a positive impact on learning.

Through the survey Jordan and Weller (2018b) applied, they identified that academics participate in online networking aiming for the development of a personal learning network, to promote the professional self, seek and foster publications, and advance careers. The aspect of promoting publications is causing tension with publishers as the case of Elsevier against ResearchGate shows (Chawla, 2017).

Social networking among teachers and other education professionals is relevant due to the need for collaboration among them. Lane and Sweeny (2019) indicate that people tend to form bonds through propinquity (those who are near them) and homophily (those who are like them). Since long ago (Granovetter, 1973, in Lane & Sweeny, 2019), we know that in an individual's social network, there are strong ties (durable and with whom people place more demands) and weak ties (typically formed with more distant associates, they require less effort to maintain and allow access to a broad variety of resources). Both are important and desirable, and their usefulness depends on certain conditions: weak ties are best when the information to be shared is highly codified and does not depend on context.

Ubiquity represents another advantage: "Social media are increasingly visible in higher education settings as instructors look to technology to mediate and enhance their instruction as well as promote active learning for students" (Ebrahimpour et al., 2016, p. 135).

Networks such as Edmodo provide additional advantages, since they allow cooperative work among teachers and students in a secure environment; with a user-friendly interface, they provide Android and iOS compatibility and serve many different language users (Ursavaş & Reisoglu, 2017).

The project MoodleNet released its platform under the premise "Connecting and empowering educators worldwide." MoodleNet provides educators with a "new open social media platform for educators initially focused on the collaborative curation of collections of open resources. MoodleNet will be an integral part of the Moodle ecosystem, sustainably empowering communities of educators to share and learn from each other to improve the quality of education" (Belshaw, 2018, n/p). This project stresses the proprietary basis of most of the social networks employed nowadays by educators. Project MoodleNet also differs from other ESN in representing a decentralized network system, which encourages more diverse participation, more efficient through locally informed decisions, and privacy, a safe space for interaction and data storage.

Another important use is professional learning. "Teachers are turning to social media for a range of education-focused resources and networks to collaborate and curate, when and with whom they want" (Trust, Carpenter, & Krupta, 2016 in Prestridge, 2019, p. 143). There is a new conceptualization of professional development, for teachers are thinking of it as a boundless and self-generating activity conducted through collaboration and independent inquiry. Professional development and professional learning are terms that have been used as synonyms, although the latter can be a consequence of the former. Therefore, the author conceives professional learning as an informal type of learning that is not required by regulatory bodies but is always accessible and turns out to be rewarding for the teacher. The use of Twitter, for example, enables teachers to connect with others because of their educational philosophy or content areas, compensating in this way for the fact that they do not find that sort of colleagues in their schools. Prestridge even has developed a typology of users of social media for this end. In it, there are two purposes: a people orientation and a content orientation. From the first one, teachers share, collaborate, support, and put ideas forward, but they also lurk or try to understand who's who in the content domain. From the content orientation, they collect, share, and build resources, but they also just take and do not contribute. She presents a typology of consumers: info consumers who find knowledge, ideas, and resources from social media, who are passive participants and practically do not contribute to their networks, and look for what their peers may consider valuable, rather than self-realization. The info networker looks for resources and has the goal of sharing them with others. They do not contribute with substantive knowledge and they retweet more than share their materials. The self-seeking contributor posts knowledge, ideas, and materials with specific reasons such as getting feedback or validation, asking questions about new tools, and asking for technical support or about educational innovations. This contributor has both people and content orientation. The last category is the vocationalist who engages in social media with professional learning purposes, wants to be a member of a community of learners, and desires to contribute to the body of knowledge. The vocationalist initiates new threads, leads chats, and tests new ideas, providing constant feedback to the community.

Mora, Signes, De Miguel, and Gilart (2015) demonstrated that "social networks have more capacity to spread information than educational web platforms. Moreover, ESN are developed in a context of freedom of expression intrinsically linked to Internet freedom. In that context, users can write opinions or comments which are not liked by the staff of schools. Social networks allow freedom of expression which could not be produced in the educational context. This feature can be exploited to enrich the educational process and improve the quality of their achievement" (p. 894).

Trust and Horrocks (2019) present several sources to acknowledge the wide range of applications these communities have: critical reflection and dialogue to knowledge exchange and problem-solving, as well as teacher agency, collaboration, and professional growth.

5.2.4 Challenges of ESN and Virtual Academic CoP

The use of academic networking has pros and cons. The literature review of Martín et al. (2016) suggests that using a commercial social network directly may not be the best option: adaptations must be made to hinder the negative effects such as engagement in noneducational activities while being connected. Examples of ESN are the TENcompetence project (Wilson, Sharples, & Griffith, 2008 in Martín et al., 2016) and ClipIt (Llinás et al., 2014 in Martín et al., 2016). ESN provide the opportunity to interact outside the class and reduce confidentiality and security risks.

Social networks have had a strong influence on the opportunity to relate to others and the way these interactions happen, although they have not been deeply explored yet in the field of education (García-Saiz, Palazuelos, & Zorrilla, 2014).

Manca and Ranieri's (2016a) review of the literature regarding the use of Facebook as a learning environment finds that this social network is seen as a bridge between everyday life and the educational sphere. Significantly, this can become a burden as Veletsianos and Kimmons find: "The SNS experience of faculty members culminates in a tension between personal connection and professional responsibility. As faculty attempt to negotiate their participation on social networking sites, they encounter issues of establishing boundaries, maintaining appropriate and meaningful connections with others, structuring participation for perceived presentation to others, and using their time efficiently" (2013, p. 45).

As Saedy, Rajaee, Jamshidi, and Jamshidi (2014, p. 1) express: "However, existing providers today, typically are simple websites that provide one or two services without presenting a complete social network that can connect users based on their actual needs." Since social networking is common among teachers (e.g., Tezer & Yıldız, 2017), it may constitute an interesting resource for academic work.

Perez and Brady (2018) found in their systematic literature review that the biggest challenge for academic staff to incorporate social media in their work is the lack of institutional support: even if they are motivated to use it, they must follow their universities' practices and regulations. Other challenges were pedagogical issues, time investment, and privacy concerns.

Although technology has been a game-changer, more than 20 years after the first online social network "SixDegrees.com" was developed (Gohel, 2015), their use as an educational resource has typically been limited to interaction among students or teachers and students who share learning and assessment resources and socialize (Martín et al., 2016).

A final challenge is that this area is still not fully understood due to a lack of studies. Jordan and Weller (2018a) analyzed a large dataset and they found that teachers are not using ESN mainly for reasons related to time and efficacy. Veletsianos and Stewart (2016) have investigated different forms of self-disclosure in the use of social media by scholars, and some other approaches observe the cultural aspects that may impact the participation of teachers, for instance, in a vACoP in China (Hou, 2015). These are steps forward to fulfill gaps in this area as Veletsianos previously observed: "scholars' experiences and practices on social media and online networks are not well understood and the evidence describing their experiences is limited and fragmented" (2016, p. 2).

5.3   A Brief Description of B@UNAM's Program and the Need for an ESN

Due to its unique characteristics and curricular review process, a brief description of B@UNAM is presented. The new curricular map and the map of thoughts are also included.

5.3.1   B@UNAM: Characteristics, Curricular Review, and New Pathways

B@UNAM is a public virtual high school program that serves Spanish-speaking students in Mexico and abroad. It belongs to UNAM, Mexico's largest public university. B@UNAM was launched in 2007 and received a national award, from ANUIES, for its innovative curricular design (Villatoro, Aznavwrian, & Vadillo, 2013). It has only 24 courses (as opposed to about 49 in most high school programs) covering all of the contents this educational level must provide; its courses are interdisciplinary; there is a skills map that is followed throughout all learning activities; learners take one course at a time, and there are online teachers and counselors who provide a close follow-up for each learner. In 2017 a trajectory model was included: at the beginning of each week, based on the results of a diagnostic exam, each learner is provided with a personalized track of contents. Students who require prerequisite knowledge go to the bronze path and get a remediation course, and then he returns to the silver path (which presents the expected contents). Gifted students follow the gold path, with activities and contents that correspond to advanced placement courses (Vadillo, 2017).

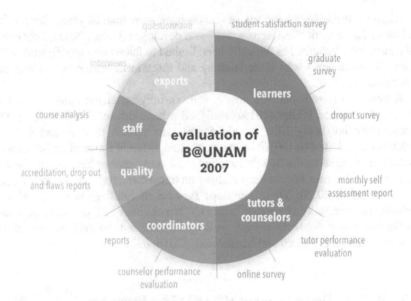

Fig. 5.1 B@UNAM evaluation model

During 2017–2018, an evaluation of the program was conducted. In Fig. 5.1, the sources of information that were included to identify strengths and weaknesses as well as opportunities to improve it are shown. The guidelines of the program's redesign were defined based on the findings of this evaluation process, as well as on a thorough analysis of literature concerning trends and prospective studies of the high school level.

5.3.2 The New Curriculum

The new B@UNAM's curricular matrix is shown in Fig. 5.2, and its main characteristics are the following:

- The admission process requires completing three prerequisite courses.
- All 24 formal courses are transdisciplinary, with authentic academic activities aimed to foster deep learning.
- Learning activities foster throughout the program the development of five types of thinking: (a) computational, (b) mathematical, (c) social, (d) humanistic, and (e) scientific.
- Students take one course at a time, with a 30-h commitment per week, during a period of 4 weeks to complete all required academic activities.

The main characteristics of its new version are as follows: courses are transdisciplinary, there is an intelligent tutoring system to answer frequent questions and to build individual tracks according to each learner's needs, and learning activities

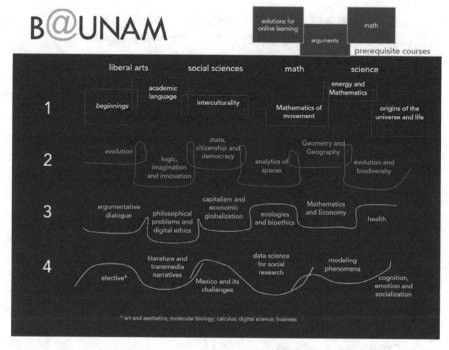

Fig. 5.2 B@UNAM's curriculum

throughout the program foster the development of five types of thinking (humanistic, social, computational, mathematical, and scientific), as seen in the map of Fig. 5.3. An operational definition of each skill guides developers while designing learning activities.

5.3.3 Personalized Tracks

Following our learning engineering approach, an intelligent tutoring system will be included to coach learners and to answer frequent questions about content. Also, a multilevel course design will be included, based on the diagnostic evaluations at the beginning of each section of every course unit, combined with analytical information processes to identify individual tracks according to each learner's needs. Three different trajectories will be considered:

- Bronze: it is followed by students with poor knowledge of previous subjects required to understand the current topic. For example, see the left-hand side of Fig. 5.4; if a student does not know what a cellular membrane is and its properties, then he/she will not be able to understand the immune system; therefore, he/she should take a remedial short course on the cellular membrane and then go back to the core of its current course.

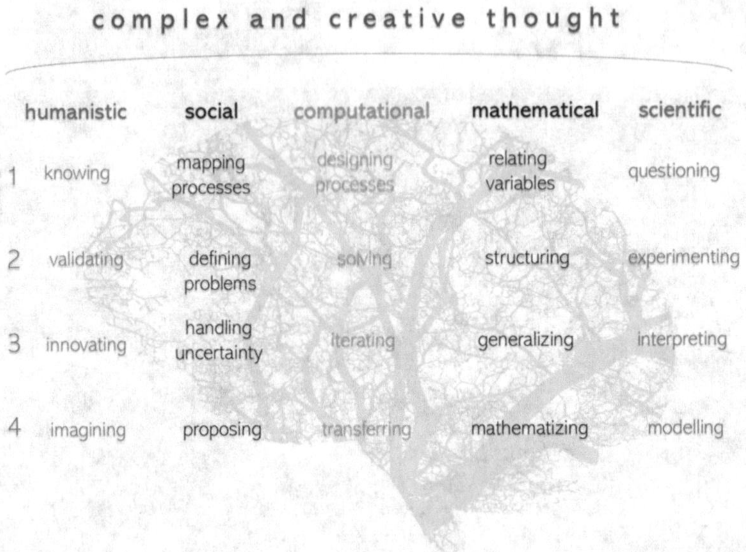

complex and creative thought
humanistic
social
computational
mathematical
scientific
1
knowing
mapping processes
designing processes
relating variables
questioning
2
validating
defining problems
solving
structuring
experimenting
3
innovating
handling uncertainty
iterating
generalizing
interpreting
4
imagining
proposing
transferring
mathematizing
modelling

Fig. 5.3 Types of thinking included in the redesign of B@UNAM

unit 2 Inmune system
overarching case: HIV
function: relation
level: cell
why is there no vaccine for HIV?
head transplant
written / other format
theoretical practical
diagnostic test
communication and cell recognition
blood: physical properties
figurative elements
function
inmune system
diagnostic test
blood groups and transfussion
vaccines
transplants
prevention policies
another disease: chikungunya, ebola
cell membrane
how to identify blood groups
antigen,antibodies histocom-patibility
clips:
Influenza A (H3N2) vaccine
zika and its effects on NCS

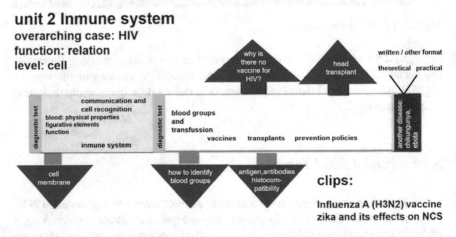

Fig. 5.4 B@UNAM's multilevel design

- Silver: for students going according to the academic program.
- Gold: it is designed for advanced students that already know some of the current unit contents. For example, see the right-hand side of Fig. 5.4; if a student already knows about blood groups, transfusions, vaccines, transplants, and prevention policies, then he/she will be given a research project to address the question of why is there not yet a vaccine for AIDS?

In the end, all students must comply with the final part of the course to complete it.

5.4   Initial Experience: Academic-Match

The first attempt to design an ESN is presented, and the conceptual framework and its main characteristics are described in detail. Additionally, some ethical issues are discussed.

5.4.1   Planning Academic-Match

Martín and her colleagues (2016) came up with several recommendations for the development of ESN: content must match the personal learning interests of users, published material must be reliable and from a trusted source, and visual elements must be included, especially for students. The inclusion of microblogs with a tree structure constitutes another advantage (Krejci & Siqueira, 2014).

Manca and Ranieri (2016b) underline the importance of virtual spaces where educators can share teaching practices and have professional development opportunities.

An ESN for teachers is relevant since there is an important gap in the literature on academic social networks. For example, "little is known about how and under what conditions teachers form social network ties" (Lane & Sweeny, 2019, p. 80). Academic outcomes among teachers have been briefly described in cases such as Tapped In (Suthers & Dwyer, 2015), eTwinning (Kearney & Gras-Velázquez, 2018), and, with a deeper detail in the case of the Center for Collaborative Online International Learning, COIL, at State University of New York (Moore & Simon, 2017; Rubin & Guth, 2017), where even the administrator's experience is documented regarding the creation of the globally networked learning environment (GNLE) for the COIL Center initiative (Jansen, 2017).

One of the few published examples of ESN as a tool for curricular design is the experience by Cochrane, Guinibert, Simeti, Brannigan, and Kala (2015), where students take the role of experts and contribute to the curricular redesign of a program, following the idea of empowerment of learners in Ihanainen and Gallagher's manifesto, cited in Cochrane's work.

5.4.2   Academic-Match: A Conceptual Proposal

Here, the design of the platform used in this study is described in detail with an emphasis on its characteristics and some ethical and security issues.

5.4.2.1   Objectives, Design Restrictions, and Basic Premises

The main objective of this study was to validate the use of an ESN as a means to accomplish academic work among teachers. The specific objectives were:

- To test the validity of a set of predefined fundamental premises about teachers' capacities.
- To identify the functionalities that best serve the academic needs of the project.
- To identify technical issues that may hinder interaction among teachers in social networks.
- To search for possible design flaws in the ESN site.

The design restrictions we established were:

- The platform that would be called *match académico*, *matcha* for short, should be as simple and attractive as possible. It should feel like an informal space where teachers are free to express their opinions and ideas, investing only a small amount of time. Matching should appear promptly, and it should be relevant for the users.
- Teachers should be matched according to criteria related to their contribution on academic projects: knowledge domain, type of innovator, and leadership style.
- At least during the first trial, *matcha* is meant only to be used as a collective intelligence resource of our 1600+ certified teachers. In the future, other uses such as crowd voting and crowd creation may be established and students might also use it to connect with peers and teachers, for academic tasks. It is also intended to function as a networked improvement community such as the one at Carnegie Foundation (Means, 2018).
- It would be a white label or private label social network that would allow teachers to build a mini-community for academic interaction. All types of participants in Prestridge's typology (2019) should find themselves comfortable and rewarded while using it.
- In this version, it would only have a messaging option, but, in further versions, collaborative workspaces should be included.

The project was aimed at B@UNAM's faculty who have served as academic tutors since at least 2015. Based on the teacher's training and experience, the following premises were established:

- They have developed digital skills for teaching online.
- They possess the know-how to conduct collaborative tasks online.
- They can to work using any type of device.

5.4.2.2   Characteristics of Academic-Match

The platform was developed in late 2018 by the staff of UNAM's online high school (B@UNAM). It is called *Academic-match* (or *matcha*, for short), and its purpose is to facilitate the integration of efficient teams that need to collaborate in academic projects. In the beginning, it serves online teachers and counselors at B@UNAM (see Fig. 5.5); however, in the future, it will be open to learners participating in class projects.

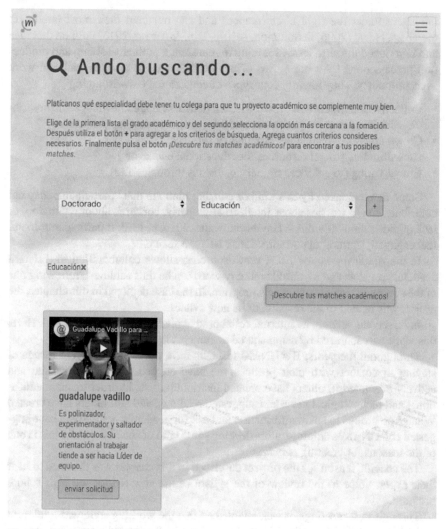

Fig. 5.5 B@UNAM's *matcha* platform

It offers a context where teachers make contact with peers to collaborate on academic-specific projects or just to share resources and ideas about their areas of expertise. In its initial version, it considers a profile that:

- Portrays individuals' type of leadership.
- Describes their most and least developed facets of innovation.
- Includes personal and professional information they want to share, through a brief video.

It is meant to be a crowdsourcing tool to develop projects, work on educational ideas, and share learning resources. Surowiecki, in 2004, wrote that groups are very intelligent under the right circumstances and can outsmart their most intelligent members. Howe (2006, in Solomon, Ariffin, Din, & Anwar, 2013) built on that idea and introduced the term crowdsourcing to represent a problem-solving and production process done by a big crowd of people that react to an open call and work asynchronously. They identify four types of educational crowdsourcing:

- Collective intelligence or crowd wisdom (teachers share knowledge and contribute with ideas to solve problems, make forecasts, or develop projects).
- Crowd creation (to generate products or services).
- Crowdfunding (to get resources for educational endeavors).
- Crowd voting (to get ideas or make decisions through votes).

Some dangers faced by this kind of collaboration are that "participants may not contribute due to worries about intellectual property, concerns about reputation, or lack of motivation" (Paulin & Haythornthwaite, 2016, p. 132). If participants do not feel engaged, learning opportunities may become scarce.

In the *Academic-match* ESN, crowdsourcing allows collaboration in different academic projects and a transdisciplinary work can be thus achieved, following one of the central principles of the new program. In the case depicted in this chapter, the project was related to the revision of its new syllabi.

In its second version, an internal collaborative platform will be developed. Here, this work was achieved by using shared documents in Google Drive.

Throughout the years, B@UNAM teachers have developed different means of staying in contact with their peers: some have created a personal website and actively promoted it, others have created distribution lists in order to share educational and technological news and suggested readings, and still others have created WhatsApp groups to share information and ask for academic help. Some have suggested B@UNAM's administrators develop some kind of social network: this is one of the reasons for creating *matcha*.

The specific reason for the project described in this chapter was the need to hear their expert voice in the review of the syllabi of the new redesigned high school program.

Through staff meetings, it was established that the working style was a relevant variable to be considered for all matches to have well-adjusted teams. A questionnaire, inspired in the test of Blake and Mouton (1964), was developed to work with those teachers that are project-oriented and who promote positive working morale.

Another dimension that was prioritized is the type of innovation facets of the teams derived from the matching process. It was important that, from a creative point of view of the curricular design process, syllabi would be analyzed from a variety of perspectives. Therefore, a test that identifies the prominent as well as the least developed faces of innovation (Vadillo, 2014) was used, according to the taxonomy proposed by Kelley and Littman (2005). It enables matching teachers with different predominant faces of innovation: *learning personas* (anthropologist, experimenter, or cross-pollinator), *organizing personas* (hurdler, collaborator, or director), and *building personas* (experience architect, set decorator, storyteller, or caregiver). Building a team with as many different perspectives as possible allows for a wider creative perspective. This is an innovative approach because there is a tendency to work in social networks only with colleagues belonging to a unique discipline: "This seems to obstruct the development of an open culture among academics, preventing cross-disciplinarity and reinforcing subject matter enclosures where experts from a specific field converse with those in the same field" (Manca & Ranieri, 2017, p. 134). In the academic social network proposed here, the matches look for complementary abilities and interests, encouraging a transdisciplinary approach.

5.4.2.3 Ethical and Security Issues

To communicate between colleagues to crowdsource the critique of the syllabi, a private communication mode was preferred since participants were to work with unpublished institutional documents that were not yet approved by the university accreditation committees. Therefore, even though open educational networking is a tendency nowadays, a private/institutional setting was chosen as the best option for this exercise.

Also, because privacy concerns among academics arise from the revision of the literature, "protection of privacy is the number one concern of scholars when using social media tools" (Gruzd, Staves, & Wilk, 2012, p. 2349). Teachers who participated in this study agreed to the privacy conditions and objectives of the project.

5.5 Method

The sample and procedure that were followed are presented here. Besides, the two products been sought are also described.

5.5.1 Sample

A call for participation in the project was sent via email to teachers who had received an excellent evaluation from students.

The available time to develop and create the ESN was scarce. The initial structure was detailed in October 2018 and it was tested among teachers and launched in January 2019. In February 2019 the possibility not only of messaging but also of working collaboratively using shared documents in Google Drive was also included through a Google Classroom.

Further developments are scheduled for the second semester of 2019: the inclusion of a module for teachers' collaborative work and a student module to foster peer tutoring. Later in 2019, these modules should allow the development of a third module to foster collaborative work among students and teachers. The time commitment and schedule were presented, and 12 teachers answered that they were willing to critique a syllabus.

5.5.2 Procedure

The procedure followed in our method is comprised by six stages. The first three represent a sort of a timeline for preparing the platform infrastructure. The last three activities relate to the teachers participation in the project.

- Planning phase. Based on the main and specific objectives of the project, as well as on the design restrictions and basic premises about teachers' technological knowledge levels, the layout of the site and its functionalities were defined, and a timetable was established.
- Development phase. Following the products and processes of the planning phase, the development of the platform was completed following a lean process using Trello software to complete specific tasks. The team was composed of three academic developers, a software developer, and a graphic designer.
- Beta phase. Once the platform was developed, it was tested among staff members of B@UNAM. Changes were made according to the suggestions and issues that appeared.
- Launch. After a successful trial among staff members, the platform was launched. Committed and prestigious teachers were invited to contribute via a crowdsourcing experience to the curricular revision of a syllabus of B@UNAM.
- Revision of syllabi. After teachers completed their profile in *Academic-match*, they were matched with peers who represented the disciplines that contribute to the corresponding course. Via email, they were invited to work with their team in a Google Classroom. Google documents were the tool selected for teachers to read and comment on the new syllabi. A group was set up and teachers were registered with student roles. The requirements were that participants had to have a Gmail account to be identified by name in each comment. Google documents allow three participation modes: editing, suggesting, and viewing. In this scenario teachers were allowed only to comment or suggest changes; they could not edit the document directly (editing mode). Four syllabi were selected, and copies were prepared in the classroom. Teachers were able to comment and provide insights to their peers' contributions using a shared document.
- The shared document was a noneditable syllabus where each teacher inserted comments (identification of flaws, commendations, suggestions, and questions) and could chat with the other members of the team. This process lasted for 3 days.

The first product we looked for was the profile of each teacher, in terms of the test results and their video clip. In *matcha*, participants create a profile that portrays their type of leadership, describes their most and least developed facets of innovation, and includes personal and professional information they want to share, through a brief video.

A second product was the critique of some syllabi that had been built by experts in the disciplines. Each course includes and incorporated the suggestions of very prestigious professors and researchers. This task was completed using, as mentioned above, shared documents in a Google drive, aggregated in a Google Classroom.

5.6   Results

In this section, results are presented in the following categories: general data, a test of premises, functionalities, and interaction.

5.6.1   General Data

A total of 313 teachers were invited through an email message to register and help test the new social teacher network for B@UNAM. All of them taught B@UNAM courses during 2018. A total of 117 (37.4%) were registered (30 men and 87 women). The majority were female teachers, which reflects the general distribution by sex at our program. As can be observed in Table 5.1, 10 male teachers and 29 female teachers completed the first test they were presented with (facets of innovation) and only 22 teachers completed both (7% out of the total registered participants).

As can be observed in Table 5.1, more men completed the leadership test (23.3%) than women (18.4%), and in both groups exactly a third answered the facets of the innovation questionnaire. In terms of leadership style, almost all of those who completed that test (88%) came up with a balanced profile, while being both socially and goal oriented.

There were no math teachers who completed the leadership test and only two of the participating teachers of that area answered the innovation test, as can be seen in Table 5.2. The area that had more representation was social sciences followed by liberal arts.

The distribution of selected teachers, according to the area of expertise, appears in Table 5.3. As can be seen in Table 5.3, out of 25 teachers who completed both tests, only 12 were related to the corresponding syllabus. They represent the areas of liberal arts, social sciences, and science since none of the ten math teachers who registered in *Academic-match* completed the tests. Therefore, we were not able to get any comments about that syllabus. The final selection considered the match between the area of expertise of teachers and disciplines involved in each syllabus. Therefore, the final sample was composed of 12 teachers. They were invited to the project by email and a phone call (in two cases they could not be reached by phone). However, we only worked with ten, for a teacher with no Gmail account was not able to participate and another one later explained that he did not receive the message because his inbox was full at the time.

Table 5.1 Completion of leadership and facets of innovation tests, by sex and professional area of expertise

	Leadership test	Innovation test	Professional area
30 men (25.6%)	7 did (23.3%)	6 did (85.7%)	Math: 0% Liberal arts: 33.3% Science: 66.7% Social: 0% No answer: 0%
		1 did not (14.3%)	Math: 0% Liberal arts: 0% Science: 100% Social: 0% No answer: 0%
	23 did not (76.7%)	4 did (17.4%)	Math: 0% Liberal arts: 25% Science: 25% Social: 50% No answer: 0%
		19 did not (82.6%)	Math: 21.1% Liberal arts: 10.5% Science: 10.5% Social: 5.3% No answer: 52.6%
87 women (74.4%)	16 did (18.4%)	16 did (100%)	Math: 0% Liberal arts: 31.25% Science: 12.5% Social: 50% No answer: 6.25%
		0 did not (0%)	Math, liberal arts, science, social, no answer: 0%
	71 did not (81.6%)	13 did (18.3%)	Math: 15.4% Liberal arts: 7.7% Science: 46.2% Social: 23.1% No answer: 7.7%
		58 did not (81.7%)	Math: 6.9% Liberal arts: 12.1% Science: 13.8% Social: 25.9% No answer: 41.4%

5.6.2   Test of Premises

Even though all participating teachers have worked as online tutors for at least 3 years, several technical problems were observed. They are related to the basic premises we had established:

1. The first premise was teachers have developed digital skills for online teaching. One of the social science teachers wrote comments below the main entry (not in

Table 5.2 The number of teachers who completed the tests

Area	Male teachers			Female teachers		
	Leadership	Innovation	Total men	Leadership	Innovation	Total women
Liberal arts	2	3	5	5	6	11
Social sciences	0	2	2	8	11	19
Science	5	5	10	2	8	10
Math	0	0	0	0	2	2
No answer	0	0	0	1	2	3
Total	7	10	17	16	29	45

Table 5.3 The number of teachers who complied with completing both tests, by area of expertise

	Completed both tests	Were related to the syllabus
Liberal arts	6	4
Social sciences	13	2
Science	6	6
Total	25	12

the collaborative document) which resulted in limited interaction and a lower level of precision in the comments and suggestions.

2. The second premise was they possess the know-how to conduct collaborative tasks online. One of the literature teachers wrote her comments, but she did not save them, so when she closed the document, the draft comments disappeared. She had to revisit the document and rewrite her comments.

3. The third premise was they can work using any type of device. Another teacher reported that she was not able to see the document and she was not able to find the way to send her comments. Through a 20 min video conference session with a member of B@UNAM's staff, she learned how to use the "comment" function. The main problem was she used a tablet instead of a desk computer. The visualization is different in laptops than on desk computers, which makes it difficult to reach the space where comments can be added.

5.6.3   Functionalities

The analysis of issues detected during this experience suggests that several functionalities must be integrated into the ESN. In general, there was a lack of helpful feedback for the users. The low completion rate of some or all of the steps to create their profile suggests that the lack of timely completion and step completion indicator is an important element. The same applies to the lack of follow-up for those who were not successful in completing their profile. Finally, once the match is established, there was no indicator as to what degree the potential coworkers satisfied the user's selection criteria.

The use of an external aggregator site (Google Classroom) has been a hurdle that could be solved if the ESN had a space to collaborate. Educational robust platforms such as eTwinning offer both public and private areas; they enable users to find teachers working in similar subjects, foster interaction and collaboration, and invite teachers to engage in professional development activities (Kearney & Gras-Velázquez, 2018). Video tutorials or even precise text instructions to use collaborative documents were missing from this experience.

5.6.4 Interaction

Due to the diverse number and characteristics of teachers working on each of the syllabi, specific analyses are presented. As expected, teams with more participants and more tech-savvy teachers reported more interactions and longer discussions.

5.6.4.1 Course: Populating, Migration, and Interculturality

This course involves social sciences (history, geography, demographics, anthropology, and sociology). It comprises from prehistory to classical antiquity, followed by European migrations and Spanish colonization of the American continent in the sixteenth century, imperialism and urbanization in the nineteenth century, and interculturality in the global era.

Results are shown in Table 5.4; only a female teacher participated with four positive comments and two comments related to suggestions. These are related to privacy concerns in an activity that requires sharing personal information and to potential technical problems when recording audio or video as part of a certain activity.

Table 5.4 Participation in the analysis of the populating, migration, and interculturality syllabus

	Formal	+	−	?	Complaint	Suggest	Interaction	Adjustments	Unimportant
W1 (4)	−	4	−	−	−	2	−	−	−
	−	4 (67%)	−	−	−	2 (33%)	−	−	−

Symbols: formal, details related to the writing of the document; +, positive appraisals; −, negative comments; ?, questions; complain, complaints; suggest, recommendations; interaction, response or comment to feedback from a peer; answer to suggest, answer to suggestions; adjustments, suggestions related to the need of adjusting the academic level of activity; unimp., unimportant comments (not related to the task). M, man; W, woman

5.6.4.2   Course: Literature and Transmedia Narrative

This course involves the study of literature classic genres and the characteristics of transmedia literature. Through the text selection, students explore a transversal axis, topics such as diversity, sustainability, complexity, creativity, and digital humanities. A transdisciplinary approach is followed so aspects of history, art, anthropology, cinematography, and English as a second language are integrated into the contents.

Table 5.5 shows that, from the total of 61 messages, it can be observed that teachers commenting on this syllabus were heavily concerned about formal issues (the three of them are from the literature and Spanish language area). W2 and W3 were focused on the correct use of capital letters and aspects of clarity in some sentences. A high level of positive feedback was observed from W1, and none of them expressed negative comments. Nevertheless, three of them posed several questions as a form of kind suggestions ("What if…" or "It may be necessary…"), plus content observations and suggestions of adjustments regarding the time dedicated to a certain topic or project. Interaction among them was mainly to clarify new terms such as "transmedia narrative" and to help each other to understand it. The "nonrelevant" column groups phatic comments and some greetings.

5.6.4.3   Course: Earth and Life Sciences II

This course explores how the geological evolution and the biological evolution interact and allow for biodiversity on our planet. It includes knowledge and methods from geography, biology, archeology, chemistry, and history of science.

As depicted in Table 5.6, the activity and interaction in this area were widespread. A total of 58 comments, questions, or suggestions were incorporated, and although the number of comments about form-related issues outnumbered the other categories, they are also valuable. There were very active teachers (M1 and M2) and very reserved teachers (M3 and M4), and both women had an intermediate number of contributions. The content-based comments were especially valuable (such as suggestions on specific topics that should be included in the course).

5.7   Discussion

From a wide perspective, this study is related to the opportunities provided by connectivity and, more specifically, to the participatory culture, personal learning environments, distributed work, and learner-generated contexts, where learning can be a relationship connecting people through social network ties (Haythornthwaite, 2019). That is, in a learning society, social networks may allow people to become socially organized centered on the interactions, mediated by learning, and not only

Table 5.5 Interaction among the three teachers who participated in the analysis of the literature and transmedia narrative syllabus

	Formal	+	?	Complaint	Content suggestion	Interaction	Adjustments	Unimportant
W1 (24)	–	9	4	–	6	3	–	2
W2 (26)	10	1	5	–	6	2	2	–
W3 (11)	8	–	2	–	–	–	–	1
	18	10	11	–	12	5	2	3
	(29.50%)	(16.39%)	(18.03%)		(19.67%)	(8.19%)	(3.27%)	(4.91%)

Symbols: formal, details related to the writing of the document; +, positive appraisals; –, negative comments; ?, questions; complain, complaints; content suggestions, interaction; adjustments, suggestions related to the need of adjusting the academic level of activity; unimp., unimportant comments (not related to the task). M, man; W, woman

Table 5.6 Interaction among the six teachers who participated in the analysis of the earth and life sciences syllabus

	Formal	+	–	?	Complaint	Suggest	Answer to suggest	Lower level	Unimportant
M1	12	–	–	1	–	2	3	2	6
M2	–	1	1	–	1	7	–	–	5
M3	1								1
M4	–	–	–	1	–	–	–	–	–
W1	–	1	–	3	–	1	2	–	–
W2	–	–	–	3	1	–	1	2	–
	13	**2**	**1**	**8**	**2**	**10**	**6**	**4**	**12**
	(22%)	(3%)	(2%)	(14%)	(3%)	(17%)	(10%)	(7%)	(21%)

Symbols: formal, details related to the writing of the document; +, positive appraisals; –, negative comments; ?, questions; complain, complaints; suggest, suggestions; answer to suggest, answer to suggestions; lower level, suggestions related to the need of lowering the academic level of the course; unimp., unimportant comments (not related to the task). M, man; W, woman

in the degree of availability of information and knowledge as commodities. Below the main findings, limitations and recommendations are presented.

This connection could be fostered through the use of network platforms such as Twitter or Facebook, but with the commercial disadvantages of a proprietary system. Siemens (2019, n/p) has recently commented on this topic saying that this kind of "[n]etworking options are optimized to preserve the attention of site visitors, not to enable broad network formation across sites."

Siemens stresses the difference between systems and networks: "It's hard to control people in networks, they have too much agency, they can do what they want. The lack of controllability makes it difficult to achieve intended outcomes in networks. When agents want a clear outcome, they turn to systems. Systems preserve power" (Siemens, 2019, n/p). Following these thoughts, the development of a network for professional collaboration should take into account how much the system will interfere with the "natural" development of the network. Also, several questions must be addressed: how to avoid the negative effects of the group phase to evolve into the network phase that promotes a healthy and productive relationship with the system? What kind of platform/space is ideal for these purposes? If the system is always more powerful than the network, could this be an advantage for educational purposes? Can we design a network space where the system intervenes not only to preserve the attention (with commercial purposes) but also to foster the formation of a network with a specific focus? Could a system be a creative restriction rather than a constraint?

5.7.1   Findings

The main objective of this experience was to validate the use of an ESN as a means to accomplish academic work among teachers, and it was fully achieved. We created a crowdsourcing and white label site and complied with all of the design restrictions we established. The critiques, comments, and suggestions enriched marginally the curricular redesign process, and the interaction that was achieved in the science group indicates that teachers may build on each other's ideas and collectively construct solutions that better the syllabi. We conclude that Academic-match may be a useful means to obtain ideas and feedback from teachers in a curricular redesign process. However, the main issue we found is that it did not function as an ESN, because the environment was not open, the sense of liberty was hindered by a specific task, and a specific way of contributing to the project and the number of participants in each group was very small.

The specific objectives were also addressed. First, the fundamental premises about teachers that were established were only partially validated. Some participants were not able to use shared documents or comply with the instructions in Google Classroom or to make the necessary adjustments to fully visualize shared documents when using tablets or cell phones and thought that there were no comments to review. This evidence implies the need for professional development, more precise instructions, and additional just-in-time resources that they may need to solve doubts and provide specific information.

Second, we identified functionalities that best serve the academic needs of the curricular redesign project. To promote a higher level of retention in *matcha*, we will include both times and step indicators to provide feedback to the user, so he/she can acknowledge how long will it take to complete the profile and can be sure that the process is finalized. We will also include follow-up emails and pop-up messages to encourage them to complete their profile when they have not succeeded in this initial stage. Probably the most important functionality missing in the first version is the lack of collaborative workspaces inside the ESN. This has been established as a top priority and will be solved soon.

The third objective was to identify technical issues that may hinder social interaction among teachers. Since several teachers were unfamiliar with the procedure to insert comments and to respond to them, social interaction was not always possible (e.g., in the case of a teacher who used a discussion forum and not the syllabus to make comments). Also, because Google Classroom and Drive do not offer the option of an embedded video chat, conversations among teachers through other media will be lost for academic projects. There is a need to embed a video chat that can save the sessions for further analysis.

The last specific objective was to discover design flaws in the ESN site. Due to the reduction in participants answering the second test, we will redesign the test buttons so both are visible, and a pop-up message will alert users to engage in the second questionnaire. To provide more precise information about the choices users have, the match percentage of each co-teacher will be available so selections can be better informed; this means a much more robust infrastructure such as the one provided by the eTwinning platform (Kearney & Gras-Velázquez, 2018).

The use of this initial version of *matcha* points to its potential use as a tool to form groups with specific characteristics in order to contribute to projects in a larger and more open space that allows for networking, i.e., an ESN: an online environment where the benefits of social media, such as a sense of liberty and of belonging, can promote collaborative work in academic tasks, like curricular reviews. The ease to create groups based on predetermined criteria (such as leadership style and facets of innovation of the participating teachers, as was the case of *matcha*) is another positive feature of this kind of network. The inclusion of performance indicators (such as feedback functionalities) and the solution of design flaws will probably improve results in the academic tasks involved in future trials of this platform.

The best asset of this experience is the fact that it addresses a need of educational institutions and teachers, for them to be able to work in a platform that allows for grouping according to specific criteria and to collaborate and interact with peers.

5.7.2 Limitations and Lessons Learned

The two most important limitations of this study are the small number of teachers who finally participated and the fact that *matcha* did not allow the development of networking, as an ESN would. Also, several functionalities were missing: a space for collaborative work, the possibility of looking for specific members among a large number of participants, the opportunity to visualize members through badges, or other means. Therefore, *Academic-match* must be considered a group tool, not a network tool.

An important lesson learned in this study was the need to make sure all teachers acquire all proper digital skills. These include skills required for online teaching, know-how to conduct collaborative tasks online, and using any type of device.

5.7.3 Recommendation for Future Practice and Research

This is one of the few reports on ESN specifically aimed at teachers to facilitate their collaboration in academic endeavors. Even though this is a first effort to use a social network platform to promote teachers collaborative work, the obtained results suggest that they can become more creative and innovative, not only by showing a proper use of digital technologies (e.g., analytics and artificial intelligence tools for including chatbots and intelligent tutoring systems) but also by participating in the design of educational and academic models (e.g., multidisciplinary courses with located and significant learning included, as well as multilevel learning pathways). However, more research should be carried out, granted the affordances of these resources to profit from collective intelligence. Furthermore, following a learning engineering approach (Dede, 2019), the outcomes of these research projects should be integrated into educational practices, in our particular case within the B@UNAM high school program.

Besides, instead of assuming that teachers, students, and administrators have developed a digital culture, a project regarding the complete digital transformation of the university should be promoted. This would foster the success of proposals like the one presented in this chapter.

5.8 Final Proposal

In Table 5.7, the advantages and disadvantages of Academic-match for the design and development of academic projects crowdsourced among teachers are described.

The aspects presented in both columns are relevant to the projects of B@ UNAM. Therefore, a need for an expanded space of interaction and networking was identified. This led to the development of the final proposal that is detailed below.

The Fig. 5.6 depicts the proposal, which tries to profit from the competitive advantages of each tool:

- ASN provide large numbers of teachers who have the experience of relating and working in an online and open platform.
- *Academic-match* has the possibility of dynamically changing the criteria and methods to select teachers who best fit the requirements of the projects to be developed. It represents an efficient grouping and organizing procedure.
- AvCoP constitutes a milieu where the teachers with a certain profile will be able to work collaboratively to achieve the goals of the project at hand. Whenever a sufficiently large number of teachers participate in it, algorithms will make the processes more efficient while fostering better interactions, more cohesive groups, or optimal pairings.

This proposal might lead us to a more robust vACoP and let us get into the observation of the emerging "island networks" and "core-periphery networks" as Basak, Tafti, and Huang (2019) have done. They conclude that "[i]n island networks, weak

Table 5.7 *Academic-match* advantages and disadvantages

Advantages	Disadvantages
Grouping for specific purposes based on test results and other selective instruments	Loss of openness that characterizes a network
Time efficiency for grouping or for recruiting people interested in the project	Limitations on the number of potential people interested in the project due to the ceiling effect of the N that the platform includes
Time efficiency while developing the project, for structure, indicators, and instructions are preplanned	Loss of a sense of liberty for self-organizing and developing ideas. Potential sense of supervision that limits the expression of dissent and critique relating to institutional issues
	Technical limitations to form efficient groups, compared to the experiences where algorithms define part of the dynamics of the project

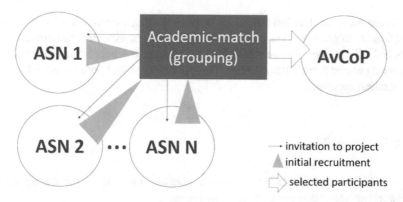

Fig. 5.6 A final proposal that integrates the advantages of academic social networks (ASN), a grouping site such as Academic-match, and an academic virtual community of practice (AvCoP)

ties create shorter paths and are important for the connectivity of the network. They have a higher edge betweenness centrality and the tie strength negatively correlates with edge betweenness centrality. In core-periphery networks, in contrast, strong ties are found in the core of the networks; they create shorter paths and are crucial for network stability. They have a higher edge betweenness centrality and the tie strength is positively correlated with edge betweenness centrality" (p. 381). Other studies (Menshikova, 2018) are using different approaches such us "overlapping communities" to find how "[o]verlapping members may obtain a range of skills and knowledge from several communities at the same time. Furthermore, a person who [is] involved in the activity of more than one community can transfer experience and information to the others, thereby providing a crossing enrichment and diverse perspectives" (p. 485).

This kind of analysis can reveal which kind of interactions are more effective in the vACoP and guide us toward a clearer view of network topology, and this proposal lets teachers feel free to engage in tasks, yet it provides a structure for organization and a space for collaboration.

5.9 Conclusions

The experience of starting an AvCoP, as complex as it seems, offers an area of enormous opportunity for academic collaboration. Our first experience presented challenges we had not foreseen due to the limited experiences we have in our academic contexts. Teachers work in isolation and sharing their expertise is not a common practice. This kind of academic community requires time and support to reach a consolidation phase.

There are potential limitations: it is possible that external ASN may impose restrictions to promote new projects, so the number of expected participants is not reached. There may also be difficulties to integrate teachers from different backgrounds coming from different ASN. This possibility would pose the need for initial integration strategies.

5.9.1 Road Map

In searching for an institutional project to give teachers a space for interacting and networking, the following steps must be taken:

1. Technical improvements to Academicmatch. There must be adjustments to include more efficient communication channels among teachers. Also, it should allow a higher degree of visibility for all participants, thus others may easily contact them based on the characteristics they might be looking for.
2. Identification of several ASN that can be invited to the planned projects. There must be channels to them that provide the possibility of dissemination of information about them among their members.
3. Connection to an AvCoP. This resource will provide a collaborative workspace.
4. Test the whole system to identify flaws and improve it. Indicators include motivation, satisfaction, and engagement in the project, products, and relations among members.

5.9.2 Potential Developments

This proposal is flexible in terms of the number of teachers that may participate in a certain project. When several institutions are involved, probably big numbers of participants will be involved. In those cases, the initial recruitment may provide thousands of teachers and the selection process may allow for hundreds or even thousands of teachers to participate in the AvCoP. In those cases that probably will be associated with public policy-making, or the analysis of national or regional educational problems, for example, decisions based on algorithms to improve the efficiency of the processes would be implemented, and networking will be activated and maintained.

However, there are other cases where the initial number of recruited teachers is limited to hundreds and the selected members for the project reaches only hundreds or even dozens. In those cases, no algorithmic input will be considered, and decisions will be taken based on observation by the project leaders, and work will be based on group interactions.

References

Basak, E., Tafti, A., & Huang, P. (2019). Network topology and tie strength in online communities of practice. In L. Aiello, C. Cherifi, H. Cherifi, R. Lambiotte, P. Lió, & L. Rocha (Eds.), *Complex networks and their applications VII* (pp. 337–387). Cham: Springer.

Belshaw, D. (2018). *MoodleNet*. White paper. Retrieved from Moodle.org website https://docs.moodle.org/dev/MoodleNet_whitepaper

Blake, R. R., & Mouton, J. S. (1964). *The managerial grid: The key to leadership excellence.* Houston, TX: Gulf Publishing Company.

Boyd, D. M., & Ellison, N. B. (2007). Social network sites: Definition, history, and scholarship. *Journal of Computer-Mediated Communication, 13*(1), 210–230.

Charalampidi, M., & Hammond, M. (2016). How do we know what is happening online? *Interactive Technology and Smart Education, 13*(4), 274–288.

Chawla, D. S. (2017). Publishers take academic networking site to court. *Science, 358*(6360), 161–161.

Cochrane, T., Guinibert, M., Simeti, C., Brannigan, R., & Kala, A. (2015). Mobile social media as a catalyst for collaborative curriculum redesign. In J. Keengwe & M. Maxfield (Eds.), *Advancing higher education with mobile learning technologies: Cases, trends, and inquiry-based methods* (pp. 1–21). Hershey, PA: Information Science Reference.

Dede, C. (2019). Improving efficiency and effectiveness through learning engineering. In C. Dede, B. Richards, & B. Sackberg (Eds.), *Learning engineering for online education* (pp. 1–14). New York: Routledge.

Dron, J., & Anderson, T. (2014). *Teaching crowds: Learning and social media.* (AU Press, Ed.). Edmonton, AB: Athabasca University Press.

Dron, J., & Anderson, T. (2015). Learning and teaching with social media. In Kinshuk & R. Huang (Eds.), *Ubiquitous learning environments and technologies* (pp. 15–29). Berlin, Heidelberg: Springer.

Ebrahimpour, A., Rajabali, F., Yazdanfar, F., Azarbad, R., Nodeh, M., Siamian, H., et al. (2016). Social network sites as educational factors. *Acta Informatica Medica, 24*(2), 134–138.

Firpo, D. R. T. Ð., Zhang, S., Olfman, L., Sirisaengtaksin, K., & Roberts, J. T. (2019). System design for an online social networking app with a notification and recommender system to build social capital in a university setting. In *Proceedings of the 52nd Hawaii international conference on system sciences*, 177–186.

García-Saiz, D., Palazuelos, C., & Zorrilla, M. (2014). Data mining and social network analysis in the educational field: An application for non-expert users. In A. Peña-Ayala (Ed.), *Educational data mining* (pp. 411–439). Cham: Springer.

Gohel, H. (2015). Design of intelligent web-based social media for data personalization. *International Journal of Innovative and Emerging Research in Engineering, 2*(1), 42–45.

Gruzd, A., Staves, K., & Wilk, A. (2012). Connected scholars: Examining the role of social media in research practices of faculty using the UTAUT model. *Computers in Human Behavior, 28*(6), 2340–2350.

Haythornthwaite, C. (2019). Learning, connectivity, and networks. *Information and Learning Science, 120*(1–2), 19–38.

Hou, H. (2015). What makes an online community of practice work? A situated study of Chinese student teachers' perceptions of online professional learning. *Teaching and Teacher Education, 46*, 6–16.

Insani, H. N., Suherdi, D., & Gustine, G. G. (2018). Undergraduate students perspectives in using Edmodo as an educational network. *English Review: Journal of English Education, 6*(2), 61–68.

James, R. (2014). ICT's participatory potential in higher education collaborations: Reality or just talk. *British Journal of Educational Technology, 45*(4), 557–570.

Jansen, J. (2017). Globally networked learning environments. An administrator's perspective. In A. Schultheis Moore & S. Simon (Eds.), *Globally networked teaching in the humanities: Theories and practices* (pp. 46–57). London: Routledge.

Jordan, K., & Weller, M. (2018a). Communication, collaboration, and identity: Factor analysis of academics' perceptions of online networking. *Research in Learning Technology, 26*, 1–13.

Jordan, K., & Weller, M. (2018b). Academics and social networking sites: Benefits, problems, and tensions in professional engagement with online networking. *Journal of Interactive Media in Education, 1*, 1–9.

Kearney, C., & Gras-Velázquez, À. (2018). *eTwinning twelve years on: Impact on teachers' practice, skills, and professional development opportunities, as reported by eTwinners*. Central Support Service of eTwinning – European Schoolnet, Brussels.

Kelley, T., & Littman, J. (2005). *The ten faces of innovation: IDEO's strategies for beating the devil's advocate & driving creativity throughout your organization*. New York: Currency/Doubleday.

Kelly, N., & Antonio, A. (2016). Teacher peer support in social network sites. *Teaching and Teacher Education, 56*, 138–149.

Klimova, B., & Poulova, P. (2015). Social networks in education. In *12th International conference on cognition and exploratory learning in digital age* (CELDA 2015), 240–246.

Krejci, R., & Siqueira, S. W. M. (2014). YouFlow microblog. *Interactive Technology and Smart Education, 11*(1), 2–14.

Lane, J. L., & Sweeny, S. P. (2019). Understanding agency and organization in early career teachers' professional tie formation. *Journal of Educational Change, 20*(1), 79–104.

Manca, S., & Ranieri, M. (2016a). Is Facebook still a suitable technology-enhanced learning environment? An updated critical review of the literature from 2012 to 2015. *Journal of Computer Assisted Learning, 32*(6), 503–528.

Manca, S., & Ranieri, M. (2016b). "Yes for sharing, no for teaching!": Social media in academic practices. *The Internet and Higher Education, 29*, 63–74.

Manca, S., & Ranieri, M. (2017). Networked scholarship and motivations for social media use in scholarly communication. *International Review of Research in Open and Distance Learning, 18*(2), 123–138.

Martín, E., Hernán-Losada, I., & Haya, P. A. (2016). Comparing social factors affecting recommender decisions in online and educational social network. *New Review of Hypermedia and Multimedia, 22*(1–2), 6–26.

Means, B. (2018). Tinkering toward a learning utopia. In C. Dede, B. Richards, & B. Sackberg (Eds.), *Learning engineering for online education* (pp. 36–52). New York: Routledge.

Menshikova, A. (2018). Evaluation of expertise in a virtual community of practice: The case of stack overflow. In D. Alexandrov, A. Boukhanovsky, A. Chugunov, Y. Kabanov, & O. Koltsova (Eds.), *Digital transformation and global society* (pp. 483–491). Cham: Springer.

Moore, A. S., & Simon, S. (2017). Globalization in the humanities and the role of collaborative online international teaching and learning. In A. Schultheis Moore & S. Simon (Eds.), *Globally networked teaching in the humanities: Theories and practices* (pp. 1–10). London: Routledge.

Mora, H., Signes, M. T., De Miguel, G., & Gilart, V. (2015). Management of social networks in the educational process. *Computers in Human Behavior, 51*, 890–895.

Nazir, M., & Brouwer, N. (2019). Community of inquiry on Facebook in a formal learning setting in higher education. *Education Sciences, 9*(1), 10.

Nistor, N., Baltes, B., Dascălu, M., Mihăilă, D., Smeaton, G., & Trăuşan-Matu, Ş. (2014). Participation in virtual academic communities of practice under the influence of technology acceptance and community factors. A learning analytics application. *Computers in Human Behavior, 34*, 339–344.

Paulin, D., & Haythornthwaite, C. (2016). Crowdsourcing the curriculum: Redefining e-learning practices through peer-generated approaches. *The Information Society, 32*(2), 130–142.

Perez, E., & Brady, M. (2018). A preliminary scoping review study of the progress of social media adoption as an educational tool by academics in higher education. *DBS Business Review, 2*, 127–153.

Porcel, C., Ching-López, A., Lefranc, G., Loia, V., & Herrera-Viedma, E. (2018). Sharing notes: An academic social network based on a personalized fuzzy linguistic recommender system. *Engineering Applications of Artificial Intelligence, 75*, 1–10.

Prestridge, S. (2019). Categorizing teachers' use of social media for their professional learning: A self-generating professional learning paradigm. *Computers & Education, 129*, 143–158.

Rabbany, R., Elatia, S., Takaffoli, M., & Zaïane, O. R. (2014). Collaborative learning of students in online discussion forums: A social network analysis perspective. In A. Peña-Ayala (Ed.), *Educational data mining* (pp. 441–466). Cham: Springer.

Rubin, J., & Guth, S. (2017). Collaborative online international learning: An emerging format for internationalizing curricula. In A. Schultheis Moore & S. Simon (Eds.), *Globally networked teaching in the humanities: Theories and practices* (pp. 15–27). London: Routledge.

Saedy, M., Rajaee, A., Jamshidi, M., & Jamshidi, N. (2014). Method and system for educational networking. In *2014 World automation congress (WAC): Emerging technologies for a new paradigm in system of systems engineering*, 1–4.

Siemens, G. (March 26, 2019). *I was wrong about networks* [web log post]. Retrieved from https://www.linkedin.com/pulse/i-wrong-networks-george-siemens/

Sliwka, A. (2003). Networking for educational innovation: A comparative analysis. In OECD (Ed.), *Networks of innovation towards new models for managing schools and systems* (pp. 49–63). Paris: OECD.

Smyrnova-Trybulska, E., Morze, N., & Kuzminska, O. (2019). Networking through scholarly communication: Case IRNet Project. In E. Smyrnova-Trybulska, P. Kommers, N. Morze, & J. Malach (Eds.), *Universities in the Networked Society* (Critical Studies of Education) (pp. 71–87). Cham: Springer.

Solomon, B., Ariffin, I., Din, M., & Anwar, R. (2013). A review of the uses of crowdsourcing in higher education. *International Journal of Asian Social Science, 3*(9), 2066–2073.

Surowiecki, J. (2004). *The wisdom of crowds: Why the many are smarter than the few and how collective wisdom shapes business, economies, societies, and nations*. New York: Doubleday & Co..

Suthers, D., & Dwyer, N. (2015). Identifying uptake, sessions, and key actors in a socio-technical network. In *Proceedings of the Annual Hawaii international conference on system sciences*, 1696–1705.

Tezer, M., & Yıldız, E. P. (2017). Frequency of internet, social network, and mobile devices use in prospective teachers from faculty of education. *TEM Journal, 6*(4), 745–751.

Trust, T., & Horrocks, B. (2019). Six key elements identified in an active and thriving blended community of practice. *TechTrends, 63*(2), 108–115.

Ursavaş, Ö. F., & Reisoglu, I. (2017). The effects of cognitive style on Edmodo users' behavior. *International Journal of Information and Learning Technology, 34*(1), 31–50.

Vadillo, G. (2014). Instrumento para identificar las facetas de la innovación. *Creatividad y Sociedad, 22*, 1–21.

Vadillo, G. (2017). Arquitectura de los cursos de B@UNAM. *Revista Mexicana de Bachillerato a Distancia, 9*(17), 26–31.

Van Waes, S., De Maeyer, S., Moolenaar, N. M., Van Petegem, P., & Van den Bossche, P. (2018). Strengthening networks: A social network intervention among higher education teachers. *Learning and Instruction, 53*, 34–49.

Veletsianos, G. (2016). *Social media in academia: Networked scholars*. New York: Routledge.

Veletsianos, G., & Kimmons, R. (2012). Networked participatory scholarship: Emergent techno-cultural pressures toward open and digital scholarship in online networks. *Computers & Education, 58*(2), 766–774.

Veletsianos, G., & Kimmons, R. (2013). Scholars and faculty members' lived experiences in online social networks. *Internet and Higher Education, 16*(1), 43–50.

Veletsianos, G., & Stewart, B. (2016). Discreet openness: Scholars' selective and intentional self-disclosures online. *Social Media and Society, 2*(3), 1–11.

Villatoro, C., Aznavwrian, L., & Vadillo, G. (2013). B@UNAM a través de sus números. *Revista Mexicana de Bachillerato a Distancia, 5*(10), 27–36.

Wenger, E. (2010). Communities of practice and social learning systems: The career of a concept. In C. Blackmore (Ed.), *Social learning systems and communities of practice* (pp. 179–198). London: Springer.

Willis, L. D., & Exley, B. (2018). Using an online social media space to engage parents in student learning in the early years: Enablers and impediments. *Digital Education Review, 33*, 87–104.

Chapter 6
Developing a Learning Network on YouTube: Analysis of Student Satisfaction with a Learner-Generated Content Activity

Daniel Belanche, Luis V. Casaló, Carlos Orús, and Alfredo Pérez-Rueda

Abstract This chapter focuses on an innovative learning project in which undergraduate marketing students create videos and post them on the official YouTube channel of the course. The project aims to apply new information technologies to university teaching, use social networks to disseminate information and knowledge, and provide students with a more active role in their learning process. We offer an overall vision to enable the appropriate application of this project and analyze students' perceptions and satisfaction with the activity; we also offer a tool (e.g., rubric) to objectively evaluate the video content. The results of the first study show that students' satisfaction with the activity is high and is affected by the emotions it generates, its perceived usefulness, and its ease of use. In a second study, we observe that the students are satisfied with the activity's evaluation rubric, because it is a simple evaluation system and serves to guarantee an objective and consistent evaluation. The description of the project and discussion of the research findings will help academics to understand the key drivers of a satisfactory experience by students. It also provides scholars with advice on how to implement educational networking in their subjects.

Keywords YouTube · Active learning · Learner-generated content · Satisfaction · Rubric · Networks

D. Belanche (✉) · C. Orús
Facultad de Economía y Empresa, Universidad de Zaragoza, Zaragoza, Spain
e-mail: belan@unizar.es; corus@unizar.es

L. V. Casaló
Facultad de Empresa y Gestión Pública, Universidad de Zaragoza, Huesca, Spain
e-mail: lcasalo@unizar.es

A. Pérez-Rueda
Facultad de Ciencias Sociales y Humanas, Universidad de Zaragoza, Teruel, Spain
e-mail: aperu@unizar.es

© Springer Nature Switzerland AG 2020
A. Peña-Ayala (ed.), *Educational Networking*, Lecture Notes in Social Networks, https://doi.org/10.1007/978-3-030-29973-6_6

Abbreviations

AVE   Average variance extracted
EHEA   European Higher Education Area
ICT   Information and communications technologies
PLS   Partial least squares
TAM   Technology acceptance model

6.1 Introduction

The digitization of information and the importance of social networks are two primary characteristics of the global society (Gutiérrez Martín & Tyner, 2012). In recent decades, new information and communications technologies (ICTs) have been increasingly present in academic institutions and educational systems (e.g., e-learning platforms, tablets, smartphones, virtual reality) (Blasco-Arcas, Buil, Hernández-Ortega, & Sese, 2013; Dündar & Akçayir, 2014; Ferrer, Belvís, & Pàmies, 2011; Ifenthaler & Schweinbenz, 2013; Merchant, Goetz, Cifuentes, Keeney-Kennicutt, & Davis, 2014). Research has found a positive link between using ICTs in the classroom and academic performance and the learning experience (Blasco-Arcas et al., 2013; Orús et al., 2016). The use of ICTs, which facilitate a combination of content, pedagogy, and technology, may encourage students' motivation and improve their learning outcomes, compared to more traditional education methods in the classroom.

In this line, the current development of Web 2.0 tools and the incorporation of "digital natives" as students are drastically changing the traditional teaching and learning paradigms (Orús et al., 2016). The use of ICTs, for example, has been found to improve interactivity in the classroom (Beauchamp & Kennewell, 2010), facilitate access to and sharing of knowledge, and foster collaboration (Pérez-Mateo, Maina, Guitert, & Romero, 2011). In addition, the incorporation of digital natives implies that students no longer want to be passive receivers of knowledge but demand continuous challenges to remain engaged in the learning process (Orús et al., 2016). In this context, using ICTs can facilitate students' active learning (Mon & Cervera, 2011; Schmid et al., 2014) and can be particularly effective to develop competencies in students that are usually less promoted in academic courses (e.g., technological and social skills) (Fralinger & Owens, 2009).

6.1.1 Active Learning

According to Shekhar, Prince, Finelli, Demonbrun, and Waters (2019), active learning is instruction that meaningfully engages students in learning through higher participation in activities. Active learning activities imply cooperation between students (Johnson, Johnson, & Smith, 1998). Initial studies prescribing active learning methods (Michael, 2006; Prince, 2004) have suggested that these methodologies are advantageous because (1) they increase the construction of meaning by the learner; (2) they help the learner to understand how to do something, rather than just assimilating facts; (3) they help transfer knowledge from general to specific contexts; (4) they provide a group learning situation, which is important since students prefer to learn with others than alone; and (5) they contribute to meaningful learning by means of articulating explanations via the learner, teachers, or peers. In addition, active learning methods are not necessarily implemented during teaching hours and may or may not involve ICTs. Active learning processes rely on user-centered learning (e.g., Wright, 2011) that involves rebalancing power in the class by reshaping the role of the teacher and the student, sharing the learning responsibility, and increasing the students' perception of value of the course content and evaluation thereof.

6.1.2   Social Media for Learning Materials

The tremendous growth of social media, which is altering the ways users obtain information (Hofacker & Belanche, 2016), favors a higher diffusion of active learning methods. These days, users rely on social media for updates about their closest networks and the latest global or local news and look for learning materials about specific issues when needed. In this way, text, image, audio, or audiovisual layouts can help disseminate knowledge in social media. Among the different formats to obtain information online, video is increasingly preferred among users (Zote, 2019). Indeed, the video format gets 21.2% more interaction in social media (e.g., likes, shares) compared to still images (Zote, 2019).

In recent years, users have gained control of the audiovisual material they want to consume and share through social media. Uploading and watching social media videos is easy, ubiquitous, and convenient. The affordability of recording and playing devices (e.g., smartphones with cameras) has contributed to the widespread use of video among both professional and amateur users. Indeed, between 2016 and 2018, video watching via the social network Instagram has quadrupled (Spisak, 2019).

Nonetheless, the leading social media for video broadcasts is YouTube, with a 96% penetration rate among US teenagers, compared to a 72% rate for Instagram (Spisak, 2019). Interestingly, social media applied to education contexts is contributing to creating communities or collaborative learning networks based on the sharing of content and the recommendation of relevant information (Lewis, Pea, & Rosen, 2010). Indeed, 70% of millennial users use YouTube to watch educational content; as a reaction, Google recently announced a US$20 million investment into its YouTube learning initiative (Cooper, 2019). Therefore, to improve the potential of this media, there is a need to generate and disseminate educative content on YouTube, taking advantage of the interactive features of the medium, such as the control of experience and bidirectional and interpersonal communication

(Gao et al., 2010). Indeed, scholars have already recognized the value of YouTube as an educational network that allows real-time or asynchronous interaction between participants of the learning process via social media (Lee, Osop, Goh, & Kelni, 2017; Zahn et al., 2014).

6.1.3 Learner-Generated Content

In the context of online knowledge sharing, the concept of learner-generated content has been extended among scholars and is becoming an increasingly used instrument to foster active learning processes. Specifically, the generation of content by students could be useful to develop relevant competencies that are frequently missing in regular courses (e.g., technological and social skills, (Fralinger & Owens, 2009)). Nevertheless, previous findings on the efficacy of learner-generated content for increasing learning are inconclusive (Orús et al., 2016). Whereas some studies support ICTs as an effective tool for active learning (e.g., Hoogerheide, Deijkers, Loyens, Heijltjes, & Van Gog, 2016; Lee & Mcloughlin, 2007; Pereira, Echeazarra, Sanz-Santamaría, & Gutiérrez, 2014), others find that the generation of educational content by students entails several difficulties. On the positive side, the activities of creating and disseminating videos by students have been shown to benefit attention and memorization of content (Clifton & Mann, 2011), learning outcomes (Kay, 2012), academic qualifications (Dupuis, Coutu, & Laneuville, 2013), and the acquisition of transversal competencies (Orús et al., 2016); in addition, students presenting content in front of a camera make an extra effort to empathize with their audience to increase others' learning (Hoogerheide et al., 2016). On the negative side, participants in these student-generated content activities often do not perceive them to have added value (Bennett, Bishop, Dalgarno, Waycott, & Kennedy, 2012), find generated content too difficult to create and express a lack of self-efficacy or interest in the activity, or try to avoid active learning activities in class (Cole, 2009). All these diverse results encourage us to further investigate how to increase students' satisfaction with active learning activities to ensure a positive impact of learner-generated content on education systems.

Furthermore, few studies analyze the subjective and/or psychological variables that affect participation in student-generated content activities. Taking into account that these types of activities have unusual characteristics in the academic sphere (students are creating the teaching content and using audiovisual and interactive tools; teachers need to objectively assess the quality of both the content and the format of these creations), there is a need to improve our understanding of the variables that affect the perceived value of these activities from the students' perspective. In addition, implementation of these initiatives requires an objective assessment of students' tasks related to their participation. In this line, the use of rubrics (i.e., grading matrixes) may contribute to facilitating this evaluation (Reddy, 2007) and to carrying it out in a more objective way (Jonsson, 2014).

6.1.4 Project Objectives

In this innovative teaching project, undergraduate students of a basic marketing course were encouraged to make videos and post them on the official YouTube channel of the course. This project is based on active learning and learner-generated content paradigms since students have to generate and understand meaningful contents, in a way that it helps them to transfer knowledge to other members of an educational network (classmates, teachers, and other users interested in the content). After 4 years, more than 1400 students have had the opportunity to participate in the activity, resulting in more than 300 videos about theoretical concepts of the course. Nevertheless, the interest in or success of the project cannot be measured only in terms of the number of participating students or the videos posted. It is important to investigate the factors that contribute to the success of the initiative by focusing on students' perceptions and intentions about their participation.

Thus, this chapter contributes to the literature by identifying key psychological factors that determine satisfactory participation by students. In this vein, our research not only includes well-grounded cognitive factors (e.g., Liao, Huang, Chen, & Huang, 2015), but also affective cues (students' positive and negative emotions related to the activity) that are frequently ignored in the previous literature in this field. We focus on students' satisfaction with their participation in this active learning activity, as well as with the rubric that was developed and used by the teachers to increase objectivity in evaluating such an innovative activity. Thus, this chapter contributes to expanding the knowledge about how to successfully introduce active learning activities in a higher education context. We also provide valuable information about the process that may guide teachers in the implementation of similar active learning activities.

Therefore, the aim of this chapter is to evaluate the project as a teaching method from the students' perspective. We analyze the process that leads to student satisfaction with an activity based on student-generated content that is thereafter posted on YouTube. Based on the results of our project, we also aim to explain the key factors that should be considered for successfully introducing such an innovation. Specifically, we try to answer the following research questions:

- What aspects improve students' satisfaction with the generation of academic content to be posted on social networks?
- What is the role of cognitive factors (e.g., ease of use, usefulness) and affective factors (e.g., positive and negative emotions derived from participation), among others?
- How could teachers evaluate the content generation as part of the students' work?
- What is the degree of student satisfaction with the evaluation criteria, and what are the main determinants thereof?

6.1.5 Chapter Structure

To answer these questions, the remainder of the chapter is structured as follows. First, we review the literature on the latest advances and challenges regarding active learning methodologies in general and the use of YouTube as a learning tool in particular. A section then briefly explains our teaching project and the context in which it was implemented, as well as the overall results of the project. Afterward, we rely on quantitative methods to assess the key factors determining the perceptions of a wide sample of students about the teaching project. Specifically, the first study analyzes with detail their expectations and final perceptions about the activity, as well as the influence of cognitive and affective factors on their satisfaction. The second study focuses on the development of an evaluation system for the activity, which is based on a rubric, and the students' perceived utility and satisfaction with the tool. After presenting the main results of the project and the two empirical studies, which answers to the research questions mentioned above, the chapter ends with a discussion of the main contributions of the project. We stress the useful knowledge generated that can be used to implement similar active learning activities in other teaching contexts. Finally, we highlight the limitations and future research lines.

6.2 Literature Review

In this section, we review the literature related to the relevant topics for this chapter. First, studies that deal with the integration of new technologies in the new educational models in higher education, which rely on the acquisition of competencies and the active role of the students in their learning process, are reviewed. The importance of developing rubrics to evaluate student-generated videos is also addressed. After that, we offer an overview of the previous literature about the specific use of YouTube as a teaching tool in a higher education context.

6.2.1 The Integration of ICTs in New Educational Models

The European Higher Education Area (EHEA) educational model relies on the acquisition of competencies by students (Cano, 2008). This new model implies that each degree should develop different specific and cross-curricular competencies in order to ensure that students acquire certain skills and knowledge. Specific competencies are focused on particular knowledge related to the fields involved in the degree, whereas cross-curricular competencies transcend a particular field and have a multidisciplinary nature (Pârvu, Ipate, & Mitran, 2014). Cross-curricular competencies are defined as "the set of intellectual, personal, and social skills that all students need to develop in order to engage in deeper learning" (British Columbia

Ministry of Education, 2013, p. 3). Cross-curricular competencies involve social and interpersonal skills, skills to manage ICTs (Pereira et al., 2014), or general academic skills such as creativity (Azevedo, Apfelthaler, & Hurst, 2012; Pârvu et al., 2014). In this way, cross-curricular competencies encourage students to look at problems from different approaches, so as to see the relationships between subjects and their own previous learning and personal experiences as members of larger social communities (British Columbia Ministry of Education, 2013).

In addition, EHEA legislation implies a change in the students' role in their learning process, evolving from "passive" to "active" learners (Orús et al., 2016). Based on the learning-by-doing concept, active learning is defined as "anything that involves students doing things and thinking about the things they are doing" (Bonwell & Eison, 1991, p. 2). Active learning requires activities that challenge the learners to perform tasks to engage in their own learning, such as games or discovery learning (van Diepen, Stefanova, & Miranowicz, 2009).

Among the many active learning activities, previous research notes the powerful learning effects of explaining to others (e.g., Fiorella & Mayer, 2014; Hoogerheide et al., 2016). "Learning by teaching involves explaining to-be-learned material with the goal of helping others learn" (Fiorella & Mayer, 2016, p. 728). Learning by teaching involves not only the actual teaching but also preparing to teach (Okita, Turkay, Kim, & Murai, 2013). Research has found that both active learning strategies stimulate cognitive processes and improve memorization and understanding of the materials, leading to better academic performance (Fiorella & Mayer, 2013; Okita et al., 2013). Learning by teaching involves several active learning strategies, such as selecting and organizing the most relevant information to include in the explanation and elaborating on the material via a coherent structure that can be understood by others (Fiorella & Mayer, 2016). The final result is the creation of learner-generated content. In our project, students engage in these learning strategies to generate content in the form of a short video that explains theoretical concepts of a marketing course.

6.2.2   Web 2.0 Tools

Web 2.0 tools facilitate the creation of learner-generated content, which may improve students' performance and learning (Lee & Mcloughlin, 2007) and increase their level of involvement (Pérez-Mateo et al., 2011). Gupta (2014) developed a framework and proposed that the use of Web 2.0 tools, such as wikis, blogs, and discussion boards, can produce effective learning (in terms of cognitive, affective, and metacognitive responses), as long as there is a fit between the characteristics of the technology and the pedagogical task. However, other authors observe certain barriers to the successful integration of Web 2.0 (e.g., lack of students' familiarity with the tools and lack of institutional support (Bennett et al., 2012)), and the use of technology may exert a negative effect on learning outcomes (e.g., students who are required to complete many discussion posts may perceive this as a tedious task) or raise certain concerns, such as those pertaining to intellectual property rights, privacy, and copyright laws (Rodriguez, 2011).

On the other hand, previous research shows that when students generate videos, it enhances their learning (Stanley & Zhang, 2018), skills (Forbes et al., 2016), cross-curricular competencies (Orús et al., 2016), and e-leadership (Chua & Chua, 2017). Levina and Arriaga (2014) indicate that in addition to cognitive needs, content creators seek affective gratification (entertainment, social recognition). Therefore, beyond the utilitarian value of the activity (improved learning, acquisition of competencies, academic performance), the student could obtain hedonic benefits, reflected in positive and negative emotions derived from his or her participation in the activity. In this sense, the value of the hedonic aspects reflects personal and subjective experiences linked to the emotions of the individual and complements the vision based solely on the utilitarian value of behavior (Babin, Darden, & Griffin, 1994). Taking into account that users appreciate the functional and hedonic value of their online behavior (Schulze, Schöler, & Skiera, 2014), there is a need for better understanding of how cognitive and affective variables impact students' satisfaction with learner-generated content activities using Web 2.0 tools.

6.2.3 The Use of YouTube in Education

Recently, the use of YouTube in higher education contexts has been increasing (Alon & Herath, 2014; Chan, 2010; Fralinger & Owens, 2009; Krauskopf, Zahn, & Hesse, 2012; Orús et al., 2016; Sherer & Shea, 2011; Torres-Ramírez, García-Domingo, Aguilera, & De La Casa, 2014; Tugrul, 2012). The integration of YouTube in the classroom is especially easy due to the popularization of digital technologies, such as internet connections, digital cameras, and mobile devices; the development of software, such as streaming and editing programs; and the wide communication potential of visual media, given that people learn more deeply when ideas are expressed in words and pictures, rather than in words alone (Mayer, 2001).

In 2018, YouTube reached 149 million users worldwide, and more than 1 billion hours of video are watched daily (Dogtiev, 2019). "Digital natives" are very familiar with these applications because they have grown up in a society characterized by new information technology (Chan, 2010). In this context, higher education centers are using their official YouTube channels to provide students with useful knowledge and enriching experiences. Students obtain satisfaction from the use of online videos for teaching (Torres-Ramírez et al., 2014; Tugrul, 2012). Following the previous literature, using YouTube to post educational videos facilitates the searching of video information about any topic, promotes student-to-student and teacher-to-student collaboration, and improves users' learning process (Domínguez Fernández & Llorente Cejudo, 2009).

Previous studies analyze the use of YouTube channels as a platform to post educational videos. Kay (2012) finds important benefits arising from video use in education, such as control over learning, improved attitudes and learning behaviors, and increased learning performance. Duncan, Yarwood-Ross, and Haigh (2013) review approximately 1500 minutes of video in 100 different YouTube educational sites. The authors conclude that lecturers should be more proactive in recommending suitable YouTube material as supplementary learning materials. Everson, Gundlach, and Miller (2013) note the importance of clearly communicating specific learning goals in order to reduce feelings of intrusiveness (e.g., bothersome, lack of worthiness). Dewitt et al. (2013) conclude that YouTube is an effective information technology in learning and teaching. Employing YouTube to teach increases accessibility to resources and improves students' attention and memory (Clifton & Mann, 2011). Prior studies also find that YouTube video use in the classroom leads students to higher levels of recall compared to an audio format (Mayer, 2001) and that students using recommended online videos obtain higher marks than those who decided not to watch them (Dupuis et al., 2013).

6.2.4 The Use of YouTube for Active Learning

Previous research focuses exclusively on the use of YouTube to support teaching materials. In most cases, students are passive spectators of the educational materials. While it is true that watching videos involves a certain degree of active learning (e.g., students analyze and evaluate the content, interpret its meaning, and answer questions raised in the audiovisual material), few studies develop learning activities using YouTube as a tool for learner-generated content (Orús et al., 2016). In one of the few exceptions, Jenkins and Dillon (2013) argue that creating and uploading videos to YouTube, and then presenting them in the classroom, build collaborative knowledge, especially in areas such as social sciences, education, and humanities. However, the authors do not examine the students' work, perceptions, and learning outcomes. Fralinger and Owens (2009) propose a similar activity resulting in higher levels of interest, better learning, achievement of objectives, and acquisition of skills or workgroup competencies. In a project carried out by Pereira et al. (2014), 29 students created and uploaded a video in Babelium (open-source software) about different diagnostic techniques. The authors show improvements in the students' competencies compared to traditional teaching methods, in terms of information search, organization, problem-solving, and technological skills. Alon and Herath (2014) carry out a project in which students develop a social media plan, which includes a YouTube video, to promote a specific country image. The authors find that students have a positive attitude toward the task, which is perceived as more productive and enjoyable compared to traditional learning experiences (e.g., lectures). Finally, in relation to a video creation project on YouTube, Zahn et al. (2014) show that the students involved in the project acquire media-related skills to present the topic for an audience, increase their knowledge, and are highly satisfied with the course concept, compared to students who do not take part of the video creation task.

However, previous studies on this topic show unclear results regarding students' overall perception about the activity. For instance, some authors suggest that students do not perceive the value of generating educational content (Bennett et al., 2012). Cole (2009) concludes that students could have concerns regarding the difficulty of the task; they sometimes lack self-confidence or interest in the activity. Therefore, more research has to be developed in order to understand the impact of ICTs on education systems from the perspective of learner-generated content.

6.2.5 Evaluating Learner-Generated Content

Evaluating student-generated content is a challenging task (Momeni, Cardie, & Diakopoulos, 2016; Whitaker, Orman, & Yarbrough, 2016). As previously stated, active learning activities involve the implementation of different learning strategies and the development of specific and cross-curricular competencies. These competencies are difficult to measure and evaluate. In addition, the evaluation of content generated by students represents a complex task for teachers, who face the challenge of separating the consideration of form-related aspects from that of content-related aspects. In this way, the use of a rubric is an objective instrument that avoids reliance on subjective evaluations of content and aesthetics (Ryan, 2017). The rubric is a tool that provides the evaluation system with objectivity, given that it serves to grade complex skills and shows the expected assessment criteria in a comprehensive list (Andrade, 2000; Stiggins, 2001). In this way, the rubric offers transparency in the evaluation system and helps students to control their performance during the activity's development (Jonsson, 2014); in addition, it allows teachers to involve students in the evaluation by means of peer-to-peer assessment (Panadero, Jonsson, & Botella, 2017).

6.3 Overall Description of the Project

This project extends across four academic years, starting in 2013. During these 4 years, we implemented the project and included modifications to improve its development, the commitment and satisfaction of the agents involved (students and teachers), and its evaluation.

In this section we present the information about the development of the teaching project. Specifically, a first subsection (Sect. 6.3.1) describes the context in which the activity is applied and the evolution that its development has taken in the four academic years in which it has been implemented. The details about the teaching methodology, including instructions, procedures, and materials, are also explained in a second subsection (Sect. 6.3.2). Finally, a third subsection (Sect. 6.3.3.) reports the overall results of the activity regarding the creation of the YouTube and several statistics and metrics about its use.

6.3.1 Context of the Project

The context of the project is a university course (Introduction to Marketing) of the Degree in Business Administration and Management at a major university in Spain. This course was selected for three reasons. First, it is a first-year compulsory course where students' learning is based on the outcomes of several activities carried out during the semester, rather than on a global exam. Thus, development of the project was consistent with the assessment policy and was perceived as appropriate by both teachers and students (Alon & Herath, 2014). Second, young college students are active consumers, and sometimes creators, of YouTube con-tent. Providing an opportunity to create a video for YouTube could be seen as more appealing for students than just creating a video. Third, marketing is a subject with a very practical nature; therefore, students require active learning strategies to understand concepts and put them into practice.

During the first year of implementation, the activity was voluntary and was not taken into account for the students' final grade. Given the success of the activity and the high degree of satisfaction of students and teachers (see Orús et al. (2016) for further details), the activity became compulsory and was included in the assessment policy of the course.

In the following years, adjustments and modifications were made to the activity, which was continuously scrutinized for avenues of improvement. Specifically, efforts were exerted to understand the students' perceptions and evaluations of the activity, with the aim of not only creating effective methodologies in terms of learning and acquisition of competencies but also on achieving their satisfaction. Furthermore, given the novelty of the activity and the complexity of its evaluation, resources were devoted to designing adequate and objective evaluation criteria.

6.3.2 Development of the Project

The development of the activity was similar for each academic year and throughout the 4 years (see Fig. 6.1). First, an introductory session to the activity was developed. This introductory session took place in the first weeks of the course. The teachers explained to the students the objectives of the project and contextualized its importance in the development of the curriculum within the degree. In this way, the project aimed at improving the students' comprehension of the theoretical concepts of the course, turning the theory into practice in such a way that the students showed a real application of the theoretical concepts, improving the learning process through a greater retention of concepts, increasing the students' involvement with the course, and developing a series of cross-curricular competencies, such as creativity, synthesis, and teamwork skills.

In the introductory session, the teachers also explained the materials and instructions for making the video activity. Students were asked to form groups of four to

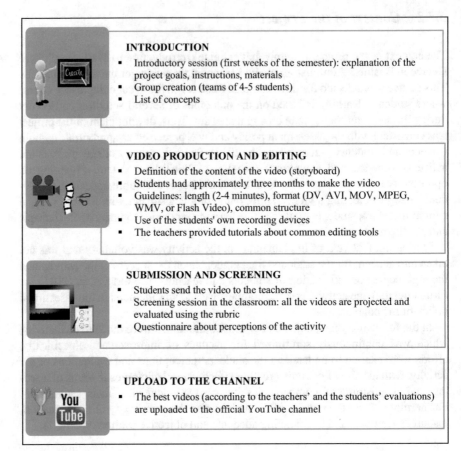

INTRODUCTION
- Introductory session (first weeks of the semester): explanation of the project goals, instructions, materials
- Group creation (teams of 4-5 students)
- List of concepts

VIDEO PRODUCTION AND EDITING
- Definition of the content of the video (storyboard)
- Students had approximately three months to make the video
- Guidelines: length (2-4 minutes), format (DV, AVI, MOV, MPEG, WMV, or Flash Video), common structure
- Use of the students' own recording devices
- The teachers provided tutorials about common editing tools

SUBMISSION AND SCREENING
- Students send the video to the teachers
- Screening session in the classroom: all the videos are projected and evaluated using the rubric
- Questionnaire about perceptions of the activity

UPLOAD TO THE CHANNEL
- The best videos (according to the teachers' and the students' evaluations) are uploaded to the official YouTube channel

Fig. 6.1 Timeline of activities

five people and create a video explaining a marketing concept based on the introduction to marketing subject. The teachers offered a list of the marketing concepts that could be explained in the video. In this way, a first orientation was given to the students about its content. Specifically, the explanation could be illustrated with a real company (e.g., commercial strategies), and different techniques were recommended, such as interviews with experts, oral presentations, documentary (combination of images, video clips, and narration), dynamic PowerPoint presentations, animations, etc.

The choice of the video's content and execution was left exclusively up to the students. However, each video was required to have the same structure: it had to start with a mask that clearly identified that it had been created by students of the degree of business management and administration at the university. In addition, the students had to identify themselves and the topic addressed in the video.

The video recording could be done with digital camcorders, digital cameras, or mobile phones with certain recording quality. The admissible formats were those allowed by YouTube: DV, AVI, MOV, MPEG, WMV, or Flash Video. The videos were edited by the students, with the supervision of teachers of the subject. To this end, the teachers developed some tutorials on the use of common editing tools (e.g., Windows Movie Maker, YouTube Video Editor), which were at the students' disposal in case of any doubts or need for clarifications.

The teachers supervised the students' work throughout the semester. First, the students had to define the content of the video and translate it into a graphic document (storyboard), which served as a guide for the working group and as a first insight into the project for the teacher. Next, the deadline for completion of the video was set (for the week before a holiday break at the end of the semester). Once the video was made, the working group sent it to the responsible teacher. Subsequently, the teacher responsible for the group evaluated the work according to the criteria developed in the rubric.

To stimulate the visualization and facilitate evaluation of the videos by the students, all the groups were asked to show their video in class during the last school days of the semester. These sessions enabled them to evaluate the videos made by their peers using the same rubric that the teachers employed. Finally, those videos rated over a minimum of quality were uploaded to the YouTube channel.

6.3.3   Overall Results of the Project

The YouTube channel was created on February 17, 2014. The channel can be accessed via the Weblink https://www.youtube.com/channel/UCXj1vWQprt0bGU SnEweGjHA (see Fig. 6.2). The channel is organized into seven reproduction lists corresponding to the six theoretical units of the course (Essentials of Marketing, Consumer Behavior, Decisions about Product, Decisions about Price, Decisions about Place, Decisions about Promotion) and one list with other content.

After 4 years of development, more than 1400 students have had the opportunity of taking part of the activity, producing more than 300 videos. However, the channel contains 89 videos. As previously stated, each year the teachers select the best videos, based on their evaluations as well as those of the students' peers; these videos are then uploaded to the channel. A few former students have requested the teachers to remove their videos from the platform due to personal concerns.

In terms of performance, the channel has more than 140 subscribers (as of January 2019). The total number of views has surpassed 66,000, and the channel can thus be considered as having a high impact in terms of spreading marketing knowledge through online networks. Specific statistics also support this contribution: 63% of the videos (56 out of 89) have more than 100 views, and 18% (16 out of 89 videos) have more than 1000 views. Interestingly, one video has reached more than 12,000 views (see Fig. 6.3). This video describes, in Spanish, the different levels of the product as identified in many marketing manuals.

Overall, considering the limited dimension of the project, these numbers reveal that the channel has performed relatively well and that the scope of the learning activity surpasses that of a regular university class.

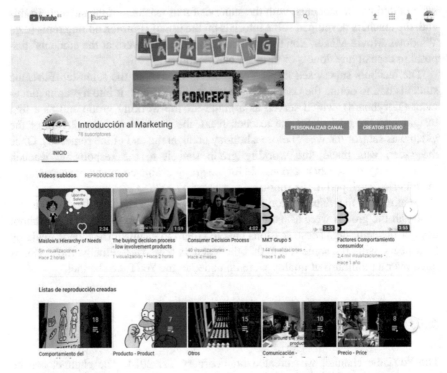

Fig. 6.2 YouTube channel front page

6.4　Study 1. Students' Satisfaction with the Activity

This study deals with the analysis of the students' expectations (before carrying out the activity) and perceptions (after concluding the activity) about cognitive and affective variables related to the activity. It is structured in three subsections. The first one (Sect. 6.4.1) deals with the proposed research model and develops a set of hypotheses according to our literature review. Next, the results of the study are presented in detail (Sect. 6.4.2). Finally, we discuss the conclusions derived from Study 1 findings (Sect. 6.4.3).

6.4.1　Research Model of Study 1

Based on the active role of the student, the project allows students to improve their knowledge about the contents of the course, as well as a series of cross-curricular competencies, such as a greater communication capacity, synthesis, or creativity (Orús et al., 2016). Students are evaluated on their ability to analyze, synthesize, solve problems, work as a team, and apply knowledge to practice.

Fig. 6.3 The most watched video on the YouTube channel

Online video has become a common format for personal expression and identity construction. Smartphones have become an extension of the youngest individual's identity. Those who, half a century ago, wrote poetry today post videos on YouTube, becoming identity icons. In this identity consolidation, there are high degrees of exhibitionism. Content creators seek affective gratification (entertainment, social recognition) (Levina & Arriaga, 2014).

As a new technology-related activity to be adopted by students, we considered the technology acceptance model (TAM, Davis, 1989) as the theoretical framework to understand the determinants of its adoption. Due to the parsimony of this model and its wide scope, the TAM is the leading model for analyzing technology adoption (Bagozzi, 2007; Venkatesh, Morris, Davis, & Davis, 2003). Indeed, this framework is very easy to apply to heterogeneous contexts and has significantly outperformed previous models in terms of explained variance (Bagozzi, 2007; Venkatesh et al., 2003).

According to the TAM, users evaluate a technology on the basis of the cost-benefit paradigm (Davis, 1989), that is, considering technology usefulness as the belief that the use of the system would enhance the individual's performance and ease of use as the belief that using the system would be free of effort (Belanche, Casaló, & Guinalíu, 2012).

In our context of analysis, students were trained and had some experience with the technology. Thus, our interest was on investigating not only adoption intentions but also their satisfaction after participating in the activity. In this regard, previous research already explained that satisfaction is the key variable in technology post-adoption processes (i.e., rather than attitudes), as far as users already hold a personal first-hand experience with the technology (Bhattacherjee, 2001). Nevertheless, considering the cognitive determinants of satisfaction with the activity (i.e., usefulness and ease of use) may be not enough to account for the real causes of student satisfaction. In this regard, the marketing literature has acknowledged the affective nature of satisfaction, such that both positive and negative emotional cues are essential to shape user's satisfaction (Oliver, 1993). In educational contexts, students have been found to take into account their positive and negative learning-related emotions, which are critical to determine the success of a learning activity (Pekrun, Goetz, Titz, & Perry, 2002). In general terms, positive (negative) emotions refer to a set of emotions felt by the user as a consequence of their use or participation in an activity (e.g., happiness or fulfillment for the positive emotions; frustration or embarrassment for the negative emotions).

Therefore, beyond the cognitive/utilitarian value of the activity, it seems interesting to examine the importance of the students' emotions related to their learning and performance of teaching activities. Specifically, in addition to usefulness and ease of use variables, the influence of positive and negative emotions related to participation in the activity was analyzed. The following hypotheses formally present the relationships proposed in the post-participation research model:

- H1: Student's perceived usefulness has a positive effect on satisfaction with the activity.
- H2: Student's perceived ease of use has a positive effect on satisfaction with the activity.
- H3: Student's positive emotions of the participation have a positive effect on satisfaction with the activity.
- H4: Student's negative emotions of the participation have a negative effect on satisfaction with the activity.

6.4.2 Method of Study 1

In order to analyze the students' expectations and perceptions about the activity and to test the proposed relationships, we used questionnaires to obtain primary information from the students. Specifically, two surveys were developed at different points during the course (prior to development of the activity and their final perceptions once the activity ended).

At the beginning of the course: After conducting the introductory session, we measured the students' expectations and anticipated emotions (both positive and negative) related to the activity (Bagozzi, Belanche, Casaló, & Flavián, 2016). Among other measurements, the expectations of interactivity were collected (Blasco-Arcas et al., 2013) as well as expectations about the ease of use and usefulness (Koufaris, 2002), since these are two key variables of technological acceptance models (e.g., Davis, 1989). To ensure anonymity during the process, the students were required to choose a nickname and to remember it for further questionnaires. This allowed us to link the students' expectations with their final perceptions. The teachers collected all the nicknames and, for the final questionnaire, projected them in class to facilitate recall by the students. To measure all the variables included in this questionnaire, 7-point Likert scales were used, where students indicated their degree of agreement or disagreement with a series of statements (from 1, "totally disagree," to 7, "totally agree"). The Appendix shows all the scale items employed in the study.

At the end of the semester: Once the activity had been completed, a final questionnaire was administered to understand the students' perceptions of the activity, as well as the emotional consequences. In addition, to evaluate the success of the activity, the students were asked about their perceived learning during the course (through the acquisition of specific and cross-curricular competencies (Orús et al., 2016), their satisfaction with the activity (Orús et al., 2016), and two behavioral intentions: intention to carry out the activity in future courses and intention to recommend the activity to peers (Belanche, Casaló, & Flavián, 2010). As in the first questionnaire, we used 7-point Likert scales.

For the two questionnaires, we used scales validated in previous studies and adapted to our research context. This helped us to guarantee the content validity of the measures – that is, the degree to which items correctly represent the theoretical content of the construct. In addition, we tested face validity – that is, whether these items are judged to be appropriate to measure the targeted construct (Casaló, Flavián, & Guinalíu, 2008). In this way, we followed Zaichkowsky's (1985) method and asked a panel of ten experts in education and marketing to classify each item as "clearly representative," "somewhat representative," or "not representative" of the construct of interest. All items reached a high level of consensus among the experts, being considered as clearly representative, and were thus retained (Lichtenstein, Netemeyer, & Burton, 1990).

Once the information had been collected, the effectiveness of the project was evaluated based on three analyses: comparison of expectations and evaluations, final perceptions of the students, and empirical model of student satisfaction with the development of the activity.

6.4.3   Results of Study 1

This sub-subsection presents the results of the empirical study that was carried out in order to analyze the students' expectations and perceptions about the activity. As previously stated, the data for this study was gathered through two questionnaires implemented at two different times (after the introductory session and before starting the activity, and once the activity had been completed). Out of the 345 students who enrolled in the course (according to the university records, since many of them do not attend lessons or exams), 265 responded to the initial questionnaire and 234 to the questionnaire at the end of the course. Given that the analysis required the comparison between expectations (before carrying out the activity) and perceptions (after carrying out the activity), only responses from students who had responded to both questionnaires were contemplated. This procedure allowed us to obtain 189 valid cases. This is considered an adequate size, since it meets the sample size requirements of PLS (ten observations multiplied by the maximum value of the construct that has the highest number of indicators or the endogenous construct with the largest number of exogenous constructs, Davcik, 2014). In our case, both positive and negative emotions have 6 indicators, hence requiring a minimum sample size of 60.

Next, Sect. 6.4.3.1 examines the level of fit between the students' expectations before carrying out the activity and their final perceptions. After that, a more detailed analysis of the students' final perceptions about the activity is provided in Sect. 6.4.3.2. Finally, Sect. 6.4.3.3 shows the results of a structural equation model of satisfaction with the activity depending on cognitive and affective variables.

6.4.3.1   Comparison of Expectations and Evaluations

The first step in the analysis consisted of a comparison between the expectations that the students had about the activity and their subsequent perceptions following the experience. The students evaluated the activity in terms of ease of use, usefulness, interactivity, subjective learning, positive emotions, and negative emotions. For this, the average values of the items collected in the corresponding scales were calculated.

Table 6.1 and Fig. 6.4 provide the descriptive information related to these variables. Related samples t-tests were carried out to statistically compare the differences between the expectations and the final perceptions.

As can be observed from Table 6.1 the ease of use and the usefulness were assessed as being significantly higher following the activity compared to expectations. Similarly, the positive emotions experienced surpassed expectations, although to a lesser extent. However, negative emotions and interactivity were very similar (Table 6.1).

As a negative result, we can observe that the perception of learning was significantly lower than expected before starting the activity (Table 6.1). This result can

Table 6.1 Descriptive statistics for expectations and perceptions about the activity and t-test for related samples

Variable	Expectations		Perceptions		
	Mean	SD	Mean	SD	Related samples t-test
Usefulness	4.36	1.25	4.84	1.30	−4.272**
Ease of use	3.50	1.49	3.85	1.32	−2.985**
Interactivity	5.11	1.16	5.08	1.26	0.245
Subjective learning	5.30	0.94	4.90	0.96	5.117**
Positive emotions	4.23	1.11	4.48	1.23	−2.455*
Negative emotions	3.03	1.33	3.12	1.45	−0.889

Note: **$p < 0.01$; *$p < 0.05$

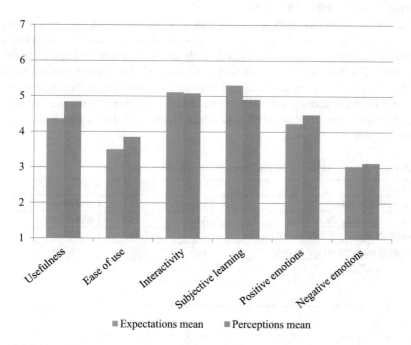

■ Expectations mean   ■ Perceptions mean

Fig. 6.4 Mean differences between expectations and perceptions about the activity

be attributed to several factors. First, at the beginning of the semester, the students may have perceived the activity as a way of "breaking the routine" of conventional teaching methodologies, thus overvaluing the novelty of the project.

In addition, at this point in the course (when information about expectations was collected), students had a smaller workload and may have felt more optimistic about their performance in the coming months.

These factors, together with an overly enthusiastic explanation of the activity by the teachers, may have contributed to inflated expectations about the learning that would be gained from the activity. On the other hand, this result may be due to the development of the activity itself. Students may have perceived the activity as less favorable than expected.

Therefore, future projects will try to delve further into this issue so as to adjust the students' expectations and final perceptions as far as possible. Despite the result, it can be said that, in general, the activity was assessed positively and exceeded the initial expectations of the students.

6.4.3.2 Descriptive Analysis of the Final Perceptions

To continue evaluating the effectiveness of the project, the following variables measured in the second questionnaire were analyzed: perception of learning, satisfaction with the activity, intention to carry out the activity again, and intention to recommend the activity to peers.

Table 6.2 shows the average values and standard deviations (SD) of each measure, as well as the results of one-sample t-tests that were carried out to analyze whether the average values were significantly above or below the medium point of the scale (4). The results confirmed that all the variables present positive values. Thus, the students were satisfied with the activity, considering that their learning derived from the activity was adequate and had favorable intentions to carry out the activity again and to recommend it to others (Table 6.2). These results are graphically displayed in Fig. 6.5.

Table 6.2 Average values of the final perceptions and one sample t-test

Variable	Mean	SD	One-sample t-test
Subjective learning	4.90	0.96	12.870**
Satisfaction with the activity	4.87	1.11	10.679**
Intention to carry out the activity again	4.37	1.54	3.245**
Intent to recommend the activity	4.53	1.30	5.532**

Note: **$p < 0.01$

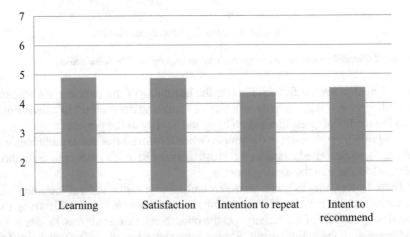

Fig. 6.5 Average values of the final perceptions about the activity

6.4.3.3   Model of Satisfaction with the Activity

Finally, we developed a structural equation model to test the hypotheses of our framework, that is, to check the influence of students' perceptions of ease of use and usefulness and their resulting feelings about the activity (positive and negative) on the level of satisfaction with the activity. The study of satisfaction is fundamental since it is the variable that includes the global evaluation of all aspects related to the activity (Severt, 2002). Thus, the model enables us to understand in depth which variables need to be stressed to achieve student satisfaction. Specifically, the data was analyzed first to confirm the validity of the measurement model and subsequently to identify the relationships between variables. This study used, in accordance with the growing trends of teaching marketing research (e.g., Orús et al., 2016), the partial least squares (PLS) regression estimation procedure. This procedure is especially useful when the phenomenon under research is relatively new (Roldán & Sánchez-Franco, 2012), as is the case with the use of YouTube in university education. The software used was Smart PLS 2.0 (Ringle, 2018).

First, validation of the measurement instruments was carried out. Specifically, we confirmed that the Cronbach alpha values of each scale were higher than the 0.65 threshold (Steenkamp & Geyskens, 2006) for usefulness ($\alpha = 0.936$), ease of use ($\alpha = 0.852$), positive emotions ($\alpha = 0.924$), negative emotions ($\alpha = 0.915$), and satisfaction ($\alpha = 0.925$). The variable "positive emotions" was considered a first-order factor, as well as "negative emotions," following the original research from which these scales were obtained (Bagozzi et al., 2016). Then, we assessed the composite validity (ρ_c), average variance extracted (AVE), and correlations between the scales, as depicted in Table 6.3. A confirmatory factor analysis found that all items loaded on their respective constructs above the minimum of 0.7 required by the literature (Henseler, Ringle, & Sinkovics, 2009). As shown in Table 6.3, all the multi-item variables had composite reliability values above 0.9, which indicates the strong internal consistency of the constructs. In addition, the AVE of these constructs was in all cases higher than 0.7, ensuring the convergent validity of the scales. That is, all the items that compose a determined scale converge on only one construct (Belanche, Casaló, & Flavián, 2012). Finally, to evaluate the discriminant validity – i.e., whether a determined construct is significantly distinct from other constructs that are not theoretically related to it (e.g., Belanche, Casaló, & Flavián, 2012) – the

Table 6.3 Composite reliability, AVEs, and correlations between the measurement instruments used in the model on student's satisfaction with the activity

	ρ_c	AVE	1	2	3	4	5
1. Positive emotions	0.942	0.801	**0.895**				
2. Negative emotions	0.925	0.756	0.147	**0.869**			
3. Ease of use	0.907	0.766	0.251	0.070	**0.875**		
4. Usefulness	0.959	0.887	0.666	0.066	0.162	**0.942**	
5. Satisfaction	0.942	0.732	0.737	0.231	0.311	0.620	**0.856**

Note: The elements on the diagonal (in bold) are the square root of the AVE. The elements below the diagonal are the correlations between constructs

Table 6.4 Summary of results of the structural model on student's satisfaction with the activity

Factors	Dependent variable	
	Satisfaction with the activity	
	β	T
Usefulness	0.241	2.992**
Ease of use	0.131	2.337*
Positive emotions	0.525	6.966**
Negative emotions	−0.129	2.457*
R^2	0.607	

Note: **$p < 0.01$; *$p < 0.05$

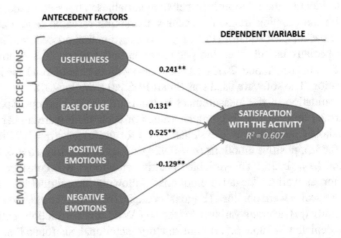

Fig. 6.6 Impact of perceptions and emotions on student's satisfaction with the activity

square root of the AVE was assessed and was found to be higher than the correlation between each pair of variables (Fornell & Larcker, 1981) (Table 6.3). In other words, we checked that each construct shared more variance with its measures than with the other constructs in the model (Real, Leal, & Roldán, 2006; Wiertz & De Ruyter, 2007).

Once the measures had been validated, we estimated the proposed effects and their significance using PLS with the recommended bootstrap of 500 iterations (Chin, 1998). The estimated coefficients and their significance are shown in Table 6.4. The analysis reveals that positive emotions, perceived usefulness, and ease of use exerted a significant effect on student satisfaction with the activity. On the contrary, negative emotions had a negative influence on satisfaction (Table 6.4). This explains our dependent variable, student satisfaction with the activity, to a large extent ($R^2 = 0.607$). Figure 6.6 shows graphically that these results, both cognitive (perceptions of usefulness and ease of use) and affective (positive and negative emotions), contribute significantly to student's satisfaction with the activity.

The results suggest the importance of emphasizing the usefulness and ease of use of the activity and focusing on the associated positive emotions while trying to reduce the negative ones. However, it is interesting to note that positive emotions influenced satisfaction to a greater extent compared to the other factors. Additionally, usefulness is presented as a more relevant factor compared to ease of use, which is consistent with the proposals of technological acceptance models (e.g., Davis, 1989).

6.4.4   Discussion of Study 1

The first study analyzes the key drivers of students' satisfaction with their participation in this active learning activity. Specifically, this study includes not only cognitive factors traditionally identified in the literature on adoption of innovations (e.g., ease of use) but also affective factors influencing this process (positive and negative emotions related to participation in the activity).

Results regarding the perceptions of students before and after their participation explain the impact of the project in several respects. In general, the comparison reveals that perceptions about the activity after they had actually experienced it were in general very positive. In particular, after participating in the project, students perceived that it was more useful and easy than initially expected. This result suggests that, from the students' perspective, the benefits (value) of the activity surpassed the costs (difficulties) to a greater extent than they had anticipated before their participation. Similarly, the level of positive emotions derived from their participation was greater after participation than they had expected, indicating that the students considered their participation as a positive affective experience.

However, the levels of subjective learning after participation were lower than had been expected at the beginning of the activity. This finding can be explained by the fact that the students' expectations were very high (probably too high) before participation; they may have initially perceived the project as an interesting social media-driven activity for young people compared to the usual classroom lessons. In this regard, although high expectations could help motivate students, in the future teachers should adjust students' expectations at the beginning of the project in order to avoid disproportionate enthusiasm about their participation in the activity.

In any case, the fundamental indicators of activity performance were very satisfactory. Analyses of the students' responses indicated high levels of learning derived from the activity, satisfaction with the activity, and intentions to carry out the activity again and to recommend it to others. These findings suggest that the project successfully reached its goals. In other words, the time and resources invested by the teachers (e.g., in designing the project and developing an evaluation system specifically for it) were appreciated and welcomed by the students. Thus, the development of such an innovative project is worthwhile and valued by students; this prompts us to continue with its development, also for other groups or subjects.

The model analyzing students' satisfaction with the activity presents interesting results. As previously mentioned, satisfaction with the activity was high, and it was influenced by both affective variables (emotions) and cognitive variables (perceived usefulness and ease of use). In particular, positive emotions related to participation (e.g., happiness, fulfillment) had the greatest satisfaction impact, followed by perceptions of usefulness and ease of use. In contrast, negative emotions (e.g., frustration, embarrassment) negatively influenced students' satisfaction with the activity.

Therefore, affective cues are as useful as cognitive cues in predicting students' satisfaction. Indeed, the combination of both cognitive and affective determinants explains a great amount of variance of satisfaction as the dependent variable. Thus, complementing already established knowledge about the influence of cognitive variables (Liao et al., 2015), these findings reveal that emotional aspects (both positive and negative ones) play a relevant role in students' overall assessment of their active experience in educational networks.

6.5   Study 2. Students' Satisfaction with an Evaluation Rubric

Study 2 focuses on the development of a rubric to evaluate the students' work, as well as an examination of the students' perceived utility satisfaction with this objective assessment tool. This section is structured as follows: first we present the proposed rubric (Sect. 6.5.1). Then, in the results Sect. 6.5.2, we check the usefulness and satisfaction of the rubric from the students' perspective. Finally, we discuss the main findings of this second study (Sect. 6.5.3).

6.5.1   Presentation of the Evaluation Rubric for the Project

In the introductory session about the activity, the teachers presented the rubric as the instrument for evaluating the video. To develop the rubric, several sources were consulted that deal with the creation of rubrics and show examples according to the competencies intended to be achieved.[1] Following the previous literature, we considered several constructs to assess the content created by the students based on the video's content (e.g., adequacy, consistency, organization) and format (e.g., structure, design, creativity) (Pérez-Mateo et al., 2011).

Table 6.5 presents the rubric and the weight given to the different evaluation categories. As can be observed, each criterion included a brief explanation (where needed). In addition, evaluation of the content was considered more important than evaluation of the format (70% and 30%, respectively). The evaluator (teacher or peers) gave a score to each category (from 0 to 10), and the final evaluation was calculated as the weighted average value (see Fig. 6.7).

After the videos had been shown in class, during which the students evaluated their peers' videos, a survey was administered with the aim of understanding their

---

[1] http://rubistar.4teachers.org

Table 6.5 Assessment rubric for student-generated audiovisual material

Category	Weight (%)
The content of the video is appropriate to the subject *(The video explains a concept related to the subject)*	10
The content of the video is relevant to the subject	10
The content of the video properly explains a marketing concept *(The content of the video allows me to understand the concept clearly)*	10
The content of the video is interesting	10
The content of the video is original *(The video contains information that is unique and different from the rest)*	10
The content of the video is rigorous *(The video uses convincing, high-quality arguments that allow me to understand the concept)*	10
The content of the video is clear *(The information contained in the video is easy to understand)*	10
The format of the video is visually attractive	6
The video is entertaining	6
The content of the video is in a clear order *(The video is easy to follow and understand)*	6
The video combines different resources *(The video uses resources such as text, music, images, animations, narration…)*	6
The video is creative	6

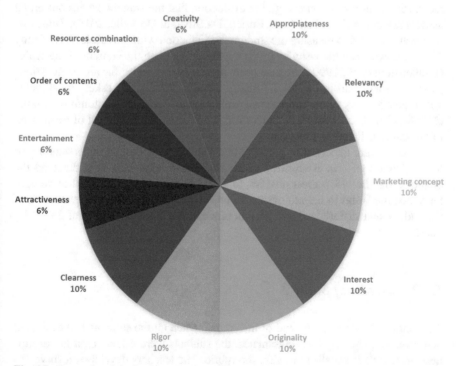

Fig. 6.7 Category weight for the evaluation rubric

Table 6.6 Measurement scales, descriptive statistics, and one-sample t-tests about the usefulness of, and satisfaction with, the assessment rubric

Usefulness	Mean	SD	One-sample t-test
I think that the rubric facilitates understanding of the evaluation criteria established by the teacher	4.85	1.17	13.320**
I think that the rubric informs me of the weighting of evaluable aspects in relation to the total score	4.91	1.16	14.399**
I think that the rubric helps me to understand the qualities that the video must have	5.14	1.15	18.116**
I believe that use of the rubric helps to guarantee objective and homogeneous evaluation by the teacher	4.94	1.27	13.615**
I think that the rubric allows us to self-evaluate	5.08	1.22	16.288**
I think that the rubric is a simple system by which to evaluate	5.06	1.24	15.700**
Satisfaction	**Mean**	**SD**	**T**
I am satisfied with the rubric as a video evaluation tool	5.03	1.29	14.697**

Note: **$p < 0.01$

perceptions regarding the usefulness of the rubric and their satisfaction with it. The decision to consider perceived usefulness is based on the belief that it consistently affects individual behavior (Bhattacherjee, 2001). Six items were developed to evaluate the usefulness of the rubric in relation to different aspects. On the other hand, satisfaction represents a global evaluation that the student carries out on all aspects related to the rubric (e.g., Eurico, Da Silva, & Do Valle, 2015). Thus, one item was included to measure the students' satisfaction with the rubric. As in Study 1, we followed Zaichkowsky's (1985) method to validate the content of the scales (Lichtenstein et al., 1990). We also used 7-point Likert scales for all the questions (from 1, "totally disagree," to 7, "totally agree"). Finally, we asked the students about the adequacy of the rubric as an evaluation tool with a dichotomous variable (1, "the rubric is adequate"; 0, "the rubric is not adequate"). The list of items used in this study is listed in Table 6.6.

Students were expected to perceive that the rubric facilitated understanding of the evaluation criteria established in the subject, to help them to understand the qualities that the video must possess, to allow them to verify the level of competence acquired, and to help to ensure an objective and homogeneous evaluation of the audiovisual materials. Through this process, we obtained a total of 337 valid questionnaires.

6.5.2 Results of Study 2

As mentioned before, at the end of the course, when all the students had evaluated and been evaluated by their classmates, the students were asked about the usefulness of, and their satisfaction with, the rubric. The teachers developed a survey to

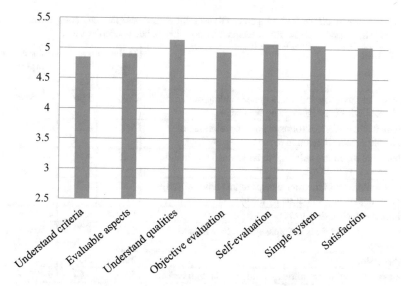

Fig. 6.8 Average values of the perceptions about the usefulness and satisfaction with the assessment rubric

determine the usefulness of the rubric as an objective evaluation instrument and the students' satisfaction with it. As indicated above, a total of 337 valid responses were collected. The data was analyzed through the statistical software IBM SPSS v22.0. First, the average values and standard deviations (SD) of each indicator related to the usefulness and satisfaction with the rubric and the results the one-sample t-tests to analyze if these values were significantly above the medium point of the scale (4) are displayed in Table 6.6. The average values are also graphically represented in Fig. 6.8. The results of the one-sample t-tests show that students were satisfied with the rubric and considered it useful.

In the same way, it was observed that 81.6% of the students considered that the rubric was adequate. To more specifically identify the differences between students who considered the rubric adequate and those who did not, a series of t-tests for independent samples were carried out. Table 6.7 shows the average values for these two groups, regarding each specific item, and the results of the t tests. As expected, students who considered the rubric adequate indicated higher levels of satisfaction and usefulness compared to those who did not consider the rubric to be an adequate evaluation instrument. These mean differences are illustrated in Fig. 6.9. However, it is noteworthy that even students who did not consider the rubric adequate evaluated the usefulness and satisfaction by assigning values around the central point of the scale, which suggests that rather than perceiving the rubric in a negative way, they were neutral toward it.

Finally, we carried out a linear regression model to understand which aspects of usefulness had the greatest influence on student satisfaction with the rubric. This estimation procedure was chosen due to the nature of the variables (they are not

Table 6.7 Mean differences in perceived usefulness and satisfaction; comparison between students who consider the rubric as adequate or not - independent samples t-test

Usefulness	Is the rubric adequate?		Independent samples t-test
	Yes	No	
I think that the rubric facilitates understanding of the evaluation criteria established by the teacher	5.00	3.96	6.308**
I think that the rubric informs me of the weighting of evaluable aspects in relation to the total score	5.05	4.22	4.996**
I think that the rubric helps me to understand the qualities that the video must have	5.28	4.35	5.598**
I believe that use of the rubric helps to guarantee objective and homogeneous evaluation by the teacher	5.17	3.73	8.390**
I think that the rubric allows us to self-evaluate	5.20	4.42	4.345**
I think that the rubric is a simple system by which to evaluate	5.23	4.09	6.503**
Satisfaction	**Yes**	**No**	**t**
I am satisfied with the rubric as a video evaluation tool	5.29	3.67	9.408**

Note: $**p < 0.01$

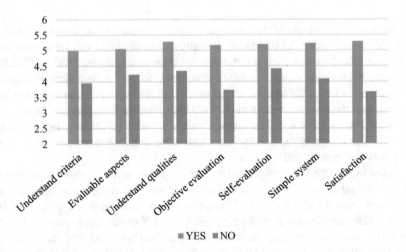

Fig. 6.9 Mean differences in perceived usefulness and satisfaction; comparison between students who consider the rubric as adequate (blue bars) or not (orange bars)

multi-item constructs). The regression coefficients and their significance are shown in Table 6.8. As can be observed from Table 6.8, satisfaction was particularly influenced because the rubric served to guarantee an objective and consistent evaluation by the teacher and because it was perceived as a simple evaluation system.

The fact that the rubric facilitates self-assessment, enhances understanding of the qualities that the videos should have, and facilitates understanding of the evaluation criteria also had positive effects on satisfaction with the rubric. It is important to note that these relationships explain satisfaction with the rubric to a large extent ($R^2 = 0.558$).

Table 6.8 Influence of usefulness on satisfaction with the rubric: results of the linear regression model

Item	Effect (β)
I think that the rubric facilitates understanding of the evaluation criteria established by the teacher	0.102*
I think that the rubric informs me of the weighting of evaluable aspects in relation to the total score	0.041
I think that the rubric helps me to understand the qualities that the video must have	0.123**
I believe that use of the rubric helps to guarantee objective and homogeneous evaluation by the teacher	0.272***
I think that the rubric allows us to self-evaluate	0.137**
I think that the rubric is a simple system by which to evaluate	0.239***

Note: *$p < 0.1$; **$p < 0.05$; ***$p < 0.01$

6.5.3 Discussion of Study 2

In the second study, we explain the development of an evaluation rubric. This evaluation system is part of a trend in the current university system where, in many cases, traditional exams have been replaced by the assessment of task development, oral presentations, etc. (Jonsson, 2014). These new evaluation systems require new assessment tools, such as rubrics, that provide objectivity and fairness (Jonsson, 2014; Reddy, 2007).

In this regard, the main finding of our second study is that students perceived the rubric as useful and were satisfied with it as an evaluation instrument. The results indicate that students perceived the rubric as an objective and homogenous evaluation tool that aids in self-evaluation in a simple way; indeed, these were the factors that had the greatest contribution to students' satisfaction with the rubric.

In addition, the rubric became a form of guidance that was valued by the students as it helped them understand the qualities that their video had to have and the weighting of the evaluable aspects. In general, the results regarding students' perceptions of the rubric show that the majority (81.6%) considered it fit for purpose and were satisfied with it. Even those students who were not satisfied with the rubric as evaluation system admitted that it was a helpful tool for understanding the evaluation criteria.

6.6 General Discussion

This chapter aims to deepen understanding of active learning processes by focusing on a project by which university students of an introductory course in marketing created didactic videos to be shared via YouTube. This proposal considers that the technological development (e.g., wide availability of recording devices) and the incredible growth of social media video formats represent an opportunity to expand learner-generated content and to refresh the learning systems in higher education.

The following subsections discuss the main contributions of this chapter to the field as well as some recommendations for teachers who are willing to apply this project in their courses (Sect. 6.6.1). The last Sect. 6.6.2 explains the limitations and further research lines to continue advancing on these kind of projects.

6.6.1 Development of the Project

The overall performance of the YouTube video channel for learning purposes in terms of number of views and impact on the audience seems to have been successful. Nonetheless, after four academic courses developing this project, we carried out two studies to assess students' satisfaction with their participation in the activity and its evaluation. Considering the findings of both studies, we conclude that the results derived from the innovative teaching project are satisfactory and promising.

These results confirm and expand previous findings about the benefits of active learning methodologies (Fiorella & Mayer, 2016; Okita et al., 2013). We found that students' participation in an active learning activity driven by ICT not only increases their curricular competencies (Blasco-Arcas et al., 2013; Pereira et al., 2014) but also enhances student's satisfaction, which could be critical from a managerial approach of higher education institutions (Severt, 2002). Indeed, implementing successful technology instruments that enhance students' satisfaction has been considered recently as an effective strategy to orientate higher education institutions toward the market (Stickney, Bento, Aggarwal, & Adlakha, 2019). In this regard, we accounted not only for the effect of cognitive utilitarian factors traditionally analyzed in the literature (Davis, 1989) but also for the influence of affective outcomes. The importance of positive and negative emotions has been stressed in previous studies focusing on student's perceptions toward their learning (Pekrun et al., 2002), but has been neglected in the study of student's perceptions about their participation in a technology-driven educational network. In this line, our work contributes to better understand the drivers for implementing and succeeding in a project relying on online video platforms, and specifically YouTube, as the main tool to develop an educational network. This trend is increasingly followed in many institutions of different education levels (Lee et al., 2017).

In addition, taking into account the high student satisfaction, the results of our project suggest that the activity and the rubric are effective (possibly because this methodology is more suitable for today's university students' demands compared to traditional lectures), so we recommend the application of this activity and wider use of the rubric in other subjects and degree courses. This result is in line with previous works which encourage the use of clear and easy-to-understand rubrics as a way to increase the transparency of the assessment (Jonsson, 2014) and orientate students' thinking and learning processes (Andrade, 2000).

Teachers who are willing to apply this kind of project in their courses should take into account that (1) students' participation in active learning activities not only increases their knowledge acquisition but also enhances their cross-curricular competences (e.g., use of ITCs, creativity (Orús et al., 2016)); (2) the development of these initiatives involves a lot of time and effort on the part of the teachers, who should seek to improve the project along with several academic courses; (3) when properly implemented, students are satisfied with the initiative and indicate an intention to continue using it, but teachers should be careful not to "overinflate" expectations; (4) students' satisfaction with their participation not only depends on cognitive or utilitarian aspects but also on the emotions related to their participation; and (5) use of a rubric may be appropriate because it increases the objectivity, fairness, and amount of information that students need to be satisfied with the evaluation system of such an innovative assignment.

6.6.2   Limitations and Further Research Lines

Our project presents several limitations that need to be addressed. First, although the project relies on active learning methodologies, some of the steps in the process of video management were not performed by students. In particular, after creating the videos, students sent their content to teachers who, after the evaluation period, uploaded the videos to the YouTube channel. Thus, students did not upload the videos themselves, as this stage of the process was managed by the teachers. Further studies could explore whether giving students full responsibility for managing the YouTube channel improves the positive impact on their learning process.

In addition, we focused on just one course in marketing, and only used YouTube, as the prototypical and most frequently used social media platform for video sharing (Spisak, 2019). Although YouTube standards and openness suggest that this could be the best platform through which to spread knowledge via online videos, further research should be conducted employing other platforms (e.g., Vimeo, thematic platforms). Similarly, the project should be implemented in other university subjects to test its efficacy across groups and courses. Nonetheless, this chapter focuses on a specific case and thus the findings cannot be generalized to the broader population of students. Other areas that may introduce bias, such as students' level of involvement or lack of participation (Orús et al., 2016), would deserve further attention in future research.

From a methodological approach, we relied on surveys as quantitative methods which are very common in studies on education innovation (e.g., Blasco-Arcas et al., 2013; Stanley & Zhang, 2018). Surveys are very useful to evaluate the teaching method from the students' perspective when there are many participants and allow the comparison of responses and the estimation of relationship parameters with precision. Nevertheless, the use of qualitative methods, such as in-depth interviews or focus groups, or mixed methods, combining both quantitative and qualitative techniques, may help to deepen our understanding of participant's perceptions and reactions toward the activity (e.g., Shekhar et al., 2019). The study of each specific interaction between participants (e.g., comments, likes) could also be useful to evaluate the project as a tool for improving educational networking (Lee et al., 2017).

Finally, our findings reveal that students' emotions toward the creation of learner-generated content clearly influenced their satisfaction with the project. In this vein, further research should be carried out to better assess how to increase positive emotions and to avoid negative emotions related to students' participation in this kind of initiatives. In this way, teachers could help to increase students' satisfaction with the project, as receiving buy-in from students is crucial to guarantee the successful implementation of active learning activities.

Appendix: Items Used in the Questionnaires of Study 1

Usefulness of the activity (Koufaris, 2002)
I think that carrying out the YouTube activity…
… has been useful to me
… has been advantageous to me
… has been beneficial to me

Ease of use of the activity (Koufaris, 2002)
I think that carrying out the YouTube activity …
… has been easy to do
… has implied little effort to do
… has been simple to do

Interactivity (Blasco-Arcas et al., 2013)
I think that carrying out the YouTube activity …
… has facilitated interaction with my peers (classmates)
… has given me the opportunity to discuss with my peers (classmates)
… has facilitated dialogue with my peers (classmates)
… has allowed me to exchange information with my peers (classmates)

Learning outcomes (Orús et al., 2016)
I think that carrying out the YouTube activity …
… has improved my comprehension of the theoretical concepts of the course
… has improved my learning of the course
… has improved my knowledge of marketing
… has improved my knowledge of marketing strategies carried out by companies
… has had a positive impact on my final grade of the course
… has helped me to improve my ability to work in group
… has helped me to elaborate and present ideas publically
… has improved my skills in video editing tools
… has improved my skills in using new Information and Communication Technologies
… has improved my ability to summarize
… has improved my creativity skills

Emotions derived from the activity (Bagozzi et al., 2016)
Carrying out the YouTube activity has made me feel …
… excited (enthusiastic)
… delighted
… happy
… fulfilled

... proud
... good in general
... embarrassed
... worried
... anxious (stressed)
... uncomfortable
... frustrated
... bad in general

Satisfaction with the activity (Orús et al., 2016)
Carrying out this activity has been interesting.
I think that the development this activity has been correct.
I am satisfied with the development of the activity.
In general terms, I am satisfied with this activity.

Behavioral intentions (Belanche, Casaló & Flavián, 2010)
If I had to carry out a similar activity voluntarily in the future, I would have the intention to do it.
I think that I will recommend other students to carry out this activity.
Note: For the sake of simplicity, we only provide the items for the final questionnaire regarding perceptions of usefulness, ease of use, interactivity, and subjective learning. The same items were used in the first questionnaire, at the beginning of the semester, to measure the students' expectations toward the activity. Instead of using "I think that carrying out the YouTube activity has been... (present perfect tense)," we worded "I expect the YouTube activity to be... (present simple tense)."

References

Alon, I., & Herath, R. K. (2014). Teaching international business via social media projects. *Journal of Teaching in International Business, 25*(1), 44–59.

Andrade, H. (2000). Using rubrics to promote thinking and learning. *Educational Leadership, 57*(5), 13–18.

Azevedo, A., Apfelthaler, G., & Hurst, D. (2012). Competency development in business graduates: An industry-driven approach for examining the alignment of undergraduate business education with industry requirements. *The International Journal of Management Education, 10*(1), 12–28.

Babin, B. J., Darden, W. R., & Griffin, M. (1994). Work and/or fun: Measuring hedonic and utilitarian shopping value. *Journal of Consumer Research, 20*(4), 644–656.

Bagozzi, R. P. (2007). The legacy of the technology acceptance model and a proposal for a paradigm shift. *Journal of the Association for Information Systems, 8*(4), 244–254.

Bagozzi, R. P., Belanche, D., Casaló, L. V., & Flavián, C. (2016). The role of anticipated emotions in purchase intentions. *Psychology & Marketing, 33*(8), 629–645.

Beauchamp, G., & Kennewell, S. (2010). Interactivity in the classroom and its impact on learning. *Computers & Education, 54*(3), 759–766.

Belanche, D., Casaló, L. V., & Flavián, C. (2010). Providing online public services successfully: The role of confirmation of citizens' expectations. *International Review on Public and Nonprofit Marketing, 7*(2), 167–184.

Belanche, D., Casaló, L. V., & Flavián, C. (2012). Integrating trust and personal values into the Technology Acceptance Model: The case of e-government services adoption. *Cuadernos de Economía y Dirección de la Empresa, 15*(4), 192–204.

Belanche, D., Casaló, L. V., & Guinalíu, M. (2012). Website usability, consumer satisfaction and the intention to use a website: The moderating effect of perceived risk. *Journal of Retailing and Consumer Services, 19*(1), 124–132.

Bennett, S., Bishop, A., Dalgarno, B., Waycott, J., & Kennedy, G. (2012). Implementing Web 2.0 technologies in higher education: A collective case study. *Computers & Education, 59*(2), 524–534.

Bhattacherjee, A. (2001). Understanding information systems continuance: An expectation-confirmation model. *MIS Quarterly, 25*(3), 351–370.

Blasco-Arcas, L., Buil, I., Hernández-Ortega, B., & Sese, F. J. (2013). Using clickers in class. The role of interactivity, active collaborative learning and engagement in learning performance. *Computers & Education, 62*, 102–110.

Bonwell, C., & Eison, J. A. (1991). *Active learning: Creating excitement in the classroom.* Green Mountain Falls, CO: Eric Publications.

British Columbia Ministry of Education. (2013). *Defining cross-curricular competencies. Transforming curriculum and assessment. Ministry of Education.* Technical Report, British Columbia Ministry of Education.

Cano, E. (2008). La Evaluación por competencias en la Educación Superior. *Profesorado: Revista de Currículum y Formación del Profesorado, 12*(3), 11–27.

Casaló, L. V., Flavián, C., & Guinalíu, M. (2008). The role of satisfaction and website usability in developing customer loyalty and positive word-of-mouth in the e-banking services. *International Journal of Bank Marketing, 26*(6), 399–417.

Chan, Y. M. (2010). Video instructions as support for beyond classroom learning. *Procedia-Social and Behavioral Sciences, 9*, 1313–1318.

Chin, W. W. (1998). The partial least squares approach to structural equation modeling. *Modern Methods for Business Research, 295*(2), 295–336.

Chua, Y. P., & Chua, Y. P. (2017). How are e-leadership practices in implementing a school virtual learning environment enhanced? A grounded model study. *Computers & Education, 109*, 109–121.

Clifton, A., & Mann, C. (2011). Can YouTube enhance student nurse learning? *Nurse Education Today, 31*(4), 311–313.

Cole, M. (2009). Using Wiki technology to support student engagement: Lessons from the trenches. *Computers & Education, 52*(1), 141–146.

Cooper, P. (2019, January 22). *YouTube stats that matter to marketers in 2019* [Online exclusive]. Hootsuite. Retrieved from: https://blog.hootsuite.com/youtube-stats-marketers/

Davcik, N. S. (2014). The use and misuse of structural equation modeling in management research: A review and critique. *Journal of Advances in Management Research, 11*(1), 47–81.

Davis, F. D. (1989). Perceived usefulness, perceived ease of use, and user acceptance of information technology. *MIS Quarterly, 13*(3), 319–340.

DeWitt, D., Alias, N., Siraj, S., Yaakub, M. Y., Ayob, J., & Ishak, R. (2013). The potential of Youtube for teaching and learning in the performing arts. *Procedia-Social and Behavioral Sciences, 103*, 1118–1126.

Dogtiev, A. (2019, January 7). *YouTube revenue and usage statistics* [Online exclusive]. Business of Apps. Retrieved from: http://www.businessofapps.com/data/youtube-statistics/

Domínguez Fernández, G., & Llorente Cejudo, M. D. C. (2009). La educación social y la web 2.0: Nuevos espacios de innovación e interacción social en el espacio europeo de educación superior. *Pixel-Bit. Revista de Medios y Educación, 35*, 105–114.

Duncan, I., Yarwood-Ross, L., & Haigh, C. (2013). YouTube as a source of clinical skills education. *Nurse Education Today, 33*(12), 1576–1580.

Dündar, H., & Akçayir, M. (2014). Implementing tablet PCs in schools: Students' attitudes and opinions. *Computers in Human Behavior, 32*, 40–46.

Dupuis, J., Coutu, J., & Laneuville, O. (2013). Application of linear mixed-effect models for the analysis of exam scores: Online video associated with higher scores for undergraduate students with lower grades. *Computers & Education, 66*, 64–73.

Eurico, S. T., da Silva, J. A. M., & do Valle, P. O. (2015). A model of graduates' satisfaction and loyalty in tourism higher education: The role of employability. *Journal of Hospitality, Leisure, Sport & Tourism Education, 16*, 30–42.

Everson, M., Gundlach, E., & Miller, J. (2013). Social media and the introductory statistics course. *Computers in Human Behavior, 29*(5), 69–81.

Ferrer, F., Belvís, E., & Pàmies, J. (2011). Tablet PCs, academic results and educational inequalities. *Computers & Education, 56*(1), 280–288.

Fiorella, L., & Mayer, R. E. (2013). The relative benefits of learning by teaching and teaching expectancy. *Contemporary Educational Psychology, 38*(4), 281–288.

Fiorella, L., & Mayer, R. E. (2014). Role of expectations and explanations in learning by teaching. *Contemporary Educational Psychology, 39*(2), 75–85.

Fiorella, L., & Mayer, R. E. (2016). Eight ways to promote generative learning. *Educational Psychology Review, 28*(4), 717–741.

Forbes, H., Oprescu, F. I., Downer, T., Phillips, N. M., McTier, L., Lord, B., et al. (2016). Use of videos to support teaching and learning of clinical skills in nursing education: A review. *Nurse Education Today, 42*, 53–56.

Fornell, C., & Larcker, D. F. (1981). Evaluating structural equation models with unobservable variables and measurement error. *Journal of Marketing Research, 18*(1), 39–50.

Fralinger, B., & Owens, R. (2009). YouTube as a learning tool. *Journal of College Teaching & Learning, 6*(8), 15–28.

Gao, H., Hu, J., Wilson, C., Li, Z., Chen, Y., & Zhao, B. Y. (2010). Detecting and characterizing social spam campaigns. In *Proceedings of the 10th ACM SIGCOMM conference on Internet measurement* (pp. 35–47). New York: ACM.

Gupta, S. (2014). Choosing Web 2.0 tools for instruction: An extension of task-technology fit. *International Journal of Information and Communication Technology Education, 10*(2), 25–35.

Gutiérrez Martín, A., & Tyner, K. (2012). Educación Para Los Medios, Alfabetización Mediática y Competencia Digital. *Comunicar, 19*(38), 31–39.

Henseler, J., Ringle, C. M., & Sinkovics, R. R. (2009). The use of partial least squares path modeling in international marketing. In R. R. Sinkovics & P. N. Ghauri (Eds.), *New challenges to international marketing* (pp. 277–319). Bingley: Emerald Group Publishing Limited.

Hofacker, C. F., & Belanche, D. (2016). Eight social media challenges for marketing managers. *Spanish Journal of Marketing-ESIC, 20*(2), 73–80.

Hoogerheide, V., Deijkers, L., Loyens, S. M., Heijltjes, A., & van Gog, T. (2016). Gaining from explaining: Learning improves from explaining to fictitious others on video, not from writing to them. *Contemporary Educational Psychology, 44*, 95–106.

Ifenthaler, D., & Schweinbenz, V. (2013). The acceptance of Tablet-PCs in classroom instruction: The teachers' perspectives. *Computers in Human Behavior, 29*(3), 525–534.

Jenkins, J. J., & Dillon, P. J. (2013). Learning through YouTube. In S. P. Ferris & H. A. Wilder (Eds.), *The plugged-in professor* (pp. 81–89). Oxford: Chandos Publishing.

Johnson, D. W., Johnson, R. T., & Smith, K. A. (1998). *Active learning: Cooperation in the college classroom*. Edina, MN: Interaction Book Company.

Jonsson, A. (2014). Rubrics as a way of providing transparency in assessment. *Assessment & Evaluation in Higher Education, 39*(7), 840–852.

Kay, R. H. (2012). Exploring the use of video podcasts in education: A comprehensive review of the literature. *Computers in Human Behavior, 28*(3), 820–831.

Koufaris, M. (2002). Applying the technology acceptance model and flow theory to online consumer behavior. *Information Systems Research, 13*(2), 205–223.

Krauskopf, K., Zahn, C., & Hesse, F. W. (2012). Leveraging the affordances of Youtube: The role of pedagogical knowledge and mental models of technology functions for lesson planning with technology. *Computers & Education, 58*(4), 1194–1206.

Lee, C. S., Osop, H., Goh, D. H. L., & Kelni, G. (2017). Making sense of comments on YouTube educational videos: A self-directed learning perspective. *Online Information Review, 41*(5), 611–625.

Lee, M. J., & McLoughlin, C. (2007). Teaching and learning in the Web 2.0 era: Empowering students through learner-generated content. *International Journal of Instructional Technology and Distance Learning, 4*(10), 21–34.

Levina, N., & Arriaga, M. (2014). Distinction and status production on user-generated content platforms: Using Bourdieu's theory of cultural production to understand social dynamics in online fields. *Information Systems Research, 25*(3), 468–488.

Lewis, S., Pea, R., & Rosen, J. (2010). Beyond participation to co-creation of meaning: Mobile social media in generative learning communities. *Social Science Information, 49*(3), 351–369.

Liao, Y. W., Huang, Y. M., Chen, H. C., & Huang, S. H. (2015). Exploring the antecedents of collaborative learning performance over social networking sites in a ubiquitous learning context. *Computers in Human Behavior, 43*, 313–323.

Lichtenstein, D. R., Netemeyer, R. G., & Burton, S. (1990). Distinguishing coupon proneness from value consciousness: An acquisition-transaction utility theory perspective. *Journal of Marketing, 54*(3), 54–67.

Mayer, R. E. (2001). *Multimedia learning*. Cambridge: University Press.

Merchant, Z., Goetz, E. T., Cifuentes, L., Keeney-Kennicutt, W., & Davis, T. J. (2014). Effectiveness of virtual reality-based instruction on students' learning outcomes in K-12 and higher education: A meta-analysis. *Computers & Education, 70*, 29–40.

Michael, J. (2006). Where's the evidence that active learning works? *Advances in Physiology Education, 30*(4), 159–167.

Momeni, E., Cardie, C., & Diakopoulos, N. (2016). A survey on assessment and ranking methodologies for user-generated content on the web. *ACM Computing Surveys (CSUR), 48*(3), 41.54.

Mon, F. M. E., & Cervera, M. G. (2011). El nuevo paradigma de aprendizaje y las nuevas tecnologías. *REDU: Revista de Docencia Universitaria, 9*(3), 55–73.

Okita, S. Y., Turkay, S., Kim, M., & Murai, Y. (2013). Learning by teaching with virtual peers and the effects of technological design choices on learning. *Computers & Education, 63*, 176–196.

Oliver, R. L. (1993). Cognitive, affective, and attribute bases of the satisfaction response. *Journal of Consumer Research, 20*(3), 418–430.

Orús, C., Barlés, M. J., Belanche, D., Casaló, L., Fraj, E., & Gurrea, R. (2016). The effects of learner-generated videos for YouTube on learning outcomes and satisfaction. *Computers & Education, 95*, 254–269.

Panadero, E., Jonsson, A., & Botella, J. (2017). Effects of self-assessment on self-regulated learning and self-efficacy: Four meta-analyses. *Educational Research Review, 22*, 74–98.

Pârvu, I., Ipate, D. M., & Mitran, P. C. (2014). Identification of employability skills-starting point for the curriculum design process. *Economics, Management and Financial Markets, 9*(1), 237–246.

Pekrun, R., Goetz, T., Titz, W., & Perry, R. P. (2002). Academic emotions in students' self-regulated learning and achievement: A program of qualitative and quantitative research. *Educational Psychologist, 37*(2), 91–105.

Pereira, J., Echeazarra, L., Sanz-Santamaría, S., & Gutiérrez, J. (2014). Student-generated online videos to develop cross-curricular and curricular competencies in Nursing Studies. *Computers in Human Behavior, 31*, 580–590.

Pérez-Mateo, M., Maina, M. F., Guitert, M., & Romero, M. (2011). Learner generated content: Quality criteria in online collaborative learning. *European Journal of Open, Distance and E-Learning, 14*(2), 1–12.

Prince, M. (2004). Does active learning work? A review of the research. *Journal of Engineering Education, 93*(3), 223–231.

Real, J. C., Leal, A., & Roldán, J. L. (2006). Information technology as a determinant of organizational learning and technological distinctive competencies. *Industrial Marketing Management, 35*(4), 505–521.

Reddy, M. Y. (2007). Rubrics and the enhancement of student learning. *Educate, 7*(1), 3–17.

Ringle, C. M. (2018). *SmartPLS 2.0 (M3)*. Retrieved from: http://www.smartpls.de

Rodriguez, J. E. (2011). Social media use in higher education: Key areas to consider for educators. *Journal of Online Learning and Teaching, 7*, 539–550.

Roldán, J. L., & Sánchez-Franco, M. J. (2012). Variance-based structural equation modeling: Guidelines for using partial least squares in information systems research. In M. Mora, O. Gelman, A. L. Steenkamp, & M. Raisinghani (Eds.), *Research methodologies, innovations and philosophies in software systems engineering and information systems* (pp. 193–221). Hershey, PA: Information Science Reference.

Ryan, B. M. (2017). A review of protocols in higher education; How my experience made me question the process. *Higher Education Studies, 7*(4), 71–73.

Schmid, R. F., Bernard, R. M., Borokhovski, E., Tamim, R. M., Abrami, P. C., Surkes, M. A., et al. (2014). The effects of technology use in postsecondary education: A meta-analysis of classroom applications. *Computers & Education, 72*, 271–291.

Schulze, C., Schöler, L., & Skiera, B. (2014). Not all fun and games: Viral marketing for utilitarian products. *Journal of Marketing, 78*(1), 1–19.

Severt, D. E. (2002). *The customer's path to loyalty: A partial test of the relationships of prior experience, justice, and customer satisfaction* (Doctoral dissertation, Virginia Tech).

Shekhar, P., Prince, M., Finelli, C., Demonbrun, M., & Waters, C. (2019). Integrating quantitative and qualitative research methods to examine student resistance to active learning. *European Journal of Engineering Education, 44*(1–2), 6–18.

Sherer, P., & Shea, T. (2011). Using online video to support student learning and engagement. *College Teaching, 59*(2), 56–59.

Spisak, K. (2019, January 2). *Social media trends & statistics* [Online exclusive]. Business to Community. Retrieved from: https://www.business2community.com/social-media/2019–social-media-trends-statistics-02156179

Stanley, D., & Zhang, Y. (2018). Student-produced videos can enhance engagement and learning in the online environment. *Online Learning, 22*(2), 5–26.

Steenkamp, J. B., & Geyskens, I. (2006). How country characteristics affect the perceived value of a website. *Journal of Marketing, 70*(3), 136–150.

Stickney, L. T., Bento, R. F., Aggarwal, A., & Adlakha, V. (2019). Online higher education: Faculty satisfaction and its antecedents. *Journal of Management Education*, Early cite. https://doi.org/10.1177/1052562919845022

Stiggins, R. J. (2001). *Student-involved classroom assessment.* Upper Saddle River, NJ: Merrill Prentice Hall.

Torres-Ramírez, M., García-Domingo, B., Aguilera, J., & De La Casa, J. (2014). Video-sharing educational tool applied to the teaching in renewable energy subjects. *Computers & Education, 73*, 160–177.

Tugrul, T. O. (2012). Student perceptions of an educational technology tool: Video recordings of project presentations. *Procedia-Social and Behavioral Sciences, 64*, 133–140.

van Diepen, N. M., Stefanova, E., & Miranowicz, M. (2009). Mastering skills using ICT: An active learning approach. In *Research, reflections and innovations in integrating ICT in education* (pp. 226–233). Badajoz, Spain: Formatex.

Venkatesh, V., Morris, M. G., Davis, G. B., & Davis, F. D. (2003). User acceptance of information technology: Toward a unified view. *MIS Quarterly, 27*(3), 425–478.

Whitaker, J. A., Orman, E. K., & Yarbrough, C. (2016). The effect of selected parameters on perceptions of a music education video posted on YouTube. In *International perspectives on research in music education*, 255–272.

Wiertz, C., & de Ruyter, K. (2007). Beyond the call of duty: Why customers contribute to firm-hosted commercial online communities. *Organization Studies, 28*(3), 347–376.

Wright, G. B. (2011). Student-centered learning in higher education. *International Journal of Teaching and Learning in Higher Education, 23*(1), 92–97.

Zahn, C., Schaeffeler, N., Giel, K. E., Wessel, D., Thiel, A., Zipfel, S., et al. (2014). Video clips for YouTube: Collaborative video creation as an educational concept for knowledge acquisition and attitude change related to obesity stigmatization. *Education and Information Technologies, 19*(3), 603–621.

Zaichkowsky, J. L. (1985). Measuring the involvement construct. *Journal of Consumer Research, 12*(3), 341–352.

Zote, J. (2019, February 1). *65 social media statistics to bookmark in 2019* [Online exclusive]. Sprout Social. Retrieved from: https://sproutsocial.com/insights/social-media-statistics/

Part IV
Approaches

Chapter 7
e-Assessments via Wiki and Blog Tools: Students' Perspective

Tomayess Issa

Abstract This chapter examines students' attitudes toward the use of wiki and blog tools for e-assessments in postgraduate units at an Australian university. e-assessments are being introduced in the higher education sector as it has become increasingly apparent that students are lacking the personal and professional skills required for their studies and for future employment. Given the needs of businesses, students should acquire and enhance these skills for both their studies and the workforce since organizations worldwide require their employees to have certain skills. Wiki and blog tools have become a vital part of teaching and learning, especially in higher education, as they develop students' skills, enable students to engage in independent learning, and give students access to new, innovative, and advanced information as it becomes available nationally and internationally. This chapter provides empirical evidence collected via formal and informal feedback as well as students' reflections. Data was obtained from 194 students who have been exposed to e-assessments via wiki and blog tools. It was anticipated that this study would determine whether such e-assessments assist students to develop their independent learning skills in higher education. Findings indicated that e-assessment tools enhanced students' skills and their acquisition of new cutting-edge knowledge, cultural awareness, and independent learning strategies by means of group and individual activities.

Keywords Wiki · Blog · e-assessments · Higher education · Postgraduate students · Skills · Australia

T. Issa (✉)
School of Management, Curtin University, Perth, WA, USA
e-mail: Tomayess.Issa@cbs.curtin.edu.au

© Springer Nature Switzerland AG 2020
A. Peña-Ayala (ed.), *Educational Networking*, Lecture Notes in Social Networks, https://doi.org/10.1007/978-3-030-29973-6_7

Acronyms

AI	Artificial intelligence
HCI	Human-computer interaction
GITS	Green Information Technology and Sustainability
KCM	Knowledge continuity management
KM	Knowledge management
KMIS	Knowledge Management and Intelligence Systems
SNEM	Social Networking and Education Model

7.1 Introduction

The purpose of using e-assessment approaches in higher education is to develop students' learning skills, promote strategies for independent learning, and facilitate access to new knowledge and information related to the various academic units. e-assessment plays a major role in imparting and enhancing specific skills, both personal and professional. Currently, academics in the higher education sector are increasingly making use of technology tools such as wikis and blogs for the purpose of e-assessment. The main reason for utilizing these tools in higher education, especially for assessments, is that today's students are lacking several, necessary professional and personal skills. In response to calls from organizations and businesses, academics should take some responsibility for ensuring that students are adequately equipped with the skills required by future employers. Academics have begun to use technology tools such as wikis and blogs as part of their assessment approach to impart and improve students' skills for their studies as well for the workforce in the future (Aral, 2013; Chuttur, 2009; Cole, 2009; Issa, T, Issa, & Chang, 2012; Karrer, 2008; Mahruf, Shohel, & Kirkwood, 2012).

Wikis and blog tools have been introduced in the higher education sector to advance and expand students' professional skills, namely, reading and writing, research, information and technology, critical thinking, decision-making, digital oral presentation, and drawing (i.e., concept maps), and students' personal skills including motivation, leadership, negotiation, communication, problem-solving, time management, and reflection. These tools facilitate and encourage greater interaction, participation, debate, and conversation among students and lecturers when they are utilized for various assessment tasks, class events, and activities (Issa, T, 2014). Additionally, several studies (British Medical Association, 2004; Chu, Reynolds, Tavares, Notari, & Lee, 2017; Mi & Gould, 2014; Novakovich, Miah, & Shaw, 2017; Ruge & Mccormack, 2017; Sancho-Thomas, Fuentes-Fernandez, & Fernandez-Manjon, 2009; Taraghi, Ebner, & Schaffert, 2009; Zein, 2014) have confirmed that the use of technology such as wikis and blogs will encourage students to become independent learners and will improve their academic and personal skills, particularly since it enables them to frequently access their teachers' feedback.

The wiki and blog tools were introduced in two postgraduate units, namely, Green IT and Sustainability (GITS) and Knowledge Management and Intelligence Systems (KMIS). I started teaching GITS from 2011 until now, while KMIS in 2018.

For his study the research approach was based on students' informal and formal feedback which was collected during the semester. These reports, including the formal feedback, indicated students' perceptions of their learning experience including the teaching, unit, and assessment strategies to which they were exposed. Informal feedback is a teaching and learning innovation requiring that students, during the semester, provide their anonymous feedback regarding the unit structure, layout, and assessments. This feedback assists the lecturers to improve the delivery of the unit before the end of the semester. The formal feedback (eVALUate) is collected at the end of the semester through the university's formal feedback process. It is intended to gather and report students' feedback about their learning experience and gives them the opportunity to provide anonymous evaluations of the unit and teaching approach.

From the methodological, theoretical, and practical perspectives, the findings from this study will make a significant contribution to the corpus of current literature on technology and education, especially in regard to the implementation of wiki and blog tools in higher education. In addition, this study will be useful to both academics and practitioners in higher education discipline, as it offers essential principles and guidelines for implementing wiki and blog tools in any study mode; moreover, it provides an assessment guide for wiki and blog contributions. However, one limitation of this study is that the sample comprised students in Australia only.

The study outcomes indicated that wiki and blog e-assessment gave students a better understanding of the unit including the new concepts and cutting-edge knowledge of GITS and KMIS. Wiki and blog e-assessment aims to increase communication and interaction among students as many students found this e-assessment to be motivating, exciting, and interactive. Students indicated that the incorporation of wiki and blog assessment in GITS and KMIS units improved their communication and collaboration, as well as their interpersonal, writing, reading, search/research, problem-solving, and decision-making skills, which are required for their studies and indispensable for the workplace in the future. In addition, students indicated that these assessment strategies gave them greater cultural awareness, knowledge of cutting-edge technology, and a better understanding of GITS and KMIS course material. Finally, students confirmed that e-assessments imparted and improved the personal and professional skills which, according to the literature review (Bailey & Cotlar, 1994; Griffin & Care, 2014; King, 2003; Thomas & Thomas, 2012), are being sought by future employers in business and industry.

This chapter will present an Australian perspective on the integration of wiki and blog tools in higher education. This chapter is organized as follows: (1) Introduction; (2) What Is Web 2.0 and Applications?; (3) e-Assessments Using Wiki and Blog Tools in Higher Education; (4) Research Methods and Question; (5) Unit, Assessments, and Participants; (6) Discussion of Research, Results, and New Findings; (7) Study Limitations; and (8) Conclusion.

7.2 What Is Web 2.0 and Applications?

Before to discuss the Web 2.0 applications, let us define what is Web 2.0. Web 2.0 was invented by Darcy DiNucci in 1999 and later Tim O'Reilly and Dale Dougherty originated this term in the O'Reilly Media live international Web 2.0 conference in 2004 (O'Reilly, 2007).

Web 2.0 contents are available to users locally and globally, as users can create content, information, images, and videos and allow web users (with special permission) to participate, contribute, edit, share, and delete the web contents. Web 2.0 technologies are dissimilar from Web 1.0, since Web 2.0 is very easy to create and maintain an account, and the widely Web 2.0 technology is used world-wide with an estimation of 3.5 billion users in 2019 (Statista, 2019a). According to the statistics for Jan 2019 in Australia, there are 5,000,000 Facebook active users, 15,000,000 YouTube users, and 9,000,000 Instagram active users (Cowling, 2019).

At present, Web 2.0 has a few different applications, including, namely, wiki, blog, social networking, podcasting, mashups, tagging and social bookmarking, Twitter, virtual worlds and social search engines, Instagram, WhatsApp, Snapchat, and crowdsourcing. Finally, Web 2.0 is dynamic and energetic, since it is more collaborative and cooperative from Web 1.0, as specific technologies are working on the back ends, i.e., Adobe Flex and Ajax, to simplify the user interfaces to the users. Boyd et al. (Boyd & Ellison, 2007) define social networking as a website which allow users to develop and create a public profile within a restricted system to share, connect, and communicate with other users locally and globally.

Several studies (Din, Yahya, Suzan, & Kassim, 2012; Issa, T, Isaias, & Kommers, 2016a; Issa, T & Kommers, 2013; Weaver & Morrison, 2008) indicate that social networking technology was introduced in the mid-1990s to facilitate communication, collaboration, contribution, creativity, and interaction among individuals and groups. Social networking is part of standard Internet facilities like email, browsing, and blogging. Social networking provokes new types of communication, collaboration, participation, and innovative transaction formats between consumers, stakeholders, vendors, suppliers, universities, and health departments. In 2004, Facebook was founded by Mark Zuckerberg. Facebook is a free social networking website that permits listed users to create, maintain, edit, and share information, video, and images with their contacts and networking.

Facebook is available in 37 different languages and include several features, namely, marketplace, groups, events, pages, and presence technology (Carlson, 2018; Whitty, Doodson, Creese, & Hodges, 2018). According to a recent study by Statista (Statista, 2019b) regarding the leading countries based on the number of Facebook users in Jan 2019, India claimed the first place with 300 million users, followed by the USA with 210 million users, Brazil and Indonesia with 130 million users, and, later, Mexico, Philippines, Vietnam, Thailand, Turkey, and the UK with 86, 75, 61, 50, 43, and 40 million users, respectively.

Additionally, wiki and blog technologies are becoming very popular among users, to exchange information, knowledge, images, and videos. Several studies (Choi, Ra, Jung, & Boo, 2017; Jones, 2007; Karasavvidis, 2010;

Kear, Woodthorpe, Robertson, & Hutchison, 2010) indicate that wiki is a website which allows students to create, edit, and delete pages collaboratively. Ward Cunningham is the founder and creator of the wiki, and this has led to the creation of Wiki Web (Laughton, 2011). Recently, wikis have massively enhanced the speed to which information is exchanged and revised and renewed on the web, since the main purpose behind wiki technology is knowledge sharing and many-to-many communications (Wooster, 2019). On the other hand, a blog is an online journal which allows users to add, edit, delete, and maintain their records by sharing knowledge, information, images, and video similar to a wiki (Godwin-Jones, 2003; Nardi, Schiano, Gumbrecht, & Swartz, 2004). Furthermore, blog is considered opinion sharing among users and one-to-many content (Wooster, 2019).

Finally, the major difference between wiki and blog is that in the blog, user cannot modify the original posts made by the blogger. This means blog is a more personal website compared to the wiki which is a collaborative website (Kung, 2018; Zein, 2014). Currently, wiki and blog are increasingly appealing the attention of academic to integrate these tools in the assessments and to increase students' personal and professional skills and their independent learning. To implement wiki and blog technology in the postgraduate units, the cognitive constructivist approach was used. This approach occurs in the mind of the students, with the learner making intellectual sense of the materials on their own (Chichester, Hagglund, & Edhayan, 2013; Wang, Y.-C., 2014).

7.3 e-Assessments via Wiki and Blog Tools in Higher Education

In e-assessments or known as electronic assessments, or online assessment, this means students should use the information and communications technology facility to complete an assessment. The main idea behind using e-assessment in the teaching sector is to assist students to complete a specific test or activities at a specific time. This type of assessments will allow students to enhance several skills from personal and professional, and the most important skill is time management. On the other hand, it will allow lecturers to control, easily use, and promptly receive students' work, and the lecturers will provide their feedback via the platform (Alruwais, Wills, & Wald, 2018; Daly, Pachler, Mor, & Mellar, 2010).

Web 2.0 technology is used to create and share online information and vast material among users locally and globally. Web 2.0 comprises numerous applications including wiki and blog tools. More and more academics are now beginning to integrate these tools in their assessment strategies to facilitate information sharing and encourage student creativity and collaboration via the Internet. Wiki and blog are platforms for hosting information, text, videos, images, and music. They are interactive forums since students can easily contribute material and work with their colleagues to generate new information and content (Alam & Mcloughlin, 2018; Choi et al., 2017).

Currently, wiki and blog tools are available on the Blackboard platform, and only those students enrolled in a unit have the access to these tools. Several studies (Alor-Hernández & Álvarez-Rodríguez, 2018; Barbosa Granados & Amariles Jaramillo, 2019; Dinger & Grover, 2019; Kung, 2018; Nardi et al., 2004) indicate that the use of wiki and blog tools in higher education will encourage students to express their opinions and ideas and share cutting-edge knowledge and information with their peers and form and maintain their teams.

With the cognitive constructivist approach principles in place, the social constructivist approach was taken into consideration for this study, since several studies (Burden & Williams, 1997; Kiraly, 2014; Williams & Burden, 1997) indicate that the social constructivist approach is a sociological theory of knowledge, as human growth is socially located and knowledge is created through interaction with other users. In line to applying the social constructivism theories in the education sector, the lecturer should change his/her teaching style from "people to teach" to being "facilitators of learning" (Kiraly, 2014; Richardson, 2005; Sivan, 1986). This means the lecturer should ask and test students if the necessary knowledge behind the unit was obtained via completing some e-assessment activities via wiki and blog, and students should encourage to reflect, communicate, and share his/her opinion and perspective. Adjoining and incorporation wiki and blog technologies in the postgraduate units will enable students to share and communicate knowledge with their colleagues, acquire new knowledge, and improve and develop professional and personal skills which are needed for the workforce and their study in the university.

To integrate wiki and blog activities in GITS and KMIS unit, the researcher used the Social Networking and Education Model (SNEM). SNEM was developed and created based on an intensive study of 15 countries comprising both developed and developing countries (Issa, T, Isaias, & Kommers, 2016b) (see Fig. 7.1). The study confirmed that using SN in higher education developed students' academic skills, expanded their knowledge of local and global issues, and assisted students to study and work independently; moreover, the use of SN makes students more sustainable. However, SN can also cause anxiety, loss of interest in everyday activities, health concerns, and security fears. It may prevent students from undertaking their regular activities and, finally, may hinder the development of literacy and other fundamental skills (Abdulahi, Samadi, & Gharleghi, 2014; Oberst, Wegmann, Stodt, Brand, & Chamarro, 2017). The SNEM model contains the following elements: teaching methods, learning, technology design, and psychological aspects. Each of these contains several subelements to ensure the successful adoption of SN by the higher education sector.

Teaching methods consist of four elements: learning to learn, blended learning, pedagogy, and curriculum (Driscoll, 2002; Hughes-Wiener, 1986; Lalonde & Castro, 2015; Mockler & Groundwater-Smith, 2015; Murphy, Gray, Straja, & Bogert, 2004; Oliver & Trigwell, 2005; Thrun, 1996). Figure 7.1 shows the subfactors of each of these. Learning to learn involves a set of principles and skills that will assist students to learn more effectively and to become learners for life. Blended learning combines face-to-face learning and online learning; these modes are essential in teaching and learning as a means of delivering the materials and information to students. Pedagogy and curriculum are related to teaching principles,

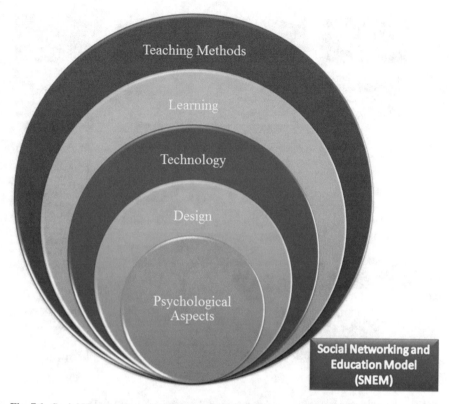

Fig. 7.1 Social Networking and Education Model (SNEM)

professional practice, leading, guiding, and methods of teaching; on the other hand, curriculum relates to assessment tasks and activities which students are required to complete successfully in order to achieve specific educational outcomes.

The learning factor emerges from learning theories, learning styles, and social learning. Learning theories are divided into two categories: connectivism and pragmatism (see Fig. 7.2). Connectivism is the theory for the digital age; it states simply that knowledge, data, and information are disseminated via a network of connections; therefore, the learning and teaching process involves the construction and traversal of those networks, especially by using the latest technology (Duke, Harper, and Johnston, 2013; Siemens, 2005). However, the pragmatism learning theory involves mainly hands-on problem-solving, teamwork, experimentation, and projects, and subsequently the outcomes will be used for decision-making (Kivinen & Ristelä, 2003; Wallace & Brooks, 2015). The learning style is based on Neil Fleming's VARK (1988) (Felder & Silverman, 1988; Fleming, 1995; Murphy et al., 2004): visual, auditory, read/write, and kinesthetic learning modes (see Fig. 7.3). Visual learners prefer to use concept maps, drawings, graphs, and flowcharts instead of texts and audio material. This type of learner relies on visual cues and can better understand the information if it is presented graphically.

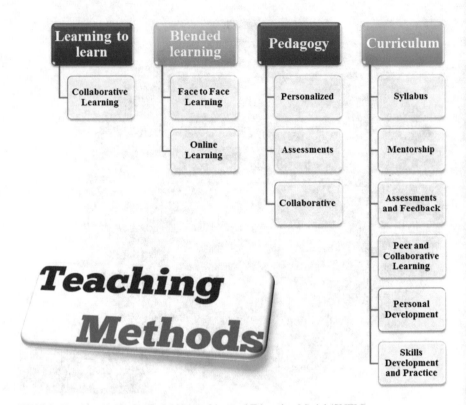

Fig. 7.2 Teaching methods: Social Networking and Education Model (SNEM)

These visual representations are tools that will help these learners to organize their ideas and understand the concepts being presented. Auditory learners prefer to listen and speak instead of taking notes from the lecturer. This group can discuss and debate with his/her classmates and lecturers as a means of understanding the concepts of the unit being studied. This group prefers to use the latest auditory technology such as MP3/MP4 audio which allows them to pause, forward, and rewind, especially those sections that they have difficulty understanding. The read/write learners prefer text, notes, and papers as a means of understanding concepts. These types of learners can interpret abstract perceptions and convey them through arguments and essays. Finally, kinesthetic learners prefer to learn through experience and practice; this means that they feel and live the experience in order to learn.

Social learning involves attention, retention, reproduction, and motivation (Bandura, 1977; Cheung, Liu, & Lee, 2015). For attention, learners need to pay attention in order to obtain the knowledge presented in the unit of study; any interruption or distraction will affect the learning process. For retention, this concept mainly concerns the process of remembering the information that the student has obtained from notes, images, models, and other materials and resources. For reproduction, students are required to reproduce the information which they

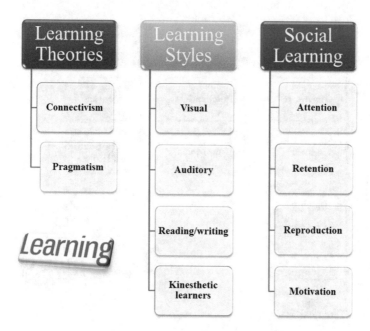

Fig. 7.3 Learning: Social Networking and Education Model (SNEM)

learned during the class, which in turn reflects their level of attention. This behavior will ensure that if students are receptive to information, their skills and observational learning will improve.

Finally, students should have the motivation to repeat their performance at the same (hopefully high) standard and be aware that appropriate performance will be given positive recognition, and conversely, substandard performance will have negative consequences. The technology factor comprises three sub-factors, namely, system, social, and security (see Fig. 7.4). These sub-factors are essential for SNEM to ensure that social networking is effective and aligned with students' needs, especially in terms of the security of data storage and the user's private information. The system factor relates to the practical implementation of social networking in higher education; universities are required to provide hardware, software, databases, Internet connectivity, and troubleshooting (support and help) for students and lecturers. The final sub-factor is social. For SN implementation to be successful, students need the latest applications and media such as Blackboard, Moodle, and other facilities.

In order to implement social networking without creating user frustration, design plays an essential role in this model, especially in terms of interface: usability, human-computer interaction (HCI), and navigation (see Fig. 7.5). Attention to these aspects of SN is vital since usability means that the interface is efficient, effective, safe, easy to learn, easy to remember, easy to use and to evaluate, practical, and visible and provides a satisfactory experience. The principles of HCI are intended to ensure that the interface is practical and visually attractive in relation to text, style,

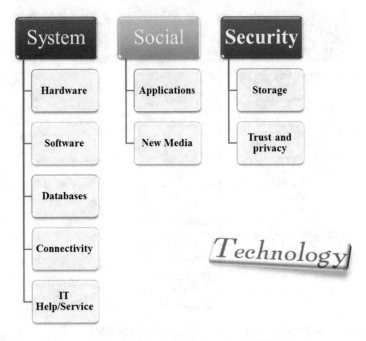

Fig. 7.4 Technology: Social Networking and Education Model (SNEM)

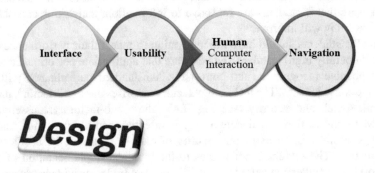

Fig. 7.5 Design: Social Networking and Education Model (SNEM)

fonts, layout, graphics, and color (Issa, T & Isaias, 2014). The navigation sub-factor aims to establish communication between the interface and navigation in the hyper-media application (Issa, T, 2008).

The final factor in SNEM is the psychological aspect (see Fig. 7.6); this factor will assist teachers to understand students' cognitive and behavioral attitudes toward the use of this technology in the learning and teaching process (Lalonde & Castro, 2015; Monks & Kawwa, 1971; Safuanov, 2004; Watson, 1963).

Furthermore, this factor will assist students to meet their study needs and require-ments, since teachers using the technology play a major role especially in terms of activities and assessments. The psychological aspects should be started from

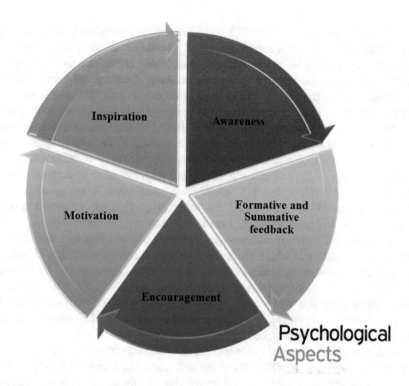

Fig. 7.6 Psychological Aspects: Social Networking and Education Model (SNEM)

awareness ending, with inspiration. According to Issa, Issa, and Kommers (2014, 30), "The constructive feedback is intended to ascertain whether students are on the right track, and to allow students to learn from their mistakes and prevent future repetitions of the same errors. Furthermore, the adoption of this approach in post-graduate and undergraduate units will improve students' confidence and motivate them to complete the assessment tasks on time, and most importantly, align with the unit objectives and aims."

The formative feedback is intended to raise students' awareness and encourage, motivate, and inspire them during the learning process, fostering their independent learning and improving their students' academic skills and students' personal skills by using the technology and, in particular, accessing their teachers' regular feed-back. Furthermore, both formative and summative feedback will encourage students to engage with the unit and will provide an exciting, memorable, and motivating experience. Motivation is the factor that assists users to move toward a goal or aim, while the purpose of encouragement is to give courage, hope, and increased confidence to users to strive to achieve their aims. The idea of awareness is to have knowledge or judgment of something, while inspiration is a process that takes place when a user sees or hears something that causes him/her to develop and generate new ideas and their personal inspirations and values (Borthwick & Gallagher-Brett, 2014; Lee & Witz, 2009; Psychology Dictionary, 2015).

The SNEM was used in the GITS and KMIS units; to develop students learning skills, especially those of communication and interaction, students must complete a set of challenging activities via wiki and blog. These activities comprise analyses and evaluations of real-life case studies, the drawing of concept maps based on unit materials, and the sharing of cutting-edge information derived from current news items relevant to the unit material. Students are required to upload their work to wiki and blog as individuals or as teams to be evaluated by the lecturer and their peers; also, the activities are presented orally to the class, again by individuals or as groups.

A mark out of 15% and 25% was allocated for students' contributions to the weekly wiki and blog, respectively. Students' wiki and blog contributions were checked by the lecturer twice a week, on Monday and Thursday afternoons. Students' submissions were graded according to quality not quantity. Providing feedback via wiki and blog is less time consuming and improves communication and collaboration between students and lecturers, as both lecturers and students provide live feedback in relation to activities during the class, and this will improve and enhance teamwork and collaboration (Biasutti & El-Deghaidy, 2015; Chu et al., 2017; Cowan & Jack, 2011; De Arriba, 2016; Matheson, 2009; Ng, 2016; Wang, L., 2016). Integrating this assessment in postgraduate units will promote students' communication skills including writing, reading, debating, written presentation and oral skills, and drawing (i.e., concept maps).

Finally, to assess students' contributions to e-assessments wiki and blog, a rubric was used comprising the following criteria: content (i.e., topic(s) is(are) covered in detail with excellent examples and knowledge of subject matter is outstanding), organization (well-presented and organized, using headings or bulleted list for group-related material), attractiveness/visual appeal (to enhance the wiki presentation, student uses excellent choice of font, color, graphics, effects, etc.), contribution to group work and discussions (student contributes to and develops the class wiki/blog, by providing her/his opinion regarding her/his peers' wiki/blog contributions), accuracy (student's observations and perspective were presented, explained, and demonstrated well), and structure and quality of writing (well structured, e.g., paragraphing, sentence structure, spacing, spelling, proofreading, no HTML errors in wiki and blog, i.e., broken links, missing images, above average standard of expression and presentation, excellent overall expression and presentation, accurate acknowledgement of sources). The informal feedback encourages students to engage with the unit and will help to provide an exciting, memorable, and motivating experience.

7.4   Research Methods and Question

Empirical data was obtained from 194 students. Both quantitative and qualitative approaches were adopted as the data was derived from formal and informal student feedback and from students' reflections. Both informal and formal feedback was collected during the semester to indicate students' perceptions of their university learning experience and included feedback about the unit and the lecturers' teaching methods.

Informal feedback is a teaching and learning innovation: during the semester, students are asked to provide anonymous feedback regarding the unit structure, layout, wiki and blog use, lecturer's teaching, assessments, and how the learning experience could be improved. This feedback is intended to assist the lecturers to enhance/improve their teaching of the unit before the end of the semester. Furthermore, through the reflective process, students offer their reflections and opinions on their learning, thereby improving their level of engagement with the unit. Students remain anonymous and the reflections are collected during week four of the semester in the class. Finally, the informal student's feedback is limited to qualitative responses.

The second method is formal feedback (eVALUate), which is collected at the end of the semester through the university's formal feedback process via an online platform. It is intended to gather and report students' feedback about their learning experience and gives them the opportunity to provide anonymous evaluations of the unit and teaching approach. The formal (eVALUate) feedback is a combination of quantitative and qualitative responses. The formal (eVALUate) is divided to teaching and unit evaluation: the unit mainly focused on learning outcomes, learning experiences, learning resources, assessment tasks, feedback, workload quality of teaching motivation, best use of the learning experiences, effective learning, and overall satisfaction, while the teaching mainly focused on knowledgeable, enthusiastic, well-organized feedback, communicates clearly, is approachable, and provides useful feedback and effective teacher.

Finally, using qualitative and quantitative approaches in this study will provide a better understanding of a research problem or issue than with one research approach alone (Gilbert, 2006; Harrison & Reilly, 2011; Maudsley, 2011; Teddlie & Tashakkori, 2009; Wiggins, 2011). This study addresses the following research question guiding this study: "How can e-assessments utilizing wiki and blog tools assist tertiary education students to develop students' learning skills and independent learning skills?"

7.5   Units, Assessments, and Participants

The GITS unit was developed subsequent to the repercussions of the global financial crisis in 2007, as businesses and users were struggling to survive, especially in the area of information technology/systems. The GITS unit aims to provide students with an understanding, knowledge, and experience of Organizational Sustainable Strategy and Green IT. As for the learning outcomes, on successful completion of this unit, students (1) demonstrate an awareness of and sensitivity to the importance of sustainable development and business strategies at a time when people and the planet must be considered in addition to profit, (2) display an understanding of the fundamentals of Green IT, (3) apply conceptual tools and frameworks to critically analyze and apply business decision-making practices and policies, and (4) translate the theories, concepts, and analytical techniques learned into practice.

Table 7.1 GITS and KMIS assessment activities and unit syllabus

Unit	Assessments	Unit syllabus
GITS	Wiki for collaborative writing 15% Reflective journals 30% Individual presentation of an IT sustainable strategy and report writing 55%	The GITS unit is mainly focused on issues relating to strategic development, IT business, sustainability tools, and green IT and other related issues
KMIS	Blog assessment 25% Reflective journal 30% Practical assessment 45%	This unit introduces knowledge management from an information systems perspective. Knowledge discovery, acquisition, learning, representation, and reasoning are covered in the unit, along with organizational learning, ontologies, semantic web, knowledge-based and expert systems, intelligent agents with e-commerce applications, and content management systems

The KMIS unit was introduced in universities after 2000, since organizations and businesses seek to organize and maintain their data. Hence, KM is considered to be a technique or method that is applied to plan, organize, motivate, and control people, processes, and systems in an organization to ensure that its knowledge-related assets are improved and effectively employed (Dalkir, 2013; Hislop, Bosua, & Helms, 2018). The learning outcomes of the KMIS unit are to (1) create a knowledge base ontology for a business application, (2) develop a small knowledge management information system for a business application using contemporary application software, (3) compare and contrast different methods and tools for knowledge management, and, finally, (4) evaluate the organizational issues associated with the implementation of knowledge management in a business. Table 7.1 shows the units' syllabuses and assessment tasks for the GITS and KMIS units.

The study participants comprised 194 postgraduate students mainly from Australia, Asia (Including India), Middle East, the USA (North and South), Russia, Mauritius, and other parts of Africa. A mixture of different nationalities and cultures is a significant characteristic of this lecturer's units, as all students are required to interact with and share their knowledge and skills, experience, and cultural perspective with their peers either during face-to-face interaction or via the technologies tools (i.e., wiki and blog). This cultural mix assists students to learn from each other by sharing knowledge, skills, and cultural perspectives, and this leads to developing self-esteem, communication skills, and self-confidence.

7.6 Discussion of Research, Results, and New Findings

This section provides the research outcomes derived from the analyses of students' informal and formal feedback. The students' informal feedback indicated that the use of wiki in the GITS unit was very interesting, motivating, exciting and drew their attention to several issues in local and global news items. Students shared the following feedback regarding the benefits of using the wiki tool in the GITS unit:

- The wiki exercises and the journals and the report have been very useful in reinforcing the concepts learned in this unit. I enjoyed the class discussions as I got to hear about new ideas and concepts regarding sustainability which I plan to take with me back home and implement them as well.
- The collaborative learning tool which is the wiki has also been helpful, because you can see other students' opinion and concepts. Together with students from different countries and different continents, I was able to learn about other countries Green IT and sustainability development.
- Wiki assisted me to improve professional skills, i.e., improved my analytical skills with respect to being sensitive and being aware of the importance of a sustainable development and business strategies. It improved my evaluation skills while ensuring that I understood the basics of Green IT and sustainability helped me gain analytical technique.
- We are working as groups in the class to complete different wiki activities related to sustainability and Green IT. I like the idea of working with a group because it allows us to search together about specific topic, communicate with each other, share our information and ideas, and indeed exchanging and discussing these ideas with the other groups in the class as well as the lecturer. All these contributions, interaction, searching, and reading helped to improve our communication skills, research skills, collaboration skills, and analyzing skills.
- This unit has many different assessments such as journals, wikis, digital presentation, and report writing. I am really changing myself not only personally but also professionally because now I have confidence about everything. When I came here that time, I am a non-English speaker and I have no English background that time, but because of the assessment, I improved my English in the field of reading, writing, and speaking; by submitting assessment, I learn so many things about English and now my English has improved, and also because of assessment requirement for references, my skill of searching in library catalogue and database has developed. I also learned many things about technology like using endnote and many more because before I never used these things in my country. By using wiki, I became aware of sustainability and Green IT by doing its exercises. I also improved my thinking skill about environment, Green IT, and the future.
- I must separately mention wiki: this is my first experience with this type of exercise, which I really liked. It allowed developing and applying my creativity when approaching different themes defined by our lecturer. The group wiki tasks during the class taught me to work under time pressure. I feel uncomfortable to complete a task in a short period, but sometimes we need to go out of our "comfort zone" to learn something new which turns out to be very useful. Presenting the wiki group work also helped to improve presentation and reasoning skills.
- The use of wiki tool to share ideas has been a great platform to obtain ideas and solutions to existing problems. The way it organizes articles, presentations, and posts helps students to pick main points and use it as a base for their academic arguments. It provides clear writing skills that can be adapted by writers who aspire to polish their skills. It enables users to structure their ideas in a summary that covers all the aspects of the article in question.

- Another important skill I learned from this unit is teamwork. This was achieved through the various wikis that were done with my team during the lecture. We had to communicate with each other effectively, listen to each other, and be non-judgmental. This was all achieved through hearing each other out and collaborating. Thus, the other subskills I gained from teamwork were communication, listening, creativity, and collaborative skills. The other skills I gained from completing this unit were conducting research effectively.
- It helps students to make improvements in creative thinking, problem-solving, decision-making, finding alternative solutions, synthesize material, group discussion, and debate. As most of the wiki activities are done during the class hours, it will help students to communicate with each other and arrange better solutions; within that limited time, it enhances their communicating and time management skills. I admit that wiki exercises enhanced my skills in the following ways: creative thinking, problem-solving, decision-making, finding alternative solutions, writing skills, research skills, teamwork, debate, time management, and communication skills.
- As challenging activities were performed in each class, by participating in those exercises, I enjoyed a lot and get boost to my writing, research, collaborative, and communication skills. I managed to participate equally in wiki exercises. Considering communication skills, I did interactions with my group mates then presented those works in the class which helped me in the improvement of my communication skills. Writing is enhanced by completing some of the writing exercises. The concept maps are created in some exercises which help me to be creative. Overall doing wiki exercises was useful and interesting as well.
- Teamwork and collaboration skills are two important ones. I got many difficulties when I worked in the first group (group 4). Although, there were four people in the group, only two of them contributed to the group's result, including activities in class or homeworks. We tried to connect them to the team for the better group's targets; however, it was difficult because they did not want to collaborate. I learned from this group activity that if I want to resolve the conflict, we must take effort to connect all team members and assign them the tasks that they would like to do with a specific deadline so that they will be happy to get the tasks done. However, if there is no success, the best approach is to just work with goodwill and active partners. With this way, the teamwork will be effective to work together.
- I admitted that during in-class activities, I truly enjoyed as well as learned skills such as communication skills during interaction with group mates and during the opportunity to speak up my views in oral discussion, public speaking skill, about some topic. Writing and presentation skills advance by completing a couple of written tasks in wikis activities. We created concept maps in wikis group tasks as well, which helps to enhance idea generation skills or more specific brainstorming. Its further benefit with my critical thinking is enhancing creativity to design presentations for wiki activities. Quick analytical skill and decision-making ability enhanced due to search and present the finding of some tasks within the class. If I am not wrong, I can say, these activities encourage me to take initiative in group task and develop my leadership skills as well.

- Wiki as a learning tool is a good avenue for us to really practice intelligent and fast research, writing, and thinking skills as they are given impromptu in class and you were only given minutes to create and then present. The classroom presentation enables us to freely impart our views on the topics and in the process hone our communication and collaboration skills. We really need to have proper communication as a team to be able to deliver. I guess in here collaboration, teamwork, and proper time management are indeed sharpened.
- The group activity helps to really get to know our classmates well, more than just classmates in class. Stronger bonds are built. For improvements, it would be good to have different groups in every activity for the group task to be able to connect also with others.
- Wikis help to enhance teamwork among the students. The collaborative work on wikis helps to communicate with other peers and get familiar with them. The wikis activities helped to know about the various cultures as all the students were from different countries and cultures. To communicate with them enables to get more knowledge and respect about other's traditions. It enhances the critical thinking and decision-making skills. These activities make the students to come up with a quick result in a short time. The wiki activities include the current topic all around the world with which I came to know the happenings in the surroundings. I got more aware about the policies of companies and their effects on nature and society. It also enables the students to look for the others work that how they can think and get more knowledge from their work.
- Overall, wikis are very effective in enhancing reading, writing, drawing mind maps, decision-making, communication, teamwork, and thinking skills.

By using word cloud generator website (https://tagul.com/), the author developed the following word cloud to emphasize students' feedback, that completing wiki exercise improve and advance students' knowledge and skills (see Fig. 7.7).

Furthermore, to confirm the study aims, the manual coding was used to identify the themes and specific skills from the qualitative data via the formal and informal students' feedback. Manual coding is an analytical process in quantitative data, i.e., students' feedback, to assess and identify the themes, i.e., skills in this study, to confirm the study's aims (Rogers, 2018; Simula, 2018).

The major benefit of using the manual (or human) coding is to allow the researcher to understand the qualitative data and text, including the jokes. The main idea behind the manual coding is to identify the specific skills which students have agreed on using in the formal and informal feedback (Creswell & Poth, 2017; Lichtman, 2012; Saldaña, 2015; Wolcott, 1994).

Based on manual coding results, Figs. 7.8, 7.10, 7.12, and 7.14 were generated. Based on the students' informal and formal feedback, the researcher used manual coding strategic by reading the qualitative data and manually assigning a code for each skill via wiki tool use for GITS unit (Saldaña, 2015). Figure 7.8 generated the students' skills based on GITS e-assessment via wiki. Generating these skills by the manual coding indicated that using the wiki tool in GITS unit assisted the students to enhance their students' learning skills and independent learning skills which achieve the study aims and goals.

Skill
Active
Wiki
Help
Idea
Group
Student
Work
Class
Use
Learn
Tool
Time
Team
Present
Each
Well
Improve
Discuss
Share
Unit
Task
Topic
Exercise
More
Very
Think
Different
Creative
Really
Good
Research
New
Communicate
Many
Person
Doing
Come
Lot
Collaborate
Sustainable
Writing
View
Make
Enhance
Map
Out
Listen
Enjoy
Post
Journal
Culture
Teamwork
Green
Develop
Concept
Knowledge

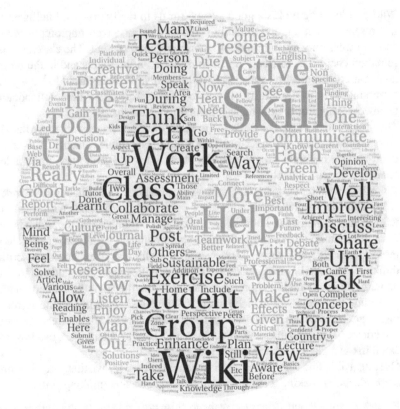

Fig. 7.7 Wiki tool use: GITS unit

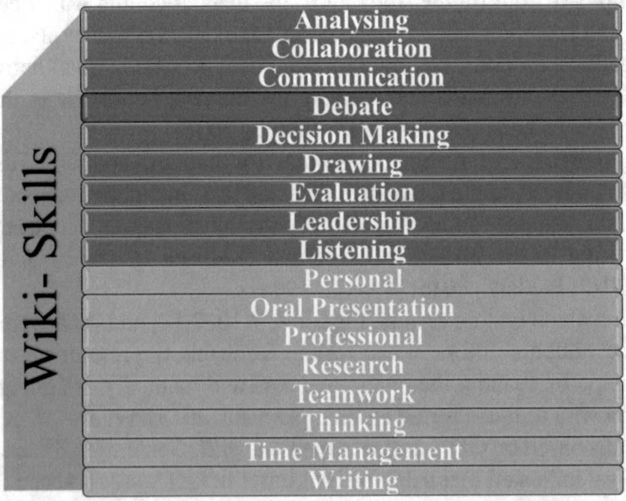

Wiki- Skills
Analysing
Collaboration
Communication
Debate
Decision Making
Drawing
Evaluation
Leadership
Listening
Personal
Oral Presentation
Professional
Research
Teamwork
Thinking
Time Management
Writing

Fig. 7.8 Wiki skills via GITS unit

Regarding the blog tool used in the KMIS unit, students' informal feedback confirmed that this tool enabled them to have a remarkable and brilliant learning experience. According to both the students and the lecturers, the blog tool assisted students to improve their personal and professional skills in communication, leadership, debating, time management, problem-solving, and decision-making, all of which are essential for their current university studies, as well as the workforce in the future. The blog tool in the KMIS unit has become very sophisticated; as students recognized and confirmed, it makes the classes more interactive, cooperative, and fun. Students shared the following comments:

- The debate was fun and forces you to argue the side of a topic that you might not necessarily agree with, while also improving our public speaking skills. I think it would also be interesting to see people's opinions in debates before specific content is taught versus after a lecture, to see how their mindset has changed based on what they have learned throughout the class.
- The blogs are another crucial part of the [KMIS] unit; we can learn about the other's student's perspective, adapt more knowledge, and improve the writing skills as well. These blogs help the students to share their knowledge, which makes our knowledge limitless because if we keep the knowledge limit to ourselves, it will be limited only. Blogs are easy to access, and anybody can see the other's blogs work and grab more information.
- Writing skill: By doing blog activity, students can improve their academic writing skill and team collaboration skill. By doing blog in team, students can learn how to collaborate with other people from different backgrounds.
- Teamwork: During the in-class activities such as group discussions and blogs, I must be a part of a team as there are some blog activities that need to be done in a group. Here, I learn things such as how to collaborate different views in a team, leadership.
- Students need to participate in group activities such as group discussion, debates in class, and blog activities with other students which are helpful to improve the communication skills and social relationship and encourage students to learn about perspectives of other students, and they can also learn about their countries, cultures and traditions, etc.
- Communication skills: The blog in this unit is done in groups. The students in the class are from different countries. They are sharing different ideas with each other to do these blogs. It enhances their communication skills, and they learn teamwork. When they are doing blogs, they learn how to corporate with the team.
- The blog assessment helps me to review what I learned from the lecture and deepen my impression of the knowledge. I also learned a new thing from practical assessment which is ontology. This is a model that helps me to demonstrate the tacit knowledge from our mind and can be easily learned by another person.
- Learned many things from the material covered in this unit in every week such as class activities, laboratory works, blog activities completed individually or within a group, and the reflective journal and practical assessments. Every week I covered distinct topics related to the knowledge management which include knowledge

management processes, knowledge capture and codification, knowledge sharing, finding knowledge, knowledge management strategy and planning, knowledge management tools and models, and artificial intelligence systems in an organization. From the written material completed in KMIS unit till now, I enhanced my technical, personal, and professional knowledge about the knowledge management and learned how to plan, organize, share, and manage the knowledge within an organization to achieve the organization objectives. Furthermore, I also learned what are the benefits of sharing knowledge with others.

- In terms of personal growth, my interpersonal skills were improved through the group blog assessment and group discussions. With the fact that this is a small class, I was able to interact well and communicate with different people with diverse culture and have their own points of view. Listening from their own stories taught me to value and appreciate life even more. Everyone in the class shared their own knowledge and experiences which I consider as the most valuable takeaway from studying this unit. Aside from learning, I have expanded my network and made new friends. Lastly, knowing that not everything ends in this unit, the bulk of learning through assessments has made me a better person, a better practitioner in the field of information systems and technology.

By using word cloud generator website (https://tagul.com/), the author developed the following word cloud to emphasize students' feedback, that completing blog exercise boost and expand students' knowledge and skills (see Fig. 7.9).

Fig. 7.9 Blog tool use: KMIS unit

BLOG SKILLS

Fig. 7.10 Blog skills via KMIS unit

Figure 7.10 generated the students' skills in KMIS unit by completing an e-assessment via the blog. Figure 7.10 was generated by using manual coding to identify the most skills which are enhanced by students.

From the above comments, it came to our attention that initiating e-assessment via wiki and blog technology led to an exceptional and remarkable experience not only for the students but for the lecturer as well. Students confirmed that using these tools in GITS and KMIS units helped them to improve and enhance their personal and professional skills, which are essential for their studies. The purpose of using wiki in the GITS unit is to improve and enhance students' skills, and students confirmed this, and they learned from each other and made friends, and the whole learning journey was enjoyable.

In the GITS unit, the lecturer initiated 13 wikis during the semester intended for group and individual activities. The wiki activities focused mainly on specific topics: Sustainability and Green Information Technology (IT) in students' countries, IT Environment Problems, Cloud Computing Applications, Mobiles, New Green IT Technologies, News behind Sustainability and Green IT, and Case Studies.

For the KMIS unit, the lecturer initiated nine blogs during the semester intended for group and individual activities. The blog activities focused mainly on the stages of the knowledge management (KM) cycle: cognitive computing, technologies used to support knowledge networks and communities of practice, social network analysis, tools used in organizational memory management, discussion of personalization and profiling approaches to model knowledge workers, discussion of the major challenges to overcome in order to implement a successful knowledge continuity management (KCM) program in an organization, debates, and team analyses of YouTube material related to KM.

The use of wiki and blog in the GITS and KMIS units encouraged students to let their voices be heard and to give their viewpoints on various topics in groups or as individuals. All the activities were posted to wiki and blog, and the lecturer provides her feedback using the wiki and blog rubric. Students also shared their views about the use of wiki and blog in the GITS unit generally and specifically to learn the unit materials, i.e., sustainability and Green IT. The KMIS students learned about various topics related to unit materials, such as KM tools and processes and artificial intelligence (AI).

Table 7.2 shows the formal feedback (eVALUate) for the GITS and KMIS units. The data obtained from the formal feedback indicated that students were pleased and satisfied with the assessment strategies including the e-assessments for wiki and blog, as quantitative results range from 92% to 100% which is considered outstanding. Finally, overall satisfaction and other aspects, namely, learning outcomes, assessment, motivation, and best learning, are highest compared to the university average. The formal feedback (see Table 7.2) confirmed the research question that students obtained the necessary cutting-edge knowledge and skills developed their independent learning by completing the GITS and KMIS units which included e-assessments.

The preliminary analyses of the collected data revealed that the e-assessments via wiki and blog gave students a better understanding and knowledge of new concepts and cutting-edge technology related to material in GITS and KMIS units. Furthermore, e-assessment managed to increase communication and interaction among students, as most students found wiki and blog assessments were motivating, exciting, and interactive.

The findings of this study have made significant contributions to the current literature on technology and education from the methodological, theoretical, and practical perspectives, especially in the development of units which will be focused on GITS and KMIS, particularly in terms of positive student outcomes related to students' knowledge, learning, and skills. In addition, this study might benefit both academics and practitioners in the field of higher education.

The e-assessment via the blog was introduced for the first time in KMIS, and the researcher taught this unit for the first time in Semester 2, 2018, despite that the students' feedback provided outstanding evidences that completing the blog activities in KMIS unit allows students to develop students' learning skills and independent learning skills. This experience proved to be unique for the academics and students involved. Moreover, it encourages other academics to include cutting-edge units similar to GITS and KMIS units in the master's degree to increase students' awareness since academics do have some responsibility toward graduate students with this consciousness and responsiveness since these students will be our future leaders.

Finally, this study addressed the study aims by enhancing students' personal and professional skills which are valued by businesses and organizations; students confirmed that they obtained the necessary knowledge behind the units, and their skills were enhanced and improved. Students offered the following comments about the facilitator:

Table 7.2 GITS and KMIS formal feedback (eVALUate)

Year	Item 1 learning outcomes		Item 4 assessment		Item 8 motivation		Item 9 best learning		Item 11 overall satisfaction	
	Unit	University average	Unit	University average	Unit	University average	Unit	University average	Unit	University average
GITS unit										
2018	92	90	100	86	100	86	100	87	100	84
2017	100	89	91	85	100	85	91	86	91	83
2016	100	89	94	85	100	85	94	87	100	84
2015	100	89	100	85	92	86	100	86	100	83
2014/S2	100	88	100	84	100	85	100	86	100	83
2014/S1	100	89	100	85	100	86	100	87	100	84
2013	100	89	100	89	100	85	100	87	100	84
2012	100	89	100	85	100	86	100	87	100	84
2011	92	89	92	85	92	84	92	86	92	84
KMIS unit										
2018	100	90	100	86	100	86	100	88	100	84

- She is the best lecturer who I have ever studied. She always gives feedback immediately to improve our works. She is a conscientious teacher and makes hard knowledge more understandable by individual and group activities.
- She understands the needs of students and the unit as well. She is really helpful and provides detailed feedback on students work so that work can be improved for future assessments. She has an in-depth knowledge of her work. She is master in her core area.
- Availability: Continuous and quick response for any query, any time. Her valuable and timely feedback helps a lot during study period, to complete the assignments on time, and with required changes. Feedback: Her feedback for assignment results or even during progress toward assignments is highly appreciable. She never said wrong to any argument presented by students, although she corrects it. It helps to keep the student's confidence to take part in in-class activities and share their view. She is one of the teachers to whom I will remember forever. Thank you so much Madam. Words end to ohh noo.
- She is a lovable and kind person who shows enormous interest in the learning of her students and in the field. She is available to give feedback, reviews, and advice 24/7. I am very thankful to her help.
- By using word cloud generator website (https://tagul.com/), the author developed the following word cloud to highlight students' reaction about the facilitator (see Fig. 7.11).

Once again, the manual coding was used to generate the facilitator skills, and Fig. 7.12 confirmed that the facilitator for GITS and KMIS units is very popular among her students, since she provided outstanding and very helpful feedback (see Fig. 7.12).

Finally, the facilitator's unique approach has given them notable distinction within the university. Students provided the following comments regarding this facilitator:

- Our lecturer is the best teacher I ever had. Our lecturer always encourages and motivates more and more the students to do their work perfectly. She is never rude or shouts on the students for lots of queries. She responses so quickly. I love the way she teaches. She is my mentor, guru, and everything.
- The lecturer's feedback is excellent as always. Always available to help when required. Engages the class in activities promoting collaboration.
- Lecturer teaching method is very effective and clear because she always tries to give her best in the class. She always communicates with students to encourage them.
- My lecturer is a wonderful teacher as she always encourages students to do the things in the right way and is very supportive to the students. She is a well-organized teacher and communicates very clearly. We can ask anything for as many times as we want. She is very cooperative. We always get immediate replies for our inquiries, and she is always happy to help before and after class if any student has any question.

Fig. 7.11 Students' feedback about the facilitator

- My lecturer is an asset to the subject, approachable, patient with answering questions, enthusiastic about the topic, passionate, and helpful in providing research direction and tools.
- My lecturer is very approachable when it comes to looking for help. Furthermore, she provides us with up-to-date information regarding sustainability and keeps us informed. She has extensive knowledge on the subject, which is important. In addition, she is inspirational and pushes to achieve good marks. She makes the teaching environment friendly and comfortable.
- My lecturer is passionate about the unit, always willing to help. She provided motivation, i.e., the best students can write a chapter in her book.
- My lecturer style of teaching is very impressive and efficient. She has knowledge about how to conduct a unit to give maximum benefits to the students.

Fig. 7.12 Students'
feedback about the
facilitator

Her honesty and deep understanding of psychology of students make her a very efficient teacher. She exactly knows the need of each student and guides them according to their needs. She realizes that she is here to solve the problems of the students. She is happy to discuss the problems which are about other units to facilitate the students.

- My lecturer is a very honest, soft spoken, understanding, kind, and caring lecturer. She carries this unit with great knowledge. I am totally inspired by her style of teaching.
- My lecturer is one of the best teachers I have never ever had. She is doing very well with teaching.
- My lecturer has a wide knowledge about Green IT and IT business. She gives a good evaluation result, which supports and challenged students with good issues to ponder on. Introducing wiki group to the class was very useful considering the workload and type of the unit.

By using word cloud generator website (https://tagul.com/), the author developed the following word cloud to accept students' opinion about the facilitator teaching style (see Fig. 7.13).

Finally, the manual coding was used once again to assess and evaluate students' feedback about the facilitator teaching style in line to develop new themes from the formal and informal feedback (Munn, Stefano, & Pelz, 2008). Figure 7.13 generated the main themes which presented the facilitator teaching style.

In conclusion, using SNEM model in this study assisted the researcher to confirm the study's aims, along with Figs. 7.7, 7.8, 7.9, and 7.10, while

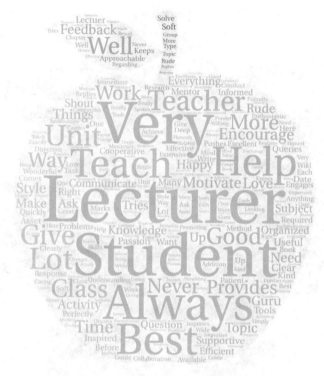

Fig. 7.13 Students' feedback about the facilitator teaching style

Figs. 7.11, 7.12, 7.13, and 7.14 identified the facilitator personality and her unique teaching style for GITS and KMIS units. Based on the formal and informal feedback, students enjoyed and prized the e-assessments and activities via the wiki and blog in GITS and KMIS units.

Finally, it was a challenge exercise for students and lecturers, as students confirmed that using wiki and blog in the units was a new challenge for them to complete activities in the class on specific time and present their findings to the class and encourage them to communicate and debate their perspective. The provided evidences from the students indicated that students enjoyed the class and obtain the necessary skills from personal and professional, and they learn the unit materials from a different perspective. As for the lecturer, she faced two challenges by using e-assessment from marking and motivating the students. The lecturer's feedback via the rubric encourages students to engage with the unit and will help students to provide an exciting, memorable, and motivating experience. As for the motivation, the lecturer encouraged her students to participate and engage in e-assessments, by completing specific activities by themselves or as a group. In the class, the lecturer plays a good role, as she checked and listened to every student contribution and discussion during the preparation stage and she provided prompt formative feedback

Fig. 7.14 Facilitator teaching style themes

to improve the student's contribution before the submission. After that, the lecturer asked every student to present their findings based on the activities and interesting research, to share their opinions and perspectives with their colleagues. This exercise encouraged and motivated students to complete their work with good quality and on time especially the shy students.

7.7 Study Limitations

This study was limited to 194 students in Australia. The motivation for conducting this study was to determine whether the use of e-assessment via wiki and blog will enhance students' independent learning by motivating them to complete all class work and assessment tasks, improving their understanding of the course material, and giving them knowledge and understanding of cutting-edge developments locally and globally. Further research with larger and more diverse units and groups of students is required in the future to strengthen the research findings.

7.8   Conclusion

This study examined and determined whether e-assessments using the wiki and blog tools promoted and enhanced students' personal and professional skills and develop independent learning skills that are required and valued by organizations and businesses. This study confirmed the research aims and helped to develop students' personal and professional skills and independent learning skills through activities such as concept maps, debates, and PowerPoint presentations in the wiki and blog on the Blackboard platform. A total of 194 students from the GITS and KMIS units confirmed that their learning techniques and skills were enhanced, and their understanding of the course material was improved through listening, collaboration, communication, cooperation, connection, and debate. Academics should be aware of the needs and requirements of employers and must take responsibility for equipping their students with the appropriate skills and knowledge enabling them to succeed in the workplace and take on leadership roles in the future. GITS and KMS students enjoyed the units, as attested to by the positive formal and informal feedback regarding the entire learning process, especially the assessments and activities. The provided evidences indicate that completing GITS and KIMIS units by postgraduate students furnishes the students with the personal and professional skills which are needed and required by the organizations worldwide, and this achieved the study's aims and objectives. In conclusion, this study fulfilled the intended aims and objectives; however, it was limited to two postgraduate units in an Australian university. In the future, the researcher will include more units from different disciplines to strengthen the research goals and findings.

References

Abdulahi, A., Samadi, B., & Gharleghi, B. (2014). A study on the negative effects of social networking sites such as Facebook among Asia Pacific University scholars in Malaysia. *International Journal of Business and Social Science, 5*(10).

Alam, S. L., & McLoughlin, C. (2018). E-citizenship skills online: A case study of faculty use of web 2.0 tools to increase active participation and learning. In M. Khosrow-Pou (Ed.), *Information and technology literacy: Concepts, methodologies, tools, and applications* (pp. 878–896). USA: IGI Global.

Alor-Hernández, G., & Álvarez-Rodríguez, J. M. (2018). Preface special issue on educational applications on the web of data: New trends and perspectives. (Elsevier). *Telematics and Informatics, 35*(3), 517–519.

Alruwais, N., Wills, G., & Wald, M. (2018). Advantages and challenges of using e-assessment. *International Journal of Information and Education Technology, 8*(1), 34–37.

Aral, S. (2013). Social media and business transformation: A framework for research. *Information Systems Research, 24*(1), 3–13.

Bailey, E. K., & Cotlar, M. (1994). Teaching via the internet. *Communication Education, 43*(2), 184–193.

Bandura, A. (1977). *Social learning theory*. New York, NY: Holt, Rinehart and Winston.

Barbosa Granados, S. H., & Amariles Jaramillo, M. L. (2019). Learning styles and the use of ICT in university students within a competency-based training model. *Journal of New Approaches in Educational Research, 8*(1), 1–6.

Biasutti, M., & EL-Deghaidy, H. (2015). Interdisciplinary project-based learning: An online wiki experience in teacher education. *Technology, Pedagogy and Education, 24*(3), 339–355.

Borthwick, K., & Gallagher-Brett, A. (2014). 'Inspiration, ideas, encouragement': Teacher development and improved use of technology in language teaching through open educational practice. *Computer Assisted Language Learning, 27*(2), 163–183.

Boyd, D. M., & Ellison, N. B. (2007). Social network sites: Definition, history, and scholarship. *Journal of Computer-Mediated Communication, 13*(1), 210–230.

British Medical Association. (2004). *Communication Skills education for doctors: An update.* UK: Retrieved from http://www.bma.org.uk/images/communication_tcm41-20207.pdf

Burden, R. L., & Williams, M. (1997). *Psychology for language teachers: A social constructivist approach*. Cambridge, MA: Cambridge University.

Carlson, M. (2018). Facebook in the news: Social media, journalism, and public responsibility following the 2016 trending topics controversy. *Digital Journalism, 6*(1), 4–20.

Cheung, C. M., Liu, I. L., & Lee, M. K. (2015). How online social interactions influence customer information contribution behavior in online social shopping communities: A social learning theory perspective. *Journal of the Association for Information Science and Technology, 66*(12), 2511–2521.

Chichester, T., Hagglund, K., & Edhayan, E. (2013). Teaching surgical residents to evaluate scholarly articles: A constructivist approach. *The American Journal of Surgery, 205*, 259.

Choi, Y.-J., Ra, J.-H., Jung, Y.-G., & Boo, Y. (2017). Coding clinic system using WIKI and unified communication in healthcare. *International Information Institute (Tokyo). Information, 20*(2A), 1043–1050.

Chu, S. K. W., Reynolds, R. B., Tavares, N. J., Notari, M., & Lee, C. W. Y. (2017). Twenty-first century skills education in Switzerland: An example of project-based learning using wiki in science education. In *21st Century Skills Development Through Inquiry-Based Learning* (pp. 61–78). Singapore: Springer.

Chuttur, M. Y. (2009). Overview of the technology acceptance model: Origins. *Developments and Future Directions Sprouts: Working Papers on Information Systems, 9*(37), 1–23.

Cole, M. (2009). Using wiki technology to support student engagement: Lessons from the trenches. *Computers and Education, 52*, 141–146.

Cowan, B. R., & Jack, M. A. (2011). Exploring the wiki user experience: The effects of training spaces on novice user usability and anxiety towards wiki editing. *Interacting with Computers, 23*(2), 117–128. https://doi.org/10.1016/j.intcom.2010.11.002

Cowling, D. (2019). *Social media statistics Australia – January 2019*.

Creswell, J. W., & Poth, C. N. (2017). *Qualitative inquiry and research design: Choosing among five approaches*. Thousand Oaks: Sage Publications.

Dalkir, K. (2013). *Knowledge management in theory and practice*. Burlington, MA: Routledge.

Daly, C., Pachler, N., Mor, Y., & Mellar, H. (2010). Exploring formative e-assessment: Using case stories and design patterns. *Assessment & Evaluation in Higher Education, 35*(5), 619–636.

de Arriba, R. (2016). Participation and collaborative learning in large class sizes: Wiki, can you help me? *Innovations in Education and Teaching International*, 1–10.

Din, N., Yahya, S., Suzan, R., & Kassim, R. (2012). Online social networking for quality of life. *Procedia - Social and Behavioral Sciences, 35*, 713–718. https://doi.org/10.1016/j.sbspro.2012.02.141

Dinger, M., & Grover, V. (2019). Revisiting web 2.0. In *Advanced methodologies and technologies in network architecture, mobile computing, and data analytics* (pp. 1777–1788). IGI Global.

Driscoll, M. (2002). Blended learning: Let's get beyond the hype. *e-Learning, 1*(4).

Duke, B., Harper, G., & Johnston, M. (2013). Connectivism as a digital age learning theory. *The International HETL Review, 2013*, 4–13.

Felder, R. M., & Silverman, L. K. (1988). Learning and teaching styles in engineering education. *Engineering Education, 78*(7), 674–681.

Fleming, N. D. (1995). *I'm different; not dumb. Modes of presentation (VARK) in the tertiary classroom.* Paper presented at the Research and Development in Higher Education, Proceedings of the 1995 Annual Conference of the Higher Education and Research Development Society of Australasia (HERDSA), HERDSA.

Gilbert, T. (2006). Mixed methods and mixed methodologies - The practical, the technical and the political. *Journal of Research in Nursing, 11*(3), 205–217.

Godwin-Jones, B. (2003). Blogs and Wikis: Environments for on-line collaboration. *Language, Learning and Technology, 7*, 12–16.

Griffin, P., & Care, E. (2014). *Assessment and teaching of 21st century skills: Methods and approach.* Springer.

Harrison, R., & Reilly, T. (2011). Mixed methods designs in marketing research. *Qualitative Market Research: An International Journal, 14*(1), 7–26.

Hislop, D., Bosua, R., & Helms, R. (2018). *Knowledge management in organizations: A critical introduction.* Oxford, UK: Oxford University Press.

Hughes-Wiener, G. (1986). The "learning how to learn" approach to cross-cultural orientation. *International Journal of Intercultural Relations, 10*(4), 485–505.

Issa, T. (2008). *Development and Evaluation of a Methodology for Developing Websites.* PhD Thesis, Curtin University, Western Australia. Retrieved from http://espace.library.curtin.edu.au/R/MTS5B8S4X3B7SBAD5RHCGECEH2FLI5DB94FCFCEALV7UT55BFM-00465?func=results-jump-full&set_entry=000060&set_number=002569&base=GEN01-ERA02

Issa, T. (2014). Learning, communication and interaction via wiki: An Australian perspective. In H. Kaur & X. Tao (Eds.), *ICTs and the millennium development goals: A United Nations perspective* (pp. 1–17). Boston, MA: Springer US. https://doi.org/10.1007/978-1-4899-7439-6_1

Issa, T., & Isaias, P. (2014). Promoting human-computer interaction and usability guidelines and principles through reflective journal assessment. *Emerging Research and Trends in Interactivity and the Human-Computer Interface, IGI Global*, 375–394. https://doi.org/10.4018/978-1-4666-4623-0.ch019

Issa, T., Isaias, P., & Kommers, P. (2016a). Social networking. In T. Issa, P. Isaias, & P. Kommers (Eds.), *Social networking and education global perspectives* (pp. 3–13). New York, NY: Springer.

Issa, T., Isaias, P., & Kommers, P. (2016b). Social networking and education model (SNEM). In T. Issa, P. Isaias, & P. Kommers (Eds.), *Social networking and education global perspectives* (pp. 323–345). New York, NY: Springer.

Issa, T., Issa, T., & Chang, V. (2012). Technology and higher education: An Australian study. *The International Journal of Learning, 18*, 223–236.

Issa, T., Issa, T., & Kommers, P. (2014). Feedback and learning support that fosters students' independent learning: An Australian case study. *The InternatiInternational Journal of Learning, 19*, 29–39.

Issa, T., & Kommers, P. (2013). Social networking for web-based communities. *International Journal of Web Based Communities, 9*(1), 5–24.

Jones, P. (2007). *When a wiki is the way: Exploring the use of a wiki in a constructively aligned learning design.* Paper presented at the ASCILITE 2007, Singapore.

Karasavvidis, I. (2010). Wiki Uses in Higher Education: Exploring barriers to successful implementation. *Interactive Learning Environments, 18*(3), 219–231.

Karrer, T. (2008). *Ten predictions for e-learning 2008: e-learning technology.* Retrieved from http://elearningtech.blogspot.com/2008/01/ten-predictions-for-elearning-2008.html

Kear, K., Woodthorpe, J., Robertson, S., & Hutchison, M. (2010). From forums to wikis: Perspectives on tools for collaboration. *The Internet and Higher Education, 13*(4), 218–225. https://doi.org/10.1016/j.iheduc.2010.05.004

King, K. P. (2003). Learning the new technologies: Strategies for success. *New Directions for Adult and Continuing Education, 2003*(98), 49–58.

Kiraly, D. (2014). *A social constructivist approach to translator education: Empowerment from theory to practice.* New York, NY: Routledge.

Kivinen, O., & Ristelä, P. (2003). From constructivism to a pragmatist conception of learning. *Oxford Review of Education, 29*(3), 363–375. https://doi.org/10.1080/03054980307442

Kung, F.-W. (2018). Assessing an innovative advanced academic writing course through blog-assisted language learning: Issues and resolutions. *Innovations in Education and Teaching International, 55*(3), 348–356.

Lalonde, M., & Castro, J. C. (2015). Amplifying youth cultural practices by engaging and developing professional identity through social media. *Youth Practices in Digital Arts and New Media: Learning in Formal and Informal Settings, 40.*

Laughton, P. (2011). The use of wikis as alternative to learning content management systems. *The Electronic Library, 29*(2), 225–235.

Lee, H., & Witz, K. G. (2009). Science teachers' inspiration for teaching socio-scientific issues: Disconnection with reform efforts. *International Journal of Science Education, 31*(7), 931–960.

Lichtman, M. (2012). *Qualitative research in education: A user's guide: A user's guide.* Thousand Oaks, CA: Sage.

Mahruf, M., Shohel, C., & Kirkwood, A. (2012). Using technology for enhancing teaching and learning in Bangladesh: Challenges and consequences. *Learning, Media and Technology, 37*(4), 414–428.

Matheson, J. (2009). *Benefits of using a Wiki.* Retrieved from http://wiki.customware.net/repository/display/wwyw/Benefits+of+using+a+wiki

Maudsley, G. (2011). Mixing it but not mixed-up: Mixed methods research in medical education (a critical narrative review). *Medical Teacher, 33*(2), 92–104.

Mi, M., & Gould, D. (2014). Wiki technology enhanced group project to promote active learning in a neuroscience course for first-year medical students: An exploratory study. *Medical Reference Services Quarterly, 33*(2), 125–135.

Mockler, N., & Groundwater-Smith, S. (2015). Curriculum, pedagogy, assessment and student voice. In *Engaging with student voice in research, education and community* (pp. 139–150). Cham, Switzerland: Springer.

Monks, T. G., & Kawwa, T. (1971). Social psychological aspects of comprehensive education. *International Review of Education, 17*(1), 66–76. https://doi.org/10.1007/BF01421371

Munn, S. M., Stefano, L., & Pelz, J. B. (2008). *Fixation-identification in dynamic scenes: Comparing an automated algorithm to manual coding.* Paper presented at the Proceedings of the 5th symposium on Applied perception in graphics and visualization.

Murphy, R. J., Gray, S. A., Straja, S. R., & Bogert, M. C. (2004). Student learning preferences and teaching implications. *Journal of Dental Education, 68*(8), 859–866.

Nardi, B. A., Schiano, D. J., Gumbrecht, M., & Swartz, L. (2004). Why we blog. *Communications of the ACM, 47*(12), 41–46.

Ng, E. M. (2016). Fostering pre-service teachers' self-regulated learning through self-and peer assessment of wiki projects. *Computers & Education, 98*, 180–191.

Novakovich, J., Miah, S., & Shaw, S. (2017). Designing curriculum to shape professional social media skills and identity in virtual communities of practice. *Computers & Education, 104*, 65–90.

O'Reilly, T. (2007). What is Web 2.0: Design patterns and business models for the next generation of software. *Communications & Strategies, 1*, 17.

Oberst, U., Wegmann, E., Stodt, B., Brand, M., & Chamarro, A. (2017). Negative consequences from heavy social networking in adolescents: The mediating role of fear of missing out. *Journal of Adolescence, 55*, 51–60.

Oliver, M., & Trigwell, K. (2005). Can 'blended learning'be redeemed? *E-learning and Digital Media, 2*(1), 17–26.

Psychology Dictionary. (2015). *What is awareness.* Retrieved from http://psychologydictionary. org/awareness/

Richardson, V. (2005). Constructivist teaching and teacher education: Theory and practice. In V. Richardson (Ed.), *Constructivist teacher education* (pp. 13–24). London, UK: Routledge.

Rogers, R. (2018). Coding and writing analytic memos on qualitative data: A review of Johnny Saldaña's the coding manual for qualitative researchers. *The Qualitative Report, 23*(4), 889–892.

Ruge, G., & McCormack, C. (2017). Building and construction students' skills development for employability–reframing assessment for learning in discipline-specific contexts. *Architectural Engineering and Design Management, 13*(5), 1–19.

Safuanov, I. S. (2004). Psychological aspects of genetic approach to teaching mathematics. *International Group for the Psychology of Mathematics Education.*

Saldaña, J. (2015). *The coding manual for qualitative researchers.* Atlanta, GA: Sage.

Sancho-Thomas, P., Fuentes-Fernandez, R., & Fernandez-Manjon, B. (2009). Learning teamwork skills in university programming courses. *Computers and Education, 53*, 517–531.

Siemens, G. (2005). Connectivism: A learning theory for the digital age. *International Journal of Instructional Technology and Distance Learning, 2*(1), 3–10.

Simula, B. L. (2018). *Book review: The coding manual for qualitative researchers.* Sage, CA/Los Angeles, CA: SAGE Publications.

Sivan, E. (1986). Motivation in social constructivist theory. *Educational Psychologist, 21*(3), 209–233.

Statista. (2019a). *Global digital population as of January 2019 (in millions).* Retrieved from https://www.statista.com/statistics/617136/digital-population-worldwide/

Statista. (2019b). *Leading countries based on number of Facebook users as of January 2019 (in millions).* Retrieved from https://www.statista.com/statistics/268136/top-15-countries-based-on-number-of-facebook-users/

Taraghi, B., Ebner, M., & Schaffert, S. (2009). *Personal learning environments for higher education: A mashup based widget concept.* Paper presented at the F. Wild, M. Kalz, M. Palmér, & D. Müler (Éd.), Mash-Up Personal Learning Environments: Proceedings of the Workshop in conjunction with the 4th European Conference on Technology-Enhanced Learning (ECTEL'09).

Teddlie, C., & Tashakkori, A. (2009). *Foundations of mixed methods research - Integrating quantitative and qualitative approaches in the social and behavioral sciences.* Thousand Oaks, CA: Sage Publisher.

Thomas, M., & Thomas, H. (2012). Using new social media and Web 2.0 technologies in business school teaching and learning. *Journal of Management Development, 31*(4), 358–367.

Thrun, S. (1996). *Learning to learn: Introduction.* Paper presented at the In Learning To Learn.

Wallace, C. S., & Brooks, L. (2015). Learning to teach elementary science in an experiential, informal context: Culture, learning, and identity. *Science Education, 99*(1), 174–198.

Wang, L. (2016). Employing Wikibook project in a linguistics course to promote peer teaching and learning. *Education and Information Technologies, 21*(2), 453–470.

Wang, Y.-C. (2014). Using wikis to facilitate interaction and collaboration among EFL learners: A social constructivist approach to language teaching. *System, 42*, 383–390. https://doi. org/10.1016/j.system.2014.01.007

Watson, A. S. (1963). Some psychological aspects of teaching professional responsibility. *Journal of Legal Education, 1*, 1–23.

Weaver, A., & Morrison, B. (2008). Social networking. *Computer, 41*(2), 97–100.

Whitty, M. T., Doodson, J., Creese, S., & Hodges, D. (2018). A picture tells a thousand words: What Facebook and Twitter images convey about our personality. *Personality and Individual Differences, 133,* 109–114.

Wiggins, B. (2011). Confronting the dilemma of mixed methods. *Journal of Theoretical and Philosophical Psychology, 31*(1), 44–60.

Williams, M., & Burden, R. L. (1997). *Psychology for language teachers: A social constructivist approach.* NY, USA: ERIC.

Wolcott, H. F. (1994). *Transforming qualitative data: Description, analysis, and interpretation.* Thousand Oaks, CA: Sage.

Wooster. (2019). *Wiki vs. Blog.* Retrieved from https://www.wooster.edu/offices/web/how/scotblogs/wiki-blog/

Zein, R. (2014). Explorative study on the ways of using blogs and wikis as teaching and learning tools in mathematics. In M. Searson & M. N. Ochoa (Eds.), *Society for Information Technology & Teacher Education International Conference 2014, held in Jacksonville, Florida, United States* (pp. 66–72). AACE. Retrieved from http://www.editlib.org/p/130712.

Chapter 8
Lurkers Versus Posters: Investigation of the Participation Behaviors in Online Learning Communities

Omid Reza Bolouki Speily, Alireza Rezvanian, Ardalan Ghasemzadeh, Ali Mohammad Saghiri, and S. Mehdi Vahidipour

Abstract Nowadays online learning communities (OLC) have been thought of as a great source of learning contents by many users in various areas such as online forums, question answering, and online social networks. Millions of posts are virally shared among different users in online social communities every day; however, different types of users deal with these posts in various ways in which, the users are divided into two categories: posters and lurkers. Lurker users join in an online community and watch other posts without performing any activities (or rarely activities). On the contrary, poster users actively participate in the creation and repost of information in online learning communities. The aim of this chapter is to investigate the lurking behaviors in educational networking or online communities in order to identify the factors influencing the participating behavior in such communities on social media. Therefore, first we introduce the role of online learning communities in educational networks and also present a bibliometric analysis. Then relevant theories and factors are reviewed, and a mathematical solution is

O. R. B. Speily · A. Ghasemzadeh
Department of Information Technology and Computer Engineering,
Urmia University of Technology, Urmia, Iran
e-mail: speily@uut.ac.ir; a.ghasemzadeh@uut.ac.ir

A. Rezvanian (✉)
Department of Computer Engineering, University of Science and Culture, Tehran, Iran

School of Computer Science, Institute for Research in Fundamental Sciences (IPM),
Tehran, Iran
e-mail: rezvanian@usc.ac.ir

A. M. Saghiri
School of Computer Science, Institute for Research in Fundamental Sciences (IPM),
Tehran, Iran
e-mail: a_m_saghiri@aut.ac.ir

S. M. Vahidipour
Faculty of Electrical and Computer Engineering, Department of Computer,
University of Kashan, Kashan, Iran
e-mail: vahidipour@kashanu.ac.ir

© Springer Nature Switzerland AG 2020
A. Peña-Ayala (ed.), *Educational Networking*, Lecture Notes in Social
Networks, https://doi.org/10.1007/978-3-030-29973-6_8

proposed to increase the participation of learners in the transmission of information through reposting. The proposed solution encourage lurking learners to participate in content reposting. Comprehensive evaluations indicated that the proposed method had significantly solved the presented challenges.

Keywords Participation behavior · Social networks · Online learning communities · Asynchronous learning · Lurkers · Posters · Information reposting

Acronyms

EN   Educational networking
ESN   Educational social networks
ICT   Information and communications technology
OLC   Online learning communities
OSN   Online social network
SN   Social network
SNA   Social network analysis
SNS   Social networking site
UTG   Uses and gratifications theory

8.1 Introduction

Conventional teaching cannot meet the increasing need of people to learn due to inadequate resources and the limitations of time and space. Therefore, individuals should find and use other methods to learn more efficiently. In addition, learning development happens in places where the learners spend most of their time (Topping, 2005). Because of their reception by Internet users, online communities and social media can be used for learning and educational purposes (Owusu, Bekoe, Otoo, & Koli, 2019; Shafipour et al., 2018; Wagner, 2011). Currently, learners not only depend on conventional learning, but also use other learning environments on Internet such as online learning communities (OLCs) to increase their learning opportunities (Ke & Hoadley, 2009; Welser, Khan, & Dickard, 2019). OLC is a kind of highly accessible and convenient learning platform that consists of educational members such as experts (i.e., teachers, instructors, trainers, and scholars) and learners (i.e., students and scholars) where learners can search what interests them and share knowledge beyond the restriction of time and space. OLCs are computer-supported public or private groups (i.e., online social networks) on the environments of web (i.e., educational online social networks) that address the learning and educational needs of their members by facilitating asynchronous learning.

Learners with special expertise help other learners needing knowledge or information (Chris Brown, 2018; Yilmaz, 2016). Despite their diversity, all these communities follow the same process: a learner posts content and other learners repost it if they like it. Finally, these posts are virally shared on the whole network. Content sharing is an essential part of OLC experience (Karaoglan Yilmaz, 2017). In addition to composing posts by themselves, learners can also rebroadcast or repost other posts of the learners that they find of particular informational value. When a learner shares another learner's post, in fact, he/she is participating in the development of a common knowledge in his own network range. It is not only a feature to display his favorite posts. Repost of interesting posts has an extensive effect on networks and spreads information by exposing a new audience to the content (Sharma & Land, 2018; Shi, Rui, & Whinston, 2014).

Participation in online community education, particularly in educational social networks (ESN), is very effective in achieving educational goals. This concept in an online learning community can be viewed as a life cycle of participation. This chapter will investigate this life cycle. The basic question of this study is to study the behavior of participation in online community learning and discovering the basic concepts that influence the level of participation in such communities and educational online social networks (OSN). Participation in an online community depends on the type of users forming such communities (Malinen, 2015; Nonnecke, Andrews, & Preece, 2006). Membership in online communities is explained by different theories, which are discussed in the following part in details (Takahashi, Fujimoto, & Yamasaki, 2003). Nevertheless, there are many users who do not participate actively in these online communities. According to the previous studies, online users are divided into two groups: lurkers (nonparticipants) and posters (participants) (Hurtubise et al., 2017; Nonnecke & Preece, 2000). In various papers, different definitions of lurkers have been provided (Amato, Moscato, Picariello, Piccialli, & Sperlí, 2018; Amichai-Hamburger et al., 2016; Kilner & Hoadley, 2005; Nonnecke & Preece, 2001).

Emphasizing on a characteristic, each of these definitions has described such users. Generally, these users join an online community or educational group consciously; however, they do not post anything. Creating a post in online communities indicates participations of the users. Regarding the level of this participation, no clear definitions have been provided in different references. In many references, the users who do not post anything in online communities are known as lurkers (Amato, Castiglione, Moscato, Picariello, & Sperlì, 2017; Nonnecke & Preece, 2001); however, a time limit has been made for this nonparticipation in some other references (Sloep & Kester, 2009). For instance, if a user did not (re)post a post in the past year, s/he is considered a lurker. In addition, there are many disagreements over the effects of such users (Nonnecke & Preece, 2000; Takahashi et al., 2003). Some studies have interpreted them as free riders who use sources free without providing communities with any benefits (Kollock & Smith, 1996). On the other hand, some research have thought of these users' behaviors as a part of the human nature and emphasized the role of communities attracting them (Schneider, Von Krogh, & Jäger, 2013; Sun, Rau, & Ma, 2014). These studies have not thought of lurkers as negative users in online communities; moreover, experiments indicate that their roles are important in identifying a group. In the majority of such communities, participation is encouraged so that users can become acquainted with the culture and common customs of a group or community before interaction. Then users can participate in them if they are willing to (Takahashi et al., 2003). Investigations also indicate that such users think of themselves as a part of an online community or group. They are included in the audience in evaluations or advertisements, too. Many methods have been introduced to attract such users (Luo, Zhang, & Qi, 2017).

First-stage use is an important indicator of OLCs' success, but the long-range success of OLCs lies on users' persistent usage. Persistent usage of OLCs can provide sufficient online learning materials and form prosperous online communication atmosphere which contributes to the long-term development of OLCs. The two fundamental interrelated challenges are access to appropriate information and shortage of participation (because of lurking behavior or low participation of some learner) in these communities. If information provided by a certain OLC can match users' information requirement, users are satisfied easily. User satisfaction plays a significant role in his/her participation and engagement (i.e., collaboration and sharing) which is vital for the OLCs' success (Rehm, Mulder, Gijselaers, & Segers, 2016). To address these challenges, this chapter proposes a method to select the effective (appropriate) posts for each learner which increases the probability of repost behavior in the OLC.

In the next section of this chapter, the literature on this subject is reviewed. Then, the proposed method is presented according to the nature of lurkers. The next section contains a detailed evaluation of the proposed method and the last section is devoted to summing up the study.

8.2   Literature Review

Nowadays, online learning communities are extensively used and have induced fundamental changes to web-based systems. Therefore, according to the review of the performed studies, this paper is devoted to the role of online communities in learning as well as reposting behavior in online communities.

Over the last decade, with the development of web-based new application software on the basis of Web 2.0, online social networks have received considerable attention. Social network services have provided this opportunity for the users to create online learning communities for educational purposes (Haythornthwaite, 2019). OLCs improve interaction, information exchange, and personal experiences between learners. The statistics reported by Duggan, Ellison, Lampe, Lenhart, and Madden (2015) show that in recent years the use of OLCs has dramatically been increased in various parts of the world. Web 2.0 environment and respective tools such as messaging, e-mail, forum, wiki, social networks, and web conference are used in developing OLCs (Datu, Yang, Valdez, & Chu, 2018;

Feng, Wong, Wong, & Hossain, 2019; Lambić, 2016). OLCs are used outside the classroom for supporting the learning and educating process. In addition, the use of social media in learning and educating purposes is about to expand. Social media capabilities such as the ability in making synchronous or asynchronous connections and tagging, posting, creating, and organizing virtual groups and resource management and sharing have provided the possibility of easy implementation of OLCS and educational purposes (Mazman & Usluel, 2010). Learning with social media takes place with greater ease due to its reception capability and mobile nature in comparison with other platforms. Nowadays, many learners use social media for sharing information and knowledge, collaborating in conducting team projects, and discussing ideas and concepts (Ahmed, Ahmad, Ahmad, & Zakaria, 2019; Dabbagh & Kitsantas, 2012; J. C. Yang, Quadir, Chen, & Miao, 2016). For example, it has been shown that learners used *Facebook* as a learning management system, and it also satisfied them (Q. Wang, Woo, Quek, Yang, & Liu, 2011), or based on Chu and Meulemans (2008), using social media is common between learners, and 90% of information exchange between the learners happens via *Facebook* and *MySpace*.

In recent years, many issues regarding information sharing in OLCS have been proposed. According to Deng and Tavares (2013), information exchange has an important role in inspiring membership and activity in OLCs, and it also has a direct relationship with the value of an OLC in the eyes of its members. In OLCs, sharing information improves the knowledge and skill of all group members. In such communities, sharing information increases the tendency of learners to participate and engage in the learning and education process (Junco, 2012). Information sharing is defined as an activity through which members exchange information, experience, and skill among themselves. Online communities are suitable for supporting interaction and sharing between the learners. Many studies have measured the impact of a special social media on information sharing and reposting behavior (Hamid, Bukhari, Ravana, Norman, & Ijab, 2016). Other studies have mostly investigated effective factors in online community information reposting. Interaction and information sharing between members is considered to be the most important activity in OLC. Therefore, studying information reposting behavior and providing a method to increase its amount are important factors in OLCs' information sharing. Not many studies have been conducted on OLCs' information sharing using information reposting behavior. The study on blogspace addresses "epidemic" interests among different blogs regarding the content cited or copied from other blogs (Adar & Adamic, 2005). By studying the cases, it estimates the relationship between two similar blogs. By relationship, it means the use of another post in the form of citation or copy. Another important point addressed in the study is the influence of a blog on another blog via a post. Leskovec, Adamic, and Huberman (2007) studied sharing small pieces of text (e.g., news) used in other articles and texts. For this purpose, a method was implemented by which the source of each piece could be specified in the network. This made it possible to study the structure of sharing in the network. The study aims to find the sources from which a post or posts are influenced.

Online communities consist of individuals who communicate with each other by exchanging messages over the Internet (Joyce & Kraut, 2006). Online communities provide a platform for individuals to exchange information about a variety of different topics such as health, recreation, professional, and technical subjects (Ridings & Gefen, 2006). These communities develop according to the needs of their creators and users. The two primary functions of online communities are information exchange and social network interactions (Ridings & Gefen, 2006). They also serve particular social functions such as facilitating public participation in democratic processes and collaborative knowledge production. Directed information exchange provides individuals with the framework to seek providing and sharing information. *TripAdvisor* is an example of one such community, which enables individuals to post and request information regarding different vacation destinations. This takes another form in the shape of social interaction groups, these specifically enable individuals to build relationships and connect with others. *Facebook* is an example of this type of network community that enables individuals to connect with one another, exchange gossip, upload pictures, and post their statuses.

Research shows that there is often more user participation in social network groups than there is in those directed at information exchange (Nonnecke & Preece, 2000). Although there is some evidence of young people starting to leave *Facebook* (Baumer et al., 2013), *Facebook* remains the largest online social network, reaching its 2.32 billion monthly active users as of the fourth quarter of 2018 (Statista, 2019). Other popular social network services include *Google+*, *Twitter*, *MySpace*, *LinkedIn*, *Pinterest*, and *YouTube*. Based on the large membership of online social networking groups, it would appear that participants are more concerned with fulfilling their own needs for affiliation and belonging, than they are in exchanging or providing information. However, both information exchange and social network online groups often take a significant amount of time to grow and develop, and initial participation in these groups is often scarce and uneven (Joyce & Kraut, 2006).

There are many terms used to describe lurkers, including nonpublic participants (Nonnecke & Preece, 2000) and "read only participants" (J.N. Williams, Van Patten, & Williams, 2004) and more negative labels, such as "abusers of common good" and "free-riders" (Kollock & Smith, 1996). Regardless of the different terms, there is a general agreement that lurkers are persistent, though silent and passive members of online communities who do not contribute to groups (Y.-L. Lee, Chen, & Jiang, 2006; Sun et al., 2014; Tagarelli & Interdonato, 2018). In contrast to lurkers, posters are active members in online discussions; hence they are generally regarded as more constructive members of online communities. A constant flow of contributors is needed in order to maintain online groups. The more active participants there are in online groups, the larger the pool of resources will be for the entire group, thus the lack of involvement among lurkers often serves as a threat to the continuity of online groups (Yeow, Johnson, & Faraj, 2006). From this perspective, it seems that lurkers should be encouraged to participate more frequently in online discussions. Although lurkers are almost invisible, it turns out that the majority of both posters and lurkers consider lurkers as part of the community. More importantly, none of the respondents to the survey showed resentment toward lurkers (Kate Merry & Simon, 2012).

8.2.1 A Bibliometric Analysis

In this section, we aim to investigate the different types of research results and address the potential trends of the key research areas related to online learning community and educational networking as the online key tools for educational purposes. Our methodology that is used in our analysis is similar to the methodology used by Rezvanian, Moradabadi, Ghavipour, Daliri Khomami, and Meybodi (2019).

The data was collected from Web of Science (WoS), an online subscription-based scientific citation indexing service that provides a comprehensive citation search that covers more than 90 million records during 1900 to present. For data collection, we searched for the ("online learning community") as the main keywords in the topic of articles belonging to WoS. The initial search resulted in 147 articles from 2001 until 2018. The search results consist of 2 Turkish language articles, 1 Portuguese language article, and 144 English language articles.

The statistical analysis is extracted from the resulted search and bibliometric analysis is performed using VOSviewer (van Eck & Waltman, 2010). VOSviewer provides network visualization on co-authorship, co-citation, and citation with respect to authors, organizations, and countries and also co-occurrence with respect to keywords.

The number of articles published during the time period 2001–2018 is shown as Fig. 8.1 which indicates the dramatically changing pattern of publications in the research community in each year from 2008 until 2018.

The top 10 institutions with highly published articles are listed in Table 8.1 in which the most articles are published by researchers from National Central University, University of Toronto and Chinese University of Hong Kong.

Besides, corresponding to each affiliation, the country in which the institution is introduced is taken out for further analysis, and this result is shown in Figure 8.2 in which institutions from the USA, China, Taiwan, and Canada have the major contributions in publication.

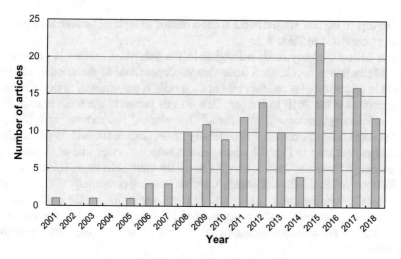

Fig. 8.1 Number of articles published during the time period 2001–2018

Table 8.1 Top 10 institutions with highly published articles with respect to number of publications and citations

Institutions	Publications	Citations
National Central University, Taiwan	5	254
University of Toronto, Canada	4	28
Chinese University of Hong Kong, Hong Kong	3	37
Sheffield Hallam University, UK	3	34
Umea University, Sweden	3	22
Ankara University, Turkey	3	6
University of Minnesota, USA	3	3
Academia Sinica, Taiwan	2	201
National Taiwan University, Taiwan	2	201
Tamkang University, Taiwan	2	201

Fig. 8.2 Top 10 countries with highly published articles

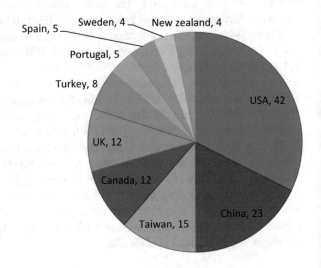

The top 10 keywords used in the articles related to topic of "online learning community" are listed in Table 8.2.

The network of keywords belonging to the published articles is presented in Fig. 8.3; in this network, each node size is proportional to the number of occurrences in all text articles and the color of each node is proportional to the publication year from blue for 2010 to red for 2018. In this network, the main keywords are "online learning community," "students," "knowledge," "education," and "Internet."

This network with respect to the density of occurrence for each keyword also is plotted as heat map in Fig. 8.4 which released the important role of "technology," "e-learning," "Internet," and "system" around the "online learning community."

As observed in the field of research, online learning community (OLC) is one of the most challenging areas of research. In these environments, the participation of members in the creation and transmission of information is vital. In the following, we look at the key areas for creating a sense of participation in the members of these communities.

Table 8.2 Top 10 keywords used in the articles

	Keyword	Frequency
1	Online learning community	66
2	Education	15
3	Social presence	12
4	e-learning	11
5	Students	9
6	Technology	8
7	Knowledge	7
8	Internet	7
9	Participation	6
10	Collaborative learning	6

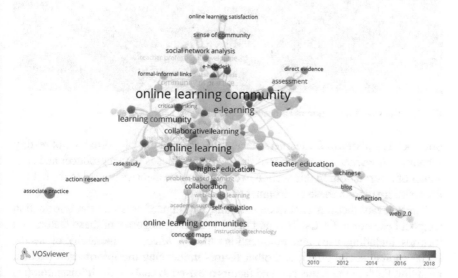

Fig. 8.3 The network of keywords belonging to the published articles

8.3   Methods

In this section, we introduce fundamental theories and factors behind the participation, and then we present the proposed method.

8.3.1   Fundamental Theories and Factors Behind the Participation

Participation of members (learners) in online learning communities is the key to the success of such environments. Members have an important role in creating educational materials and transferring them to the OLC environment. Such communities

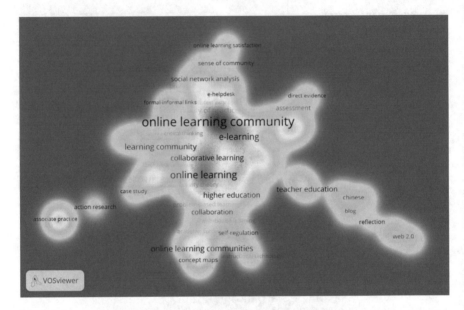

Fig. 8.4 Density visualization of co-occurrence of keyword network

often act as a platform for learners, except that roles in this platform are not readily separable. A learner as a member of these communities is both a producer and consumer of information. In this section, we will examine theories related to the field of user participation in these environments.

The relevant theories should be carefully investigated to study the information (re)post behavior in OLCs. This part deals with five categories of these factors and theories including uses and gratifications theory, *Maslow*'s hierarchy of needs, social capital, social effect, and other factors influencing the information (re)post behavior. Each of these theories and factors are used to analyze the information (re) post behavior in the OLCs.

The uses and gratifications theory (UTG) is an approach to understanding why and how people are in touch with the media to meet their needs. UGT is an audience-oriented approach to understanding social communication (Cheung, Chiu, & Lee, 2011). The relationship between audiences and the media is bilateral. Unlike the theories pertaining to the effects of the media on audiences, which investigate the effect of the performances of the media on people, UGT focuses on the effect of people's behaviors and performances on the media (Ruggiero, 2000). The main question of this theory is why people use a special medium such as an online community out of the mass media and what they use it for. In this theory, people act freely to select the media based on goals by default, and they are completely aware of their needs. The gratification of needs depends on the type of the online community. Many studies have been conducted to investigate this theory in online communities such as *Facebook* and *Wikipedia*. According to Petrič and Petrovčič (2014), these needs, which users participate in online communities to meet, include purposive values, self-discovery, entertainment values, social enhancement, and maintaining interpersonal interconnectivity. According to Nonnecke and Preece (2001), eight incentives have been introduced to use Twitter. They are self-documentation, information sharing, social interactions, entertainment, pastime, self-expression, media charm, and convenience. Therefore, these incentives are formed as expectations before use in users' minds, and users continue using an online community until there is a positive distance between the performance of such community and these expectations. User gratification is provided until individuals' expectations have a positive distance from what they acquire (Cheung et al., 2011).

In addition to gratification, content satisfaction is among the variables of learners satisfaction in the (re)post of information in OLCs, a fact which has been dealt with in this paper. Content satisfaction is based on the level of the content values of received messages (from a recipient's perspective) via an OLC. Content satisfaction leads to the purposive use of the media as educational and communicational tools. If the informational content value of messages received by learners is higher than their expectations before use, the information (re)post behaviors are more probable.

Social satisfaction is based on the relationships among users via the media. Social satisfaction is directly related to the degree of communications and exchanges among the members of an online community. For instance, *Twitter* has set up different features such as follow-up, favorite list, and repost for this purpose in order to maintain the communications among users. According to Cheung et al. (2011), the groups with many interactions among their users are more successful in satisfying their users who are more encouraged to (re)post information constantly in such groups.

The Maslow's hierarchy of needs theory was first introduced by Maslow in Personality and Motivation (Maslow, 1954). It is the foundation of many important theories in areas such as development psychology and educational management. It is one of the content theories regarding motivation. Content theories explain the quiddity of stimulated behaviors. In fact, they mainly deal with what goes on inside an individual or his surroundings and empowers his behavior. Based on Abraham Maslow's famous theory, every human has five basic needs to grow and flourish. These basic needs have been gathered together in a hierarchy. It includes a pyramid of physiological needs (required for survival), security needs (social, alimentary, occupational, and financial security), social needs (belonging to society, family, and friends), esteem needs (self-esteem, self-efficacy, success, respect from others, and self-confidence), and self-actualization needs (morality, creativity, self-stimulation, problem-solving, impartiality, and the acceptance of facts). Esteem needs are among the factors influencing the information (re)post behavior in the OLCs. For instance, a lurker may imagine he or she will not receive an appropriate feedback from others if he or she (re)posts content. The other side of this problem is also true about posters. In some cases, a user does not feel efficient enough to disseminate a post. Self-concept influences the information (re)post behavior in the OLCs. Self-efficacy has been derived from Albert Bandura's Social Cognition Theory in 1997, which refers to an individual's beliefs and judgments about his abilities to perform activities and responsibilities. On the contrary, self-esteem refers to a feeling of respect which a person has for his or her values. A person who has positive feelings for his or her values will experience a high level of self-esteem. These two concepts can be completely interpreted in specific areas. For instance, an individual may not have self-esteem in the apparel attraction; however, he or she may have a high level of self-esteem in areas related to his or her specialty. This is also true about self-efficacy. Regarding the information (re)post behavior, self-efficacy refers to an extent of reliability which an individual has to think that the post which he or she (re)posts is valuable. Previous studies indicate that self-efficacy acts as an incentive to have a positive impact on the (re)post of information in online communities. Regarding self-esteem, a learner can reach a feeling of self-esteem that he or she has an important role in an OLC and can share his or her comments with others through the (re)post of information (Alqurashi, 2016).

Social capital refers to a capacity (social quintessence) which facilitates coordination and cooperation in a community or group. In fact, social capital can be thought of as a compound concept having three dimensions such as structure, content, and functionality. Social structure is the network of social communications (Munar & Jacobsen, 2014). Social content refers to social trust and norms. Finally, social functionality is the social reaction. Many definitions have been provided for this theory, however, it is generally agreed that this capital is the effect and result of communications among different people in an OLC. According to this theory, every individual is able to use the resources existing in another individual belonging to the OLC. These resources may include useful information, personal relationships, or the capacity of an organized group.

The measurement method of social capitals individually and collectively is one of the important challenges of this area. The role of strong and weak ties in the social capital of a group is a factor influencing this area. Bonding (strong ties) and bridging (weak ties) are two important mechanisms for developing social capital. Bonding refers to the close and integrated ties among the members of homogenous groups. These emotional ties are so strong that they are usually formed among close friends and relatives. These are the strongest forms of social capital. This form of communication allows the members of these groups to support their needs in a group or network. In Hiroshi's opinion, there are four elements for a social tie: attachment, commitment, involvement, and belief. Learners expect to use other learners' information mutually in return for the participation in online communities and information (re)post (Yuan & Gay, 2006). These relationships are called mutual relationships. The more social ties there are among different groups in an OLC, the more mutual communications will emerge.

On the other hand, bridging refers to the communications among the heterogeneous members of groups in a community. It allows the members to acquire and share new information and ideas. It also broadens the radius of trust and participation. Bridge is a kind of social tie which enables two different groups to exchange information in an OLC. These bridges are very important in the flow of information, mutual understanding, and social capital, as a result.

Factors like time, topics rose in OLCs, and considerations for the technology of designing OLC can influence the (re)post behavior of learners. The (re)post of information and involvement in the topics of OLCs are time-consuming. Time is simple factor which is very important. According to Nonnecke and Preece (2001), the interactions of users increase significantly when there is no time limit. The studies conducted by Nonnecke and Preece (2000) indicated that pursuing online social topics was not a high priority in 2000; however, this may be different nowadays.

Online communities can be classified under the topics raised in them. Previous studies indicate that the type of topics in OLCs would greatly influence the information (re)post behavior of learners and the percentage of lurkers. For instance, the number of lurkers is usually smaller in online communities of the health sector in comparison with those of software support. In fact, it is because of the empathy and commitment (stronger social ties) among members in health online communities. According to X. Yang, Li, and Huang (2017), users are more willing to (re)post information in social networks than in specialized online communities.

Technology considerations can act as an important factor in the information (re) post behavior of learners and the creation of lurkers. The problems of designing an online community, difficulty of using it, paying insufficient attention to the privacy, and security of users can influence the lack of information (re)post by users in online communities. According to Cranefield, Yoong, and Huff (2011), almost 18% of lurkers stated the difficulty of designing and creating a post online as the main reason for the lack of participation in online communities. Moreover, given the difficulty of controlling and using the personal information of individuals in online communities, the privacy of users is usually violated. The fact that users can have anonymity may improve their privacy; however, this cannot be guaranteed in online communities. In fact, it is usually possible to violate the privacy and security of users in online communities.

8.4 Proposed Methods

In the previous section, we examine the various theories about learner participation and factors that influence this behavior. Participation of learners in online educational communities is made in the form of creating a post, commenting on a post, liking (or rating) a post, or reposting (sharing) a post. Reposting a post plays an important role in the transmission of information, the epidemic of information, and the increasing role of its information dissemination. This became more important in the learning environments based on social networks.

In this section, we will evaluate the proposed method to increase user participation in the transmission of information. In the studies, both groups of learners (Lurkers & Posters) are evaluated. The two main research questions in this section are as follows:

Research question 1: Is it possible to identify effective posts using the proposed mathematical method. Effective posts are the contents that have been selected based on the educational needs and learning background of learners.

Research question 2: Does the selection of effective content and the provision of more time for passive users (Which is shown in the algorithm of Table 8.6) increase the behavior of learner participation behavior?

8.4.1 Selection of k Effective Posts

In the previous sections, there were various theories and reasons for participating in OLCs. It is impossible to provide a comprehensive method for the participation of all users. In this chapter, according to the stated concepts, a method for improving participation in OLCs is proposed.

The proposed method needs no basic information on the learners because of easy-to-apply membership and using online communities. Only through the post content propagated in online communities does this method consider selecting effective posts for each learner. The learner's profile includes the learner's connections, posts created, or reposts from each learner. Due to the lack of a standard and diversity of implementation and design, in the proposed method the focus was only on the content assessment which is common in all kind of OLCs.

In order to choose the effective posts for each learner, this paper investigates three features: (1) topic similarity, (2) content similarity, and (3) learner similarity. According to the learner's profile, favorite topics of the learner can be identified by the posts they create. Learners tend to find out about the latest findings in their favorite topics. Matching posts with learners' topics of interest is very essential in OLCs. Posts that are selected in line with the learner's interests are more suitable for them. In the following section, a detailed explanation of the procedures is provided in order to measure these three features. This section explains selection method of k effective posts in OLCs and discusses its characteristics.

8.4.2 Problem Statement

Suppose that an OLC is implemented under a social network such as twitter. These groups are typically displayed as graphs of followers-followees. Learners follow other learners considering their interests and expertise. This directed graph is defined as $G = (V, E)$ where E represents the relationship between learners and V represents learners. Equation $(u, v) \in E$ shows that learner u follows learner v. If P represents the total of posts created in the whole OLC, then an online social event "post" occurs when learner u creates post p at $t \in T$ time, represented as $post(u,p,t)$. In the same manner, when learner v shares post p through learner u at time t', "reposting" occurs which can be represented as *repost* (v,u,p,t'). According to the definitions provided, the probability of repost can be defined as a function of the probability of repost $p : P \times V \times V \times T \rightarrow [1,0]$. In this function, T is the temporal domain. The probable repost of P posts by any learner from V within the temporal domain T includes the values between zero and one (zero for not posting and 1 for posting).

If σ is taken as the selection procedure of k effective posts from candidate posts (v,t) for learner v at any t time, the output $\sigma(v,t)$ is the k effective post to learner v at time t. The total candidate posts for learner v are those already created or reposted by learners set V followed by the learner v. Equation (8.1) shows the initial set of candidate posts for $t' < t$. From this stream of posts, duplicate posts already displayed for the learner in the previous time $t' < t$ should be removed.

$$Candid(v,t) = initial(c,t) - \{p \in P \mid post(v,p,t') \cup \alpha(v,t')\} \quad (8.1)$$

8.4.3 Theoretical Baseline

We have provided a heuristic method for the selection of k effective posts considering the computational capabilities and simplicity. This procedure is designed so as to be operational in online learning environments. In this section, the proposed method is introduced.

8.4.3.1 Topic Similarity

Suppose that each post and the learner have a topic vector. Post vector shows post topics. The vector of the user contains topics of interest to the user. Each topic includes a set of words $M_{topic\#} = \{w_1, w_2, \ldots, w_L\}$ for which the probability of the occurrence of L keywords is specified in the relevant topic. Then for each post p_i we define its characteristic vector with respect to categories in $M_{topic\#}$ where the membership is measured according to given threshold of its relevance. More precisely each component of the array post content topic pct_{pi} is determined as follows:

$$pct(p_i) = \left\langle pct_{w_1}, pct_{w_2}, \ldots, pct_{w_m} \right\rangle$$

$$pct_{w_i} = \begin{cases} 1 & \text{if } \exists \left\langle w_i, pr_{w_i} \right\rangle \in M_{topic\#} \ \& \ pr_{w_i \geq 0} \\ 0 & otherwise \end{cases} \quad (8.2)$$

In eq. (8.2), w_i, pr_{w_i} is a list of pairs, where the first component is the words related to the post content topic and the second one is its specific relevance degree. In the eq. (8.2), α is a fixed threshold that we set to 0.6 for example if $M_{topic1} = crypto$ $currency$, $M_{topic1} = cloud\ computing$, $M_{topic3} = Business\ Intelligence$. Let us consider the following post:

"Coinbase trading Altcoins#bitcoin"

The relevance of all post words with the topic is specified.

$$\langle \text{Coinbase}, 0.65 \rangle, \langle \text{altcoins}, 0.8 \rangle, \langle \text{bitcoin}, 0.8 \rangle$$

Then, since all of the extracted topics belongs to only one category among the fixed ones, i.e., "crypto currency," the resulting vector will contain the 1 value at index position corresponding to that category:

$$pcv = \langle 1, 0, 0 \rangle$$

For the user, we create these vectors in a similar way. To determine the similarity of the post topic, with topics of interest to the user, the Manhattan method is used according to eq. (8.3).

$$d_1(p,q) = \|p - q\|_1 = \sum_{i=1}^{n} |p_i - q_i| \tag{8.3}$$

8.4.3.2 Content Similarity

The lack of information required by learners and the lack of attention to learners' interests can be effective in non-participatory behavior. Learners tend to view posts that are similar to their interest. Therefore, in choosing posts that are effective, the similarity of post content with the posts created by the learner is considered. Each post consists of a set of words, signs, and symbols. To determine the similarity, firstly, it is necessary to perform the necessary pre-processing on the posts. Secondly, by using commonly used methods, the vector of each post is determined. To determine the similarity of a post like p with the posts collection of the learner "L" until the time t', first, the feature vectors of the user's posts must be calculated.

To do this, at first step, all the words must be extracted from all of the posts published (reposted) by the learner until time t $P_1 = \{P_{i,t}^u | l \in learners, i \in N, t \in time\}$, and at the second one, numerical values are assigned to the vector founded on the presence of the words in the text (Gou, Zhang, Chen, Kim, & Giles, 2010). One of the popular methods for this task is *tf. idf*, which transforms the text (the posts of the learners) into the feature vector based on the abundance of phrase and abundance reverse of document (De Boom, Van Canneyt, Bohez, Demeester, & Dhoedt, 2016). The calculating method of *tf. idf*$_{i,j}$ for word j in documentary i is shown in Eq. (8.4).

$$TF.IDF_{i,j} = TF_{i,j} \times IDF_j \frac{n_{i,j}}{\sum_{k=1}^{N} n_{i,k}} \log \left(\frac{N}{\left| \{i : t_j \in d_i\} \right| + 1} \right) \tag{8.4}$$

In equation (4) in the *tf* part, $n_{i,j}$ denotes the repetition number of the word j in post i and $\sum_{k=1}^{N} n_{i,k}$ is the total repetition of the words in post i. In the *idf* part, N is the total number of the posts and $|\{i : t_j \in d_i\}|$ is the number of the posts containing the word t_j. In order to solve zero denominator, it has been added with 1 (De Boom et al., 2016). The similarity between these two vectors is calculated using the angle cosine between them. The formula for calculating this criterion is shown in the equation (8.5).

$$\cos(d_1, d_2) = \frac{|d_p \cap d_q|}{|d_p| \times |d_q|} = \frac{d_p . d_q}{|d_p||d_q|} \tag{8.5}$$

Inner product expresses the angle between two vectors. If its value is equal to one, the angle will be zero and there will be maximum accordance between two vectors in terms of the similarity. If the result is zero, the angle between the vectors will be 90 degrees, so, they have minimum similarity.

8.4.3.3 Learner Similarity

People with different expertise share their posts on the network. It is very beneficial to find learners with common fields. For both learners $u, v \in V$, the degree of similarity is equal to the degree of similarity between the posts already created. The essential thing about sharing information in OLCs is to find people with the same level of information in addition to similar posts. For example, a learner who has created more than 100 posts about smartphone applications is different from someone who has just had a few posts or reposts in the same field. For either learner, the action vector can be defined (Equation 8.6). This vector contains n keywords created or reposted by the learner $v \in V$. Weighted cosine is used to determine the similarity between the vectors of learners. In this respect, the coefficients *i* (the number of keyword repeated by learner $v \in V$) are determined for *n* keywords till time *t*. Considering coefficients *i*, the level of learners' knowledge on a specific area is determined according to the number of posts made by them. The two learners are examined and taken into account for determining the similarity given the repetition of the keywords in the posts.

$$vector_u^t = i_1^t keyword_1 + \ldots + i_n^t keyword_n \tag{8.6}$$

The degree of similarity between learners **u** and **V** is equal to the value of cosine for the vectors of these two learners.

$$sim(u,v) = \cos(vector_u, vector_v, t) = \frac{vector_u^t . vector_v^t}{\left\| vector_u^t \right\| . \left\| vector_v^t \right\|} \tag{8.7}$$

In equation (8.7), $\left\|vector_v^t\right\|$ or/and $\left\|vector_u^t\right\|$ represent the value of action vector for learners v and u, respectively. If any of these values is zero, it means that the relevant learner has had no action (neither created nor reposted). In that case, the similarity between two learners is not defined. Accordingly, in collecting data, only those learners are taken into account that have at least created 10 posts or reposted. It should be noted that the action vector of a learner changes with time. In this respect, the action vector of learners at time T is used in each determination of similarity between two learners.

8.4.4   Using Logistic Regression to Predict the Probability of Reposting

Logistic regression is used to estimate the probability of post effectiveness for each learner. The binary logistic model is used to estimate the probability of a binary response (reposts happened or not) based on one or more features (content similarity, topic similarity, and learner similarity).

The goal of logistic regression in this problem is to find the best fitting model to describe the relationship between the effectiveness of the post (dependent variable, response or outcome variable) and a set of independent variables (3 features) using the sigmoid (logistics) function (Equation 8.8).

$$y_i = h\left(w^T x_i\right) = \frac{1}{1+\exp\left(-w^T x_i\right)} \tag{8.8}$$

In this equation ($y_i = 1$ for effective posts (repost happened) and $y_i = 0$ for otherwise), y_i is the prediction based on x_i inputs (value of each features). w^T is the vector of coefficients of each feature obtained from training data. In addition to classification, this method also has probable output. For example, $h_w(x_i) = 0.8$, means that the probability of post effectiveness is 80% for the sample ($h_w(x_i) = p(y_i = 1|x_i; w^T)$).

8.4.5   Probability of Post Effectiveness Based on Learner Type (Lurker and Poster)

In the proposed method, in addition to the structural characteristics of learners in the network, the attention is paid to the attitude of reposting among learners. Learners are divided into two categories of posters and lurkers based on their reposting attitude (Sloep & Kester, 2009). Lurkers are those who become a member of an online educational community but do not post, are only readers, and are not active (Nonnecke & Preece, 2000; Nonnecke & Preece, 2001). Considering this type of learners, the present study intends to reduce pure lurking behavior by setting the

Table 8.3 Repost probability determination algorithm according to the learner type (Lurkers or Posters)

00	If $t - t_u \leq \alpha \gamma_v$
01	$p\left(repost\left(v,u,p,t\right)\right) = \max\left(p_{u,v}^p, \varepsilon\right)$
02	If otherwise
03	$p(repost\,(v,u,p,t)) = \varepsilon$
04	End

time for online access to posts based on the importance of that post; that is, the posts which are more likely to be reposted by learners, are displayed longer for lurkers than for posters. The calculated probability of repost of post p of learner u by learner v at timestamp t (put formally $p(repost\,(v,u,p,t))$) is shown in Table 8.3. In algorithm (1), t_u is reposting time by learner u. t is the time of running the algorithm and γ_v is the average time interval between the posts of learner v. α is the adjusted coefficient of γ_v. At the zero line of this algorithm, by calculating the elapsed time since the repost of the posts by learner u, if time is less than the average time interval between the posts of learner $v(\gamma_v)$, there is still the probability that learner v has not seen post p. In this respect, the probability of reposting by learner v is equal to $\max\left(p_{u,v}^p, \varepsilon\right)$. Line 2 evaluates a condition in which the time elapsed since the repost of the posts by learner u is longer than the average time interval between the posts of learner v (γ_v). In this case, the learner had most likely seen the post but was unwilling to repost it. Considering this fact, it is least probable that learner v reposts the post p of learner v in timestamp $t(p(repost\,(v,u,p,t)))$. ε is equal to minimal value at 10^{-3}.

The coefficient of α provides more opportunity for lurking learners by adjusting the impact of γ_v. Normally, γ_v of lurking learners is far longer than γ_v of other learners. For ease of calculation, the coefficient α of γ_v is considered to be 1 and 2 for poster and lurker learners, respectively. In this respect, when a learner is lurking, he has more time to read and repost the posts with higher probability.

8.5 Data Collection

The data presented in this study were collected based on an unrestricted self-selected web survey (Fricker, 2008). Following the methodological procedure for posting surveys on discussion boards (Ip, Barnett, Tenerowicz, & Perry, 2010), a survey invitation was published in forum with the prior approval of the online community managers. The survey was piloted between Feb. 12 and May 12, 2017. The participants were selected from information technology and electrical engineering technophiles in popular Iranian online communities (online communities are similar to twitter environment and its members can follow other members, interested topics, etc.). Among them, only those learners are taken into accounts that had at least created 10 English posts or reposts and had more than 10 items in their list of followers and followees. In this method, qualified members were selected for a 10-day assessment period. A total of 5–20 posts a day were prepared for the volunteers' assessment. Volunteers participated in the assessment independently on a daily basis without knowing other users' opinions. These volunteers examined the posts to repost them and answered specific questions regarding each proposed feature of the posts (content similarity, topic similarity, and learner similarity). They ranked questions on features in the range of 1–5. In the 1–5 range of answers, 1 shows the lowest score of the feature in question, while 5 shows the highest score of that feature. Among the participants, lurkers are those who do not (re)post any posts publically in a certain period of time (Nonnecke & Preece, 2000) or those who rarely participate in the study (Sloep & Kester, 2009)'s. Like the method presented by Amichai-Hamburger et al. (2016), learners were asked about the last time they had (re)posted a post to identify lurkers. Based on their last (re)post in online communities, learners can select from more than 1 day, 1 month, 1 year, and never. In this study, learners who selected more than 1 year or never were considered to be lurkers.

8.6   Sample

Non-probability sampling method was used to select more than 840 participants. Participants in the survey are from social networking users in Iran who have announced that they have used online communities for educational purposes. In total, 740 participants participated in the study voluntarily and free of charge. Then, 680 participants were found acceptable. Therefore, 306 participants (45%) were selected as posters from 680 respondents, and 374 participants (55%) were selected as lurkers. The average age of lurkers was lower than posters; however, there was not a significant difference. According to statistics, 58% of respondents were male, and 42% were female. The majority of respondents were in 20–30 (34%) and 30–40 (41%) year old groups. The membership duration in online communities of 94% of respondents was more than 6 months; a fact which indicates the familiarity of most of them with the environment of an online community. Among them, 63% had academic educations (21% with master's and PhD degrees) and 37% did not have academic educations. Almost 79% of them were married, and the rest were single.

8.7   Analysis

Evaluating methodology was consisted of three phases. In the first phase, for each of the participants, k posts (as the learner wished) were displayed on a daily basis. Learners decided about the post's reposting when they observed them and they also determined when to perform the test during the day. In the second phase, according to their repost decision (previous phase), k effective and proper posts were displayed and like the first phase, learners decided about the post's reposting. In this phase, the prediction of reposting behavior was assessed. Three well-known measures, namely precision, recall, and F measure, were used to assess the predictions (Tang, Miao, Quan, Tang, & Deng, 2015; Zhang et al., 2012). These measures are used for prediction problems of binary class (posting or not reposting). Equations 8.9 to 8.11 show these measures.

$$Precision = \frac{\left|\{Predicted\ RT\} \cap \{True\ RT\}\right|}{\left|\{Predicted\ RT\}\right|} \tag{8.9}$$

$$Recall = \frac{\left|\{Predicted\ RT\} \cap \{True\ RT\}\right|}{\left|\{True\ RT\}\right|} \tag{8.10}$$

$$F - measure = 2.\frac{Precision.Recall}{Precision + Recall} \tag{8.11}$$

To evaluate the solutions for estimated features, Kendall's rank correlation coefficient was used to compare the measured values of the features (content, topic, learner similarity) with the user-expressed values (range, 1–5). This method is employed to obtain the correlation between the ranks of two quantities. If y_i is the value expressed for each feature by the user, and x_i is the proposed measured value by the proposed solution and (x_1, y_1), (x_2, y_2), …, (x_n, y_n), then the values obtained for all of the n post are considered. Each (x_i, y_i) and (x_j, y_j) pair $(i \neq j)$ is known as a concordant pair if $x_i > x_j$ then $y_i > y_j$ or if $x_i < x_j$ then $y_i < y_j$. The pair is discordant if $x_i > x_j$ then $y_i < y_j$ or if $x_i < x_j$ then $y_i > y_j$. If the two quantities are equal, the pair is neither concordant nor discordant. Kendall's correlation coefficient is calculated by equation (8.12), and varies between 1 and −1 ($-1 \leq \tau \leq 1$). If the rank of a post based on the aforementioned three features is close to its rank based on user opinions (in the questionnaire), this coefficient will be closer to 1. If the ranks differ completely, this coefficient will be closer to −1, whereas 0 shows independence of the two values.

$$\tau = \frac{2\big(Number\ of\ concordant\ pairs - Number\ of\ discordant\ pairs\big)}{n(n-1)} \tag{8.12}$$

The correlation of these features with repost behavior is studied in this paper. If the values of these features are significantly related to repost behavior, the relationship can be measured based on conventional learning. To this end, a method similar to Pearson's correlation method is employed. Since each feature has continuous values and repost behavior is a binary variable (0 or 1), Pearson's method cannot be used. Point-biserial method is utilized for such problems. If the values of each feature of each post are a continuous variable (x), and repost behavior for the post is a binary variable (y), then the point-biserial correlation coefficient is based on equation (8.13), where M_1 shows mean feature value for posts, leading to repost behavior

$y = 1$. Similarly, M_0 is the mean of features of posts not leading to repost behavior $y = 0$. Moreover, n_1 shows the number of posts in the samples except for posts leading to repost behavior ($y = 1$), and n_0 denotes the number of posts in samples which do not result in repost behavior ($y = 0$). In addition, n is the total number of samples examined, and s_n is the standard deviation for values of features of all studied post samples. This correlation coefficient varies between -1 and 1, where 1 shows maximum positive correlation between measured values and user behavior and -1 shows maximum negative correlation between measured values and user behavior. Zero (0) also shows independence of the features from user repost behavior.

$$r_{pb} = \left(\frac{M_1 - M_0}{S_n} \right) \sqrt{\frac{n_1 n_0}{n^2}} \tag{8.13}$$

In the last phase, to evaluate the increasing of information repost behavior in OLC, 5 more methods other than the proposed one were also employed: (1) latest post method, based on the time a post was created and learners visited the latest k post; (2) random k post selection method for a learner; (3) content similarity method, k posts with the highest similarity to the learner's posts; (4) topic similarity method, k post selection method with the highest measured degree of topic similarity; and (5) learner similarity method, k post selection method with the highest similarity of the posts with the learner. Furthermore, learners were categorized into lurker and posters. Each of these categories is randomly divided into six categories. During the 10-day evaluation, each of these six categories was received k post alternatively without advanced notice based on one of these six methods. Equation (8.14) shows the increasing reposting behavior evaluation metric.

$$Repost\ rate = \frac{Number\ of\ reposts}{Number\ of\ visited\ posts} \times 100 \tag{8.14}$$

8.8 Results

Table 8.4 shows the evaluation metrics of the proposed repost behavior prediction using logistic regression for lurkers and the posters. The results showed that the proposed method had precisely predicted the learner's reposting behavior. In order to assess the accuracy of choosing k effective post, common criteria were used for estimation evaluation.

Table 8.5 presents the results of analysis of Kendall's correlation coefficient for all posts in the set. Figures 8.5, 8.6, and 8.7 indicate the scatter plot of the estimated content, topic, learner similarities ([0,1]) and observed content, topic, learner similarities (range 1–5).

Table 8.6 presents the results of correlation coefficient between features and repost behaviors. These results are reflective of a positive correlation between

Table 8.4 Repost behavior prediction evaluation

	Posters			Lurkers		
	F- measure	Recall	Precision	F- measure	Recall	Precision
Proposed repost prediction	0.643	0.541	0.793	0.554	0.477	0.662

Table 8.5 Kendall's tau correlation coefficient

	Learner similarity	Topic similarity	Content similarity
τ	0.52	0.73	0.68

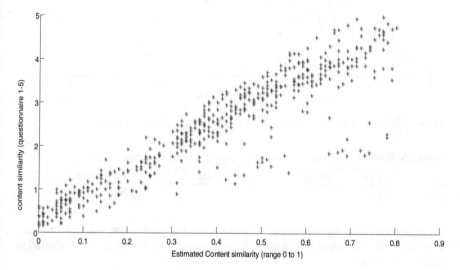

Fig. 8.5 Estimated content similarity (range 0–1) vs. content similarity (questionnaire 1–5)

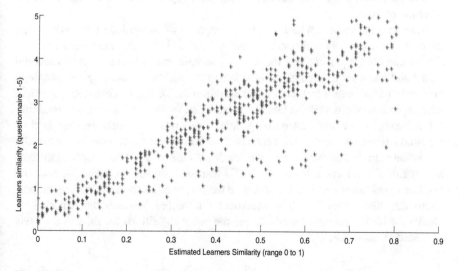

Fig. 8.6 Estimated learner similarity (range 0–1) vs. learner similarity (questionnaire 1–5)

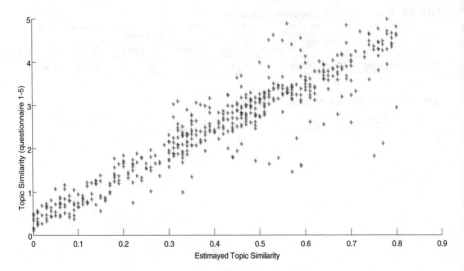

Fig. 8.7 Estimated topic similarity (range, 0 to 1) vs. topic similarity (questionnaire 1–5)

Table 8.6 Coefficient of correlation between proposed features and repost behavior

	Learner similarity	Topic similarity	Content similarity
r_{pb}	0.51	0.61	0.54

these features and repost behavior. The maximum correlation belonged to content similarity and topic similarity, whereas the minimum correlation belonged to learner similarity.

According to Table 8.7, using the proposed method the average of repost rate for each of these two categories, namely, lurkers and posters, was more than those of 5 other methods.

Also, in order to have a better comparison, a set of one-tailed t-tests were taken on ten-fold cross-validation results. Cross-validation is a technique to evaluate predictive models by partitioning the original sample into a training set to train the model and a test set to evaluate it. In k-fold cross-validation, the original sample is randomly partitioned into k equal size subsamples. Of the k subsamples, a single subsample is retained as the validation data for testing the model, and the remaining k-1 subsamples are used as training data. The cross-validation process is then repeated k times (the folds), with each of the k subsamples used exactly once as the validation data. In this experiment, we considered the result of the proposed method as algorithm A2. Hence, if the result of t-test between the proposed method and another algorithm is a negative value, then the proposed method statistically outperforms the other algorithm. The statistical differences between the experiments obtained with the proposed method, and the other algorithms for the test data sets are listed in Table 8.8.

Table 8.7 Repost rate for each of methods

Methods	Posters	Lurker
	Repost rate	Repost rate
Random	10	4
Recent	7	3
Content similarity	36	21
Topic similarity	28	17
Learner similarity	24	20
Proposed method	42	34

Table 8.8 The result of statistical t-test

	Lurker	Poster
Random	−5.342	−3.924
Recent	−34.34	−79.402
Content similarity	−16.629	−21.918
Topic similarity	−13.311	−6.456
Learner similarity	−31.034	−27.713

8.9   Discussion

Generally, the topic of increasing the repost of information has been rarely studied in education-related studies despite its great importance in social networks. In this study, in order to answer research question 1, prediction of the effectiveness of the post from the content assessment of candidate posts was emphasized regarding the variety of OLCs and their various capabilities. Features related to content similarity, topic similarity, and learner similarity were considered for the prediction of post effectiveness. In research regarding the correlation of each feature, a comprehensive assessment was done and the results of the direct effect of these features on learners' behaviors are shown in Table 8.6. The results of this study were in line with the results found in various previous research studies such as (J. Lee & Lee, 2009; Sarma & Panigrahy, 2010; W. Wang, Duan, Koul, & Sheth, 2014). In addition, the measurement methods of the presented features (content similarity, topic similarity, learner similarity) of the study were compared through the explicit feedbacks of learners in the form of a questionnaire, and their accuracies were also assessed.

The role of the proposed method in increasing repost behavior and consequently online participation (the answer to research question 2) was assessed and confirmed for both lurker and poster learners.

Regarding research question 2, both categories of learners had positive repost rates through the selection of effective posts and the display method of this study. The increase in reposts leads to an increase in the participation of learners in online society and sharing of more information. The significance of the improvement through the suggested method compared with other methods is proven in Table 8.8.

The most obvious limitation of this study was the lack of a specific standard in the implementation and design of OLCs. Therefore, the suggested method was designed based on the content of posts to make it applicable in various kinds of OLCs. Other factors such as external events (such as trends and news), context of OLC (such as health forums or technical forums), security concerns, and OLC design problems can be effective in learners' repost behavior. The investigation of these factors is ignored due to the lack of a specific standard and high-quality data. Considering at least two kinds of OLCs, future studies can investigate more effective factors in learners' repost behavior to increase the accuracy of the suggested method. In addition, the display of effective posts under adaptive navigation can be presented for learners based on an educational goal.

8.10   Conclusion

OLCs are public or private groups on the Internet that address the learning needs of their members by facilitating asynchronous learning. This paper seeks to solve the two fundamental linked challenges being access to the appropriate information and shortage of participation (because of lurking behavior or low participation of some learner) in these communities by selecting k effective posts for learner v among all candidate posts. These posts should be selected based on learner interests, characteristics, and knowledge levels to motivate the learner to repost it. This reposting behavior facilitates information sharing and increases participation in online communities. For online implementation, the proposed method is suitable to be used in all kinds of OLCs. Certain features such as content similarity, topic similarity, and learner similarity are among the major parameters taken into account in the selection of k effective posts. The proposed method focuses on the problems of online communities that are lurking learners. Comprehensive evaluations indicate that the proposed method significantly outperformed the existing methods.

Acknowledgments This research was in part supported by a grant from IPM (No. CS1398-4-222).

References

Adar, E., & Adamic, L. A. (2005). Tracking information epidemics in blogspace. *Proceedings - 2005 IEEE/WIC/ACM International Conference on Web Intelligence, WI 2005, 2005*, 207–214. https://doi.org/10.1109/WI.2005.151

Ahmed, Y. A., Ahmad, M. N., Ahmad, N., & Zakaria, N. H. (2019). Social media for knowledge-sharing: A systematic literature review. *Telematics and Informatics, 37*, 72–112. https://doi.org/10.1016/j.tele.2018.01.015

Alqurashi, E. (2016). Self-efficacy in online learning environments: A literature review. *Contemporary Issues in Education Research (Online), 9*(1), 45.

Amato, F., Castiglione, A., Moscato, V., Picariello, A., & Sperlì, G. (2017). *Detection of lurkers in online social networks* (pp. 1–15). https://doi.org/10.1007/978-3-319-69471-9_1

Amato, F., Moscato, V., Picariello, A., Piccialli, F., & Sperlí, G. (2018). Centrality in heterogeneous social networks for lurkers detection: An approach based on hypergraphs. *Concurrency and Computation: Practice and Experience, 30*(3), e4188. https://doi.org/10.1002/cpe.4188

Amichai-Hamburger, Y., Gazit, T., Bar-Ilan, J., Perez, O., Aharony, N., Bronstein, J., et al. (2016). Psychological factors behind the lack of participation in online discussions. *Computers in Human Behavior, 55*, 268–277. https://doi.org/10.1016/j.chb.2015.09.009

Baumer, E. P. S., Adams, P., Khovanskaya, V. D., Liao, T. C., Smith, M. E., Schwanda Sosik, V., et al. (2013). Limiting, leaving, and (re)lapsing. In *Proceedings of the SIGCHI Conference on Human Factors in Computing Systems - CHI '13* (p. 3257). New York, NY: ACM Press. https://doi.org/10.1145/2470654.2466446

Cheung, C. M. K., Chiu, P.-Y., & Lee, M. K. O. (2011). Online social networks: Why do students use facebook? *Computers in Human Behavior, 27*(4), 1337–1343. https://doi.org/10.1016/j.chb.2010.07.028

Chris Brown, C. L. P. (2018). *Networks for learning effective collaboration for teacher, school and system improvement.* Routledge.

Chu, M., & Meulemans, Y. N. (2008). The problems and potential of MySpace and Facebook usage in academic libraries. *Internet Reference Services Quarterly, 13*(1), 69–85. https://doi.org/10.1300/J136v13n01-04

Cranefield, J., Yoong, P., & Huff, S. (2011). Beyond lurking: The invisible follower-feeder in an online community ecosystem. *PACIS.* Retrieved from http://citeseerx.ist.psu.edu/viewdoc/download?doi=10.1.1.232.3108&rep=rep1&type=pdf

Dabbagh, N., & Kitsantas, A. (2012). Personal learning environments, social media, and self-regulated learning: A natural formula for connecting formal and informal learning. *The Internet and Higher Education, 15*, 3–8. https://doi.org/10.1016/j.iheduc.2011.06.002

Datu, J. A. D., Yang, W., Valdez, J. P. M., & Chu, S. K. W. (2018). Is facebook involvement associated with academic engagement among Filipino university students? A cross-sectional study. *Computers & Education, 125*, 246–253. https://doi.org/10.1016/j.compedu.2018.06.010

De Boom, C., Van Canneyt, S., Bohez, S., Demeester, T., & Dhoedt, B. (2016). Learning semantic similarity for very short texts. In *Proceedings - 15th IEEE International Conference on Data Mining Workshop, ICDMW 2015.* https://doi.org/10.1109/ICDMW.2015.86

Deng, L., & Tavares, N. J. (2013). From Moodle to Facebook: Exploring students' motivation and experiences in online communities. *Computers and Education, 68*, 167–176. https://doi.org/10.1016/j.compedu.2013.04.028

Duggan, M., Ellison, N. B., Lampe, C., Lenhart, A., & Madden, M. (2015). Social media update 2014. *Pew Research Center, 13*(January), 18. https://doi.org/10.1111/j.1083-6101.2007.00393.x

Feng, S., Wong, Y. K., Wong, L. Y., & Hossain, L. (2019). The internet and Facebook usage on academic distraction of college students. *Computers & Education, 134*, 41–49. https://doi.org/10.1016/j.compedu.2019.02.005

Fricker, R. D. (2008). Sampling methods for web and E-mail surveys. In *The SAGE handbook of online research methods* (pp. 195–217). London, UK: SAGE Publications. https://doi.org/10.4135/9780857020055

Gou, L., Zhang, X. (Luke), Chen, H. -H., Kim, J. -H., & Giles, C. L. (2010). Social network document ranking. *Proceedings of the 10th Annual Joint Conference on Digital Libraries - JCDL '10* (pp. 313–322). https://doi.org/10.1145/1816123.1816170

Hamid, S., Bukhari, S., Ravana, S. D., Norman, A. A., & Ijab, M. T. (2016). Role of social media in information-seeking behaviour among international students. *Aslib Journal of Information Management, 68*(5), 643–666. https://doi.org/10.1108/AJIM-03-2016-0031

Haythornthwaite, C. (2019). Learning, connectivity and networks. *Information and Learning Sciences, 120*(1/2), 19–38. https://doi.org/10.1108/ILS-06-2018-0052

Hurtubise, K., Pratte, G., Rivard, L., Berbari, J., Héguy, L., & Camden, C. (2017). Exploring engagement in a virtual community of practice in pediatric rehabilitation: Who are non-users, lurkers, and posters? *Disability and Rehabilitation, 41*, 1–8. https://doi.org/10.1080/09638288.2017.1416496

Ip, E. J., Barnett, M. J., Tenerowicz, M. J., & Perry, P. J. (2010). The touro 12-step: A systematic guide to optimizing survey research with online discussion boards. *Journal of Medical Internet Research., 12*, e16. https://doi.org/10.2196/jmir.1314

Joyce, E., & Kraut, R. E. (2006). Predicting continued participation in newsgroups. *Journal of Computer-Mediated Communication, 11*(3), 723–747. https://doi.org/10.1111/j.1083-6101.2006.00033.x

Junco, R. (2012). The relationship between frequency of Facebook use, participation in Facebook activities, and student engagement. *Computers and Education, 58*(1), 162–171. https://doi.org/10.1016/j.compedu.2011.08.004

Karaoglan Yilmaz, F. G. (2017). Social presence and transactional distance as an antecedent to knowledge sharing in virtual learning communities. *Journal of Educational Computing Research, 55*(6), 844–864. https://doi.org/10.1177/0735633116688319

Kate Merry, S., & Simon, A. (2012). Living and lurking on LiveJournal. *ASLIB Proceedings, 64*(3), 241–261. https://doi.org/10.1108/00012531211244527

Ke, F., & Hoadley, C. (2009). Evaluating online learning communities. *Educational Technology Research and Development, 57*(4), 487–510. https://doi.org/10.1007/s11423-009-9120-2

Kilner, P. G., & Hoadley, C. M. (2005). Anonymity options and professional participation in an online community of practice. *Proceedings of the 2005 Conference on Computer Support for Collaborative Learning* (pp. 272–280). https://doi.org/10.3115/1149293.1149328

Kollock, P., & Smith, M. (1996). Managing the virtual commons: Cooperation and conflict in computer communities. In *Computer-mediated communication: Linguistic, social, and cross-cultural perspectives* (pp. 109–128). Amsterdam, The Netherlands: John Benjamins. https://doi.org/10.1075/pbns.39.10kol

Lambić, D. (2016). Correlation between Facebook use for educational purposes and academic performance of students. *Computers in Human Behavior, 61*, 313–320. https://doi.org/10.1016/j.chb.2016.03.052

Lee, J., & Lee, J. (2009). Ranking user-created contents by search user's inclination in online communities. In *Search* (pp. 1215–1216). Spain: Madrid. https://doi.org/10.1145/1526709.1526935

Lee, Y. -L., Chen, F. -C., & Jiang, H. -M. (2006). Lurking as participation: A community perspective on lurkers' identity and negotiability. In *Proceedings of the 7th International Conference on Learning Sciences* (pp. 404–410). USA: Bloomington, Indiana.

Leskovec, J., Adamic, L., & Huberman, B. (2007). The dynamics of viral marketing. *ACM Transactions on the Web, 1*(1), 1–39. https://doi.org/10.1145/1232722.1232727

Luo, N., Zhang, M., & Qi, D. (2017). Effects of different interactions on students' sense of community in e-learning environment. *Computers & Education, 115*, 153–160. https://doi.org/10.1016/j.compedu.2017.08.006

Malinen, S. (2015). Understanding user participation in online communities: A systematic literature review of empirical studies. *Computers in Human Behavior, 46*, 228–238. https://doi.org/10.1016/j.chb.2015.01.004

Maslow, A. H. (1954). *Motivation and personality.* Oxford, UK: Harpers.

Mazman, S. G., & Usluel, Y. K. (2010). Modeling educational usage of Facebook. *Computers & Education, 55*(2), 444–453. https://doi.org/10.1016/j.compedu.2010.02.008

Munar, A. M., & Jacobsen, J. K. S. (2014). Motivations for sharing tourism experiences through social media. *Tourism Management, 43*, 46–54. https://doi.org/10.1016/j.tourman.2014.01.012

Nonnecke, B., Andrews, D., & Preece, J. (2006). Non-public and public online community participation: Needs, attitudes and behavior. *Electronic Commerce Research, 6*(1), 7–20. https://doi.org/10.1007/s10660-006-5985-x

Nonnecke, B., & Preece, J. (2000). Lurker demographics: Counting the silent. *Proceedings of the SIGCHI Conference on ..., 2*(1), 1–8. https://doi.org/10.1145/332040.332409

Nonnecke, B., & Preece, J. (2001). Why lurkers lurk. *AMCIS 2001 Proceedings* (pp. 1–10). Retrieved from http://aisel.aisnet.org/cgi/viewcontent.cgi?article=1733&context=amcis2001

Owusu, G. M. Y., Bekoe, R. A., Otoo, D. S., & Koli, A. P. E. (2019). Adoption of social networking sites for educational use. *Journal of Applied Research in Higher Education, 11*(1), 2–19. https://doi.org/10.1108/JARHE-04-2018-0069

Petrič, G., & Petrovčič, A. (2014). Elements of the management of norms and their effects on the sense of virtual community. *Online Information Review, 38*(3), 436–454. https://doi.org/10.1108/OIR-04-2013-0083

Rehm, M., Mulder, R. H., Gijselaers, W., & Segers, M. (2016). The impact of hierarchical positions on the type of communication within online communities of learning. *Computers in Human Behavior, 58*, 158–170. https://doi.org/10.1016/j.chb.2015.12.065

Rezvanian, A., Moradabadi, B., Ghavipour, M., Daliri Khomami, M. M., & Meybodi, M. R. (2019). *Learning automata approach for social networks* (Vol. 820). Cham, Switzerland: Springer. https://doi.org/10.1007/978-3-030-10767-3

Ridings, C. M., & Gefen, D. (2006). Virtual community attraction: Why people hang out online. *Journal of Computer-Mediated Communication, 10*(1), 00–00. https://doi.org/10.1111/j.1083-6101.2004.tb00229.x

Ruggiero, T. E. (2000). Uses and gratifications theory in the 21st century. *Mass Communication & Society, 3*(1), 3–37. https://doi.org/10.1207/S15327825MCS0301_02

Sarma, A. D., & Panigrahy, R. (2010). Ranking mechanisms in twitter-like forums categories and subject descriptors. *Human Factors, 1*, 21–30. https://doi.org/10.1145/1718487.1718491

Schneider, A., Von Krogh, G., & Jäger, P. (2013). What's coming next? Epistemic curiosity and lurking behavior in online communities. *Computers in Human Behavior, 29*(1), 293–303. https://doi.org/10.1016/j.chb.2012.09.008

Shafipour, R., Baten, R. A., Hasan, M. K., Ghoshal, G., Mateos, G., & Hoque, M. E. (2018). Buildup of speaking skills in an online learning community: A network-analytic exploration. *Palgrave Communications, 4*(1), 63. https://doi.org/10.1057/s41599-018-0116-6

Sharma, P., & Land, S. (2018). Patterns of knowledge sharing in an online affinity space for diabetes. *Educational Technology Research and Development., 67*, 247. https://doi.org/10.1007/s11423-018-9609-7

Shi, Z., Rui, H., & Whinston, A. B. (2014). Content sharing in a social broadcasting environment: Evidence from Twitter. *Mis Quarterly, 38*(1), 123–142. https://doi.org/10.2139/ssrn.2341243

Sloep, P. B., & Kester, L. (2009). From lurker to active participant. In *Learning network services for professional development* (pp. 17–26). Berlin, Germany: Springer. https://doi.org/10.1007/978-3-642-00978-5

Statista. (2019). *Number of monthly active Facebook users worldwide as of 4th quarter 2018.* Statista.Com.

Sun, N., Rau, P. P. L., & Ma, L. (2014). Understanding lurkers in online communities: A literature review. *Computers in Human Behavior, 38*, 110–117. https://doi.org/10.1016/j.chb.2014.05.022

Tagarelli, A., & Interdonato, R. (2018). *Mining lurkers in online social networks: Principles, models, and computational methods.* Cham, Switzerland: Springer. https://doi.org/10.1007/978-3-030-00229-9.

Takahashi, M., Fujimoto, M., & Yamasaki, N. (2003). The active lurker: Influence of an in-house online community on its outside environment. *Group*, 1–10. https://doi.org/10.1145/958160.958162

Tang, X., Miao, Q., Quan, Y., Tang, J., & Deng, K. (2015). Predicting individual retweet behavior by user similarity: A multi-task learning approach. *Knowledge-Based Systems, 89*, 681–688. https://doi.org/10.1016/j.knosys.2015.09.008

Topping, K. J. (2005). Trends in peer learning. *Educational Psychology, 25*(6), 631–645. https://doi.org/10.1080/01443410500345172

van Eck, N. J., & Waltman, L. (2010). Software survey: VOSviewer, a computer program for bibliometric mapping. *Scientometrics, 84*(2), 523–538. https://doi.org/10.1007/s11192-009-0146-3

Wagner, R. (2011). Social media tools for teaching and learning. *Athletic Training Education Journal, 6*(1), 51–52. Retrieved from http://csaweb109v.csa.com.ezproxy.lib.vt.edu:8080/ids70/view_record.php?id=4&recnum=41&log=from_res&SID=75pketf4eo60ftu6gurarhq7h4

Wang, Q., Woo, H. L., Quek, C. L., Yang, Y., & Liu, M. (2011). Using the Facebook group as a learning management system: An exploratory study. *British Journal of Educational Technology, 43*(3), 428–438. https://doi.org/10.1111/j.1467-8535.2011.01195.x

Wang, W., Duan, L., Koul, A., & Sheth, A. (2014). YouRank: Let user engagement rank microblog search results. In *8th Internationl AAAI Conference on Weblogs and Social Media*. Retrieved from http://knoesis.org/library/resource.php?id=1979

Welser, H. T., Khan, M. L., & Dickard, M. (2019). Digital remediation: Social support and online learning communities can help offset rural digital inequality. *Information, Communication & Society, 22*, 717–723. https://doi.org/10.1080/1369118X.2019.1566485

Williams, J. N., Van Patten, B., & Williams, J. (2004). Implicit learning of form-meaning connections. In *Form-Meaning Connections in Second Language Acquisition* (pp. 213–230). Routledge, USA: Mahwah, New Jersey.

Yang, J. C., Quadir, B., Chen, N.-S., & Miao, Q. (2016). Effects of online presence on learning performance in a blog-based online course. *The Internet and Higher Education, 30*, 11–20. https://doi.org/10.1016/j.iheduc.2016.04.002

Yang, X., Li, G., & Huang, S. S. (2017). Perceived online community support, member relations, and commitment: Differences between posters and lurkers. *Information & Management, 54*(2), 154–165. https://doi.org/10.1016/j.im.2016.05.003

Yeow, A., Johnson, S., & Faraj, S. (2006). Lurking: Legitimate or illegitimate peripheral? In *ICIS 2006* (p. 62). USA: Milwaukee.

Yilmaz, R. (2016). Knowledge sharing behaviors in e-learning community: Exploring the role of academic self-efficacy and sense of community. *Computers in Human Behavior, 63*, 373–382. https://doi.org/10.1016/j.chb.2016.05.055

Yuan, Y. C., & Gay, G. (2006). Homophily of network ties and bonding and bridging social capital in computer-mediated distributed teams. *Journal of Computer-Mediated Communication, 11*, 1062–1084. https://doi.org/10.1111/j.1083-6101.2006.00308.x

Zhang, H., Zhao, Q., Liu, H., Xiao, K., He, J., Du, X., & Chen, H. (2012). Predicting retweet behavior in Weibo social network. In *Lecture Notes in Computer Science (including subseries Lecture Notes in Artificial Intelligence and Lecture Notes in Bioinformatics)* (Vol. 7651 LNCS, pp. 737–743). https://doi.org/10.1007/978-3-642-35063-4_60

Chapter 9
Learning Spaces in Context-Aware Educational Networking Technologies in the Digital Age

Valéry Psyché, Ben K. Daniel, and Jacqueline Bourdeau

Abstract This chapter introduces the concept of learning space in the twenty-first century and considers the various contexts in which learning occurs. The increasing growth of digital educational networking technologies has contributed to the need to create and support various forms of learning spaces. These technologies have also transformed the way students engage and interact off- and online. Contemporary learning, as we know, is no longer limited to physical learning spaces. Instead, students engage in various learning spaces. Others extensively leverage opportunities afforded by social network platforms as tools for collaborative and self-directed learning. Students use social networking technologies to engage with content, connect with peers within their social networks and communicate with their teachers. With the increasing prevalence of these technologies, researchers and educators are provided with new opportunities to extend learning from physical spaces to virtual spaces and optimise pedagogical strategies that can support adaptive learning. Our goal in this chapter is to explore the extent to which understanding of learning spaces and contexts contribute to designing better student engagement and possibly better learning outcomes.

Keywords Learning space · Learning context · Context-aware educational technologies · Educational networking

V. Psyché (✉) · J. Bourdeau
TÉLUQ University, Montréal (Québec), Québec, Canada
e-mail: valery.psyche@teluq.ca; jacqueline.bourdeau@teluq.ca

B. K. Daniel
Otago University, Dunedin, New Zealand
e-mail: ben.daniel@otago.ac.nz

© Springer Nature Switzerland AG 2020
A. Peña-Ayala (ed.), *Educational Networking*, Lecture Notes in Social Networks, https://doi.org/10.1007/978-3-030-29973-6_9

Acronyms

ADDIE   Analysis-Design-Development-Implementation-Evaluation
ANR   National Research Agency
CSCL   Computer-supported collaborative learning
CAITS   Context-Aware Intelligent Tutoring System
CDL   Collaborative Distance Learning
DBR   Design-based research
FRQSC   Quebec Research Fund in Society and Culture
ITS   Intelligent tutoring system
LS   Learning spaces
MOOC   Massive open online course
TEEC   Technology Enhanced and Teaching in Context

9.1   Introduction

Many institutions of higher education are exploring innovative learning spaces to support effective engagement of students in their learning. Others have invested in institutional initiatives to redesign the physical presentation of their campuses, implement new technologies and introduce relevant pedagogical modalities.

The development of educational and social networking technologies has led to the emergence of various forms of learning spaces. These technologies have also transformed the way students engage and interact off- and online. Educational institutions continue to provide learning in physical learning spaces primarily. The impact of social networking technologies and other educational enterprise systems (e.g. learning management systems) suggests that learning is no longer limited to physical spaces (e.g. classrooms, libraries and lecture halls). Instead, students are provided with great flexibility in accessing and engaging with various learning resources available in different learning spaces and contexts. They can use repurposed platforms such as social networking ones (e.g. YouTube or Facebook) to collaborate with peers or engage in self-directed learning.

Also, students can by large freely access formal courses offered by other institutions, e.g. massive open online courses (MOOC), or in dedicated environments such as intelligent tutoring systems (ITS). Due to the widespread use of these technologies, students of the twenty-first century can readily engage with content, connect with peers within their social networks and communicate with their teachers.

With the increasing prevalence of social networks and context-aware technologies, researchers and educators are provided with new opportunities to extend the experience of learning from physical spaces to various other types of spaces while optimising pedagogical methods that can support personalisation and adaptive learning. Context-aware technologies are capable of gathering information about the organisation of the learner's learning space, the content of learning materials students engage with, the learners' characteristics (social, cognitive, etc.) and the related educational data on the learning environment. Harvesting these various forms of data provides more valuable insights into the design and optimisation of learning spaces. Research into learning spaces and learning context is critical to the design of context-aware and intelligent environments that are needed for personalisation and adaptive learning.

This chapter examines the concept of learning space in the digital age, its various types and its impact on the context where learning occurs. The chapter also looks at issues related to context modelling in the context-aware educational technologies. The chapter is conceptually grounded and sets the stage for future empirical work in this area. Further, the chapter draws upon work previously carried out in the Technology Enhanced and Teaching in Context (TEEC) (Anjou et al., 2017). The goal, in the long run, is to explore the impact of learning spaces on student's learning process and learning outcomes, with the ultimate aim of constructing a transformative theory of learning in context.

The chapter is organised as follows: first, we introduce the concept of learning space in the digital age and its impact on the learning process. We begin by examining the literature surrounding the subject. In reviewing the literature, we noticed that, even though the research community acknowledges the value of the learning space in the learning process, there is a dearth of literature on the subject, especially as it relates to learning in the twenty-first century (Sect. 9.2).

Second, we present various dimensions of the concept of learning space, taking into account the contexts in which each dimension can support learning (Sect. 9.3). This leads us to present our ideas about the modelling learning spaces and how it can be implemented to deliver context-aware services (Ejigu, Scuturici, & Brunie, 2007) such as the Context-Aware Intelligent Tutoring System, the CAITS (Forissier et al., 2013).

Third, we argue that learning space and the context in which learning occurs are intertwined (Dey, Abowd, & Salber, 2001). As such, we discuss issues surrounding context and learning space (Sect. 9.4) in order to inform the development of a context-aware learning space framework. This framework supports various educational experiments and improves adaptive and personal learning within the networked educational paradigm.

9.2   Background and Related Research

9.2.1   The Notion of Learning Space

According to Starr-Glass (2018) "learning space denotes a dedicated place (real or virtual), design by the instructor in order to enable students to meet and engage in knowledge creation". Through a purposeful design of learning spaces, the instructor can help learners to create and personalise learning.

The organisation of learning spaces is often depicted as classrooms with students seating in rows listening and taking notes while a teacher or lecturer stands in front of them and delivers knowledge or information. It is also assumed that learning mainly occurs in classrooms at a fixed time and that it is an individual activity that can be assessed using standard measures. The design of traditional learning spaces is based on the fundamental assumption that learning is mostly confined to formal space (e.g. classrooms or lecture theatres) (Thomas, 2010). Nonetheless, classrooms are not the only environments where students can acquire knowledge since much of learning takes place outside of the classroom (Milne, 2006).

This model of learning space assumes that the student's progress towards a programme of study is determined by the time spent in classrooms, the physical location of the student in the classroom and their interactions with teachers and other students. As such, the physical design, organisation of the classroom and the seating position of the student in the classroom can affect performance (Xi, Yuan, Yun Qui, & Chiang, 2017). However, the changing landscape of learning environments and students (e.g. diversity in students and learning needs and the permeation of digital technologies into learning) suggests that the traditional notion of the learning space, be it the formal lecture room, the seminar room or tutorial room, is untenable for all types of learning modalities of the twenty-first century.

It is also noted that, in the digital age, educational technologies have shifted from whiteboards to smart boards. This has shifted traditional layouts to open plan and interactive spaces and teacher-led lessons to more collaborative learning experiences.

Moreover, in highly networked learning environments, students interact in multiple modalities during their learning, which is much more complicated than the current portrayal of physical learning spaces as Erstad (2014) stated that the impact of digital technologies since the mid-1990s has implications for where and how learning might happen, whether it is online or offline and situated or distributed. It is also worth noting that while classrooms are formal learning spaces, distributed and networked learning environments can take forms of informal and non-formal learning spaces. Given all these reasons, the old classroom model is becoming archaic for both the student and the teacher.

9.2.2 The Twenty-First Century Student

The twenty-first century learners, the digital natives (Prensky, 2001), are those who grew up with new technologies like social networking ones. They were born from 1982 to the 1990s (Net Generation) and from 1995 to 2010 (Gen Z) (Brown & Lippincott, 2003). They are also known as the Google generation because they master googlism (Jansen, 2010; Biddix, 2011), or the C generation, where C stands for the twenty-first century skills, communication, collaboration, connection and creativity. Their skill in using the Internet and social networking technologies in daily life is similar to those in the educational context (CIBER group, 2008).

Typically, the characteristics of the twenty-first century students are: social, team-oriented, multitaskers, impulsives, and with short-term attention. They are hands-on with a "let us build it" approach that places increasing value on network devices (Brown & Lippincott, 2003). They do better in active learning, situated learning or learning by doing situation. Their characteristics have spread and are generalising to other generations (Prensky, 2010).

Further, learning for these generations has now become a lifelong pursuit that takes place within technological frontiers that support offline, online and blended learning (Garrison, 2011; Scott, Sorokti, & Merrell, 2016). Moreover, the way they engage with their learning is fundamentally transformed due to the permeation of various educational technologies (Daniel, 2017).

9.3   Learning Spaces in the Digital Age

In the early 2000s, discussions of learning space within many higher educational institutions largely remain constrained to three areas, namely, the classroom (where almost all learning occurs), the library and the faculty offices—where programmes are designed and students' work are graded (Temple, 2008). In parallel with the fact that the learning space remains mainly physical, institutions of higher education have come to recognise that learning happens in informal, formal and non-formal contexts. Formal learning contexts are those where teachers deliver learning activities within a formally defined curriculum (e.g. degree programme). Examples of formal spaces are a classroom or a technologically enhanced active learning classroom (Brooks, 2012). Selman, Cooke, Selman, and Dampier (1998) describe informal learning spaces as serendipitous and unsystematic (e.g. informal study groups).

In the same vein, Oblinger (2004) describes them as the results of serendipitous interactions among individuals. This type of learning is also referred to as learning outside of "designated class time" (Matthews, Andrews, & Adams, 2011), where learners are likely to acquire and accumulate knowledge skills, attitudes and insights gathered from a lifetime of experiences (Schwier, Morrison, & Daniel, 2009). Non-formal learning environments are organised as systematic educational activities that are often meant as supplementary learning support systems (e.g. learning support centres).

In the late 2000s, the appearance of networked educational and social networking technologies disrupted and transformed the nature of learning spaces leading to the consideration of various contextual factors that affect learning. During more disruptive technological forces, both students and educators owned mobile and ubiquitous technologies such as smartphones and laptops enabling them to access learning on demand, anytime and anywhere. We entered the era of the smartphone as we know it nowadays (e.g. the first iPhone was launched in June 2007).

During the 2010s, a growing body of research has called for the rethinking of learning spaces needed for the twenty-first century (Osborne, 2013; Selwyn, 2012; Temple, 2008). The contemporary forms of active learning involve the use of technologies to engage students with learning materials. The increasing growth in the use of active learning approaches in universities is evident in the transformation of physical learning spaces including the redesign of dynamic layouts that offer mobility and flexibility (Becker et al., 2017; Johnson et al., 2016).

Student-centred learning approaches aim to foster student inquiry, independent learning, collaborative working, active engagement and self-directed learning (Smith, 2017). Moreover, the ability to self-regulate learning requires an understanding of various learning contexts in which learning occurs. This is often shaped by both personal-psychological and contextual factors (Hood, Littlejohn, & Milligan, 2015). It is generally noted that research into learning space provides an opportunity to inform the development of adaptive and personalised technologies to enrich the student individual's learning need. Beyond that perspective it is possible to open the learning space in order to support the establishment of a social network for exchanges between students across the planet, leading to new educational experiences that might otherwise not be possible to achieve (Stockless, 2018). As Hung and Yuen (2010) noted, social networking technology supplements face-to-face courses and can enhance students' sense of community.

Our observation of the related literature in the field shows us that, although interesting, the studies carried out so far have focused on an archaic model of the learning space—mainly a physical classroom. Knowing that this model has evolved and continues to do so, it is necessary to consider a more appropriate model of the learning space that encompasses the current practices of the new generation of learners, teachers and instructional designers with educational and social networking technologies.

9.4 Rethinking Learning Spaces: A Conceptual Proposal

Against the above background, we argue that the modelling, the design and the organisation of learning spaces to support learners' needs in the twenty-first century should take into account the current research evidence on:

- The evolution of spaces where learning happens
- The characteristics of the twenty-first century student
- The use of social networking technologies as educational artefacts

Indeed, although the learning spaces in which learning takes place are still predominantly traditional (classroom, library, etc.), it is time to rethink these spaces in order to take into account the needs of the current generation of learners, the twenty-first century learners, and contribute to their success. Contemporary learning environments, as we know it, transcend physical spaces to virtual, (meta) cognitive and social spaces.

9.4.1   Physical Learning Spaces

Physical learning spaces such as lecture theatres, conference halls and retreat venues are valuable learning environments. They are considered part of the holistic view, one of identity and symbolic of power and prestige. However, beyond the classroom, physical learning spaces are quiet spaces or individual pods for individuals or small groups, break out spaces that could be large or small and widened corridors allowing the gathering of students away from the formal learning environments.

9.4.2   Virtual Learning Spaces

Virtual learning spaces comprise learning that is mediated both synchronously and asynchronously. In these environments, students learn to multitask and continually work outside of the classroom in spaces that promote social learning. It is viewed as an extension and enhancement of physical spaces. Wilson and Randall (2012) note that this new generation of learning spaces incorporates the use of both physical and virtual spaces.

Most studies on learning spaces have focused on physical space (Brown & Lippincott, 2003; Johnson et al., 2016). In recent years, the use of digital learning technologies alongside physical spaces has enabled synchronous and asynchronous learning activities. Therefore, the need to have different learning spaces in education is increasing, and their incorporation into teaching and learning is becoming more critical. Brown and Lippincott (2003) make it clear that, as the bare minimum, two spaces should be considered critical to learning, those of virtual and physical to integrate them into one environment. Another challenge for learning space design and learning spaces, in general, is the infrastructure necessary for their creation and maintenance. These include funding, faculty training, curriculum development and IT support.

9.4.3   Cognitive Learning Spaces

Cognitive learning spaces denote structural characteristics influencing students' cognitive processes, ranging from visual to acoustic perceptions (Arndt, 2012). Cognitive learning spaces influence learners' thought processes and learning. Cognitive learning spaces are environments that foster the creation of a specific time for writing or reflection.

9.4.4   Social Learning Spaces

Social network analysis and network visualisations can be used to explore how social interactions between learners can lead to useful learning outcomes. Learning in the context of social networks is highly self-motivated, autonomous and informal, and it forms an integral part of the higher education experience (Dabbagh & Kitsantas, 2012). Also, social networks are considered useful in developing essential skills like selecting relevant information, critically interpreting and analysing the socio-cultural context and working collaboratively and sharing knowledge (Garrison, Anderson, & Archer, 2001). However, the design of the learning space requires incorporation of strategies to create what others have referred to as Social Network Awareness (SNA). SNA is the ability of learners to identify and understand the knowledge context of peers to create a conducive environment for social engagement that is vital for knowledge construction (Lin & Lin, 2019).

Bennett (2007) asserted that "Space designs that acknowledge the social dimension of learning behaviours and that enable students to manage socialising in ways that are positive for learning are likely to encourage more time on task and more productive studying. They thereby yield a better return on the investment in physical learning spaces" (p.18). In the social aspects of learning spaces, the concepts of learning ecology (Scott et al., 2016) and learning communities (Schwier et al., 2009) are critical because they emphasise learning in a social context while recognising that learners are simultaneously involved in multiple learning settings. The social learning spaces that are capable of the blending of formal and informal learning are likely to facilitate situated and personalised learning experiences (Kovanović, Joksimović, Gašević, & Siemens, 2017). Furthermore, Oblinger (2005) pointed out that "learning is an active, collaborative, and social process that hinges on people" and that an ideal learning space is one that encourages engagement of various modalities with student peers as well as teachers.

A learning ecology is a collection of contexts—physical and virtual—that provides opportunities for learning (Scott et al., 2016). In higher education, this usually includes learning that takes place in formal, informal and non-formal contexts. Such an array of learning can take place across the institution, work settings, community and at home. Social learning spaces are instrumental in setting conditions for learning because they create a supportive environment to engage students in critical thinking and promote interactions that are richer, more gratifying and intrinsically rewarding (Garrison et al., 2001; Matthews et al., 2011).

We take the above holistic view of learning space, rethinking it along those four key dimensions (Fig. 9.1) in order to support the twenty-first century students with relevant educational networking technologies.

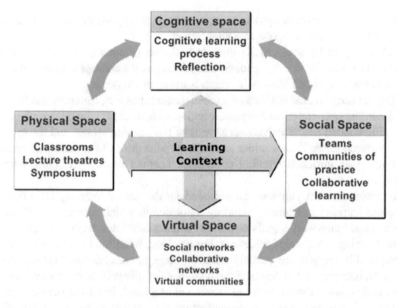

Fig. 9.1 A holistic view of learning space

9.5   Weaving Learning Space with Context Awareness

9.5.1   The Notion of Context

In artificial intelligence, the notion of context appeared in the 1990s (McCarthy, 1993), but it was not until the early 2000s where this area of research gained interests among researchers in ubiquitous and pervasive computing, mainly focusing on geolocation technologies, where the spatial and temporal dimension of the context became traceable.

Dey et al. (2001), in their definition, which is still the most widely accepted definition of the context, stated that "context as any information that can be used to characterize the situation of an entity, where an entity is a person, place or object that is considered relevant to the interaction between a user and an application".

Bazire and Brézillon (2005) have analysed 150 definitions of the context. From this study results the most cited model of context among research. It represents the components of a situation and the relations between them. According to Bazire and Brézillon (2005), "A situation could be defined by a user, an item in a particular environment, and eventually an observer". In their model of context, context and environment are separate but related.

Since 2007, CONTEXT symposia have opened the study of this concept to education where space takes a central place. Thus, according to Bettini et al. (2010), "space is an important context in many context-aware applications, and most context definitions mention space as a vital factor".

According to Miraoui (2018), context is the set of circumstances or facts that surround a particular event or situation. In education, the concept describes the various and possible circumstances in which learning might occur.

The learning context consists of students, cultural aspect, teachers, lesson plans, etc. Learning space refers to an episodic and specific moment where learning occurs. Learning context describes various aspects of the learning spaces and the relationship between them. It also refers to students' perceptions of a particular space in optimising an enriched learning experience accrued from a particular learning space.

Learning context can also be regarded as the set of learning activities and expected learning outcomes, including the space in which learning itself occurs. Context and space work together, whereby context facilitates the process of learning while learning space helps achieve it. For instance, Brézillon (2011) indicated that in physical learning space, participants use a large part of contextual information to translate, interpret and understand the utterances of others by using contextual cues like mimics, voice modulation and movement of a hand. Brezillon stresses that the efficacy of an expert system is dependent on the acquisition of knowledge of the problem being solved, the proposed solution, the situated context in which the problem prevails and the extension of the solution.

Learning context is critical to the determination of learning outcomes. It provides us with an understanding of the current context of the learner, their state of knowledge, their learning trajectories within the learning space, the relevant and suitable learning materials needed for them to succeed and the moments and space in which they learn best.

9.5.2 The Concept of Context in Education

The notion of context in education describes the various circumstances in which learning might occur, where the learning context consists of students, culture, teachers, lesson plans, etc. Context can also be regarded as the set of learning activities, expected learning outcomes, including the space in which learning itself occurs.

Context enables both teachers and students to rethink about the design of teaching and learning and the constraints of the learning spaces (Alterator & Deed, 2013). The affordances of a context must be perceived by an individual who must also have the abilities to interact with these attributes. Although openness can disrupt teaching conventions, it is the social activity of the inhabitants that define the possibilities of learning space (Lefebvre, 1991).

The emergence of massive open online courses (MOOC) in 2008 and the subsequent possibility of accessing extensive data about student interactions in online learning situations triggered more interests in understanding context and learning.

Recently, studies have emerged on didactic contextualisation (symposium on didactic contextualisation, 2011), in teaching in context (Forissier et al., 2013; Psyché et al., 2018). Bourdeau et al. (2015) stressed the importance of the external context in networked collaborative learning. According to these authors, the external context of a learning situation is influenced by environmental factors that have subsequent impacts on the learning process (Anjou et al., 2017). They used this hypothesis to guide the development of context-aware tools inside a Technology Enhanced Teaching in Context (TEEC) project. Among them, a calculator that computes the nature of context (MazCalc), some context-aware authoring services and a context-aware intelligent system (CAITS) are envisioned.

The TEEC project sought to create models of context and context-aware tools for students to discover the importance of context when learning in domains such as biology, geology, French, sustainable development, and social economics. The research was motivated by the observation that science learning tends to happen in various contexts (Bourdeau, 2017) and that significance of context in intelligent tutoring systems (ITS) was not fully understood since ITS concentrated on modelling the domain, the learner and the tutoring (Woolf, 2010). The TEEC project introduced the significance of context and its role in science learning by encouraging learners to explore and discover moments of learning within a particular context, likely to lead to conceptual understanding.

9.6 Research Motivation

In the present study, we explore the concept of learning space in the digital age and its impact on the learning process.

We know that research has associated the design and organisation of learning spaces for course design with student achievement, mastery and retention (Oblinger, 2005; Ramsay, Guo, & Pursel, 2017). Moreover, recently, research has been extended to explore the impact of learning spaces on student learning outcomes (Griffith, Vercellotti, & Folkers, 2019). Also, research on interactive learning space classrooms has reported that instructors and students find them engaging, and engagement is expected to increase learning outcomes (Vercellotti, 2018). We hypothesise that the design and organisation of learning spaces in the digital age have the same positive impact on the learning process than the traditional learning spaces. However, until now, researches have focused mainly on physical or hybrid learning spaces. Our research focuses on various other forms of learning spaces.

Based on the above, we have set up some broad research directions relating to:

1. Measuring the impact of our research on a student learning: measuring the impact of learning space on a student learning.
2. Developing models, tools and services derived from research results: learning space context-aware models (e.g. ontologies), tools (e.g. a CAITS, a Context-Aware Intelligent Tutoring System, etc.) and services (e.g. context-aware authoring services) for optimal student learning.

9.7 Pilot Study

9.7.1 The TEEC Project as a Test Bed for Experimenting Learning Spaces

In this section, we introduce the TEEC[1] (Technology Enhanced and tEaching in Context) project as a test bed for implementing learning spaces. The development of the project provides a pedagogical structure and a technological environment for specifying, implementing and testing the concept of learning space for Collaborative Distance Learning (CDL) using network technologies. Subsections are project description, Learning scenario, context-aware system architecture proposal and the design-based research (DBR) methodology. The TEEC pedagogical structure and its technological environment are considered a valuable hypothesis for deploying the ideas of learning spaces exposed above in this chapter.

9.7.2 Project Description

The TEEC project is an international project aiming at providing the foundations as well as the experimentation of context-aware systems to support collaborative learning at a distance using network technologies. The foundations are the following: Given an object of study, conceptual change can occur if the clash between two learning contexts is strong, highlighted and socially mediated. This model is called the CLASH. Moreover, in order to provide a scientific foundation for context selection, the MazCalc context gap calculator has been developed (Anjou et al., 2017). The model allows the specification of parameters and values, and the computation of the gap between two contexts, both global and parameter-specific. Should the result be too low, the learning design can be modified accordingly to obtain a more significant learning design using the DBR methodology. Similar projects are promoted among others in Europe through its etwinning initiative (https://www.etwinning.net/en/pub/about.htm).

The TEEC team experimented with only two contexts since adding a third one adds more complexity. The team conducted design experiments with various domains and topics as well as various educational levels. Domains and topics include biology, geology, environmental science, social science and history and language science. Educational levels go from primary education to university education. Plus, the experimentations were conducted among partners using the same language (French), despite apparent but productive disparities. Future work should extend these foundations to partners not sharing the same language, which again means adding another level of complexity.

---

[1] https://teec.teluq.ca/en/

9.7.3   Learning Scenario

A learning scenario in the TEEC project using the MazCalc contains the following steps:

- Establish a potential collaboration between partners in a domain and at a specific educational level while considering existing curricula, teaching and evaluation practices and technologies available on both sides. Inquiry learning is recommended with learners who have to conduct some field-based research, collect data and analyse them under the guidance of the teacher.
- Select an object of study (topic) with sufficient potential in contrast of contexts as well as the feasibility of a field study on both sides.
- Select themes (subtopics) relevant to the topic and the contexts as well as the feasibility of a field study on both sides.
- Identify parameters and their potential values for calculating the context with MazCalc.
- Enter the parameters and the values into the MazCalc software, and get the results to adjust the selection of the object of study and its context parameters to obtain an accurate prediction of productive learning through "a clash of contexts" and a conceptual change.
- Design the learning scenario according to instructional design principles following the ADDIE (Analysis-Design-Development-Implementation-Evaluation) methodology: specify actors, activities, resources, time and space.
- Actor means a set of roles that can be played by a person or another (students can have the role of teaching other students). Teaching and learning activities are selected based on their relevance to support learning (inquiry learning where learners conduct field-based research, collect data and analyse them under the guidance of teachers).

Time and space indicate the synchronous versus asynchronous communication activities and the field-based versus the classroom or lab activities. The two learning scenarios that inspire the TEEC basic scenario are Aronson's Jigsaw scenario (Aronson et al., 1978) and Schwartz's LEGACY inquiry cycle (Schwartz, Lin, Brophy, & Bransford, 1999). As a result, the pattern for a TEEC learning scenario comprises five main components, as illustrated in Fig. 9.2.

9.7.4   The Case of Geothermal

The case of geothermal is the best-documented experimentation of the TEEC project so far, as described by Anjou et al. (2017, in English) and Anjou (2018, in French) in her PhD dissertation. It has experimented at two levels: university level with future science teachers and primary school level.

The following disparities produce the contrast of contexts: Geothermy in Guadeloupe means the production of electricity powered by high-temperature water

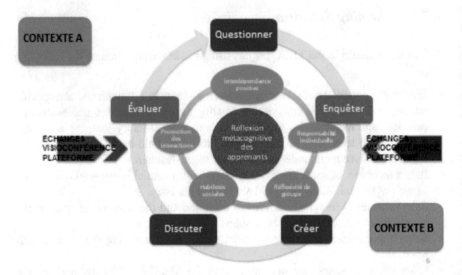

Fig. 9.2 The TEEC learning scenario

from deep layers in a recently formed geological structure in the Caribbean area. In Quebec, it means drilling rocks at a shallow level in the world's oldest geological structure to get a low-temperature increase for warming individual homes with a heat pump. Results of these two experiments indicate that the combination of the CLASH model and various contexts that involved the jigsaw inquiry learning scenario, where learners engage in the inquiry process, was productive, particularly achieving the objectives of the study (geothermal).

9.8 The Context-Aware System Architecture Proposal

The context-aware system architecture in the TEEC project includes an intelligent tutoring system. We called it the Context-Aware Intelligent Tutoring System (CAITS) architecture. It is original in that it incorporates two more models into the traditional ITS architecture: Contexts and Computer-Supported Collaborative Learning (CSCL) (Fig. 9.3).

According to Woolf (2010), an underlying ITS architecture is composed of four components: a domain model, a learner model, an instructional model and a user interface model.

Model; CSTM, context-sensitive tutoring model; CSLM, context-sensitive learner model; CSLS, context-sensitive learning scenario; Database CEM, Context Effect Manager Board (MazCalc Query interface + Parameter Visualization screen board + Calibration Tools.)

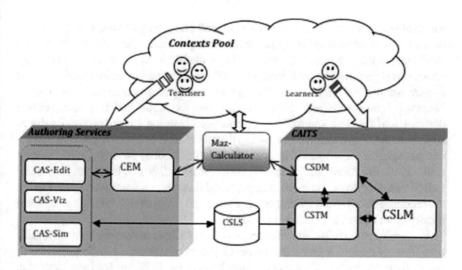

Fig. 9.3 CAITS architecture. Legend: CSDM, context-sensitive domain

VanLehn recently suggested adding CSCL to this architecture (VanLehn, 2016) and to configure the step and task loops according to its requirements. However, it is not clear how a learner could interact with her learning space while performing activities alternatively as an individual, a team or a group member.

As early as 2011, Isotani et al. underlined the need for taking CSCL into account in an ITS architecture. In their systematic review of Authoring Tools for Designing Intelligent Tutoring Systems, Beemer, Spoon, He, Fan, and Levine (2018) suggested that the next generation of ITS Authoring Systems incorporates collaborative learning. Sottilare et al. (2018) proposed an authoring framework for Designing Adaptive Instruction for Teams, and Fletcher et al. (2018) proposed a Shared Mental Models in Support of Adaptive Instruction for Teams Using the GIFT Tutoring Architecture. However, none of them takes into account the context in which teams can live and how different or even contrasted it can be.

The TEEC architecture not only provides a context model for each team but also calculates the contrast between two contexts for the same object of study, predicts the productive effect for learning on both sides and governs the instructional scenario with its actors, activities and resources.

9.9 Methods

9.9.1 What Is Design-Based Research?

The design-based research methodology (thereafter called DBR) has its roots in the pioneering work by Brown (Alterator & Deed, 2013) under the name of design experiments, in an effort to reduce the gap between lab research and in situ research and to allow a process by which both theory and practice can evolve together, based on a design process. It relies upon a cybernetic principle where the result of each loop changes the behaviour of a system. DBR then evolved into a full methodology which was claimed by the DBR Research Collective and published in the *Educational Researcher Journal* (Anjou et al., 2017,) sustained by an article in the *Journal of the Learning Sciences* (Arndt, 2012), another one by Wang and Hannafin entitled design-based research and technology-enhanced learning environments (Bazire & Brézillon, 2005) and another one by Herrington et al. It has been applied in numerous pedagogical innovations (Bazire & Brézillon, 2005; Bennett, 2007; Bettini et al., 2010; Bourdeau, 2017; Branch, 2009; Brézillon, 2011). DBR can be characterised as a microsystemic methodology, based on system science principles, mainly the feedback loops mechanisms, or iterations, and the goal of comprising the complexity of an authentic situation to study it.

In contrast to experimental research, it does not aim to isolate nor control. Unlike participatory design, it promotes the development of theoretical knowledge simultaneously to the design of artefacts. To our knowledge, DBR has not been applied to study the design of collaborative learning, nor the design of context-aware learning environments nor to Learning Spaces.

In the TEEC project, in order to achieve our objectives and test the hypotheses, while working on many fronts at the same time, we needed a methodology that would: (1) allow us to tackle the design of several components at the same time, (2) be concerned both with theory and practice, (3) account for the complexity of the learning situation, (4) respect the authenticity of the learning tasks, (5) allow us to produce results repeatedly along the development of the project, (6) allow us to test not only the hypotheses but also the components. The DBR methodology proved to be the best candidate for our project, even though it had not been applied yet to the study of context in learning, nor to telecollaborative learning, nor the design of a context-aware learning environment (Bourdeau, 2017).

The generic DBR process was instantiated in the following way: four components and three feedback loops for each iteration (see Fig. 9.4). The four components are: context modelling, instructional scenario, experimentation with data collection and results and lessons learned. An iteration means using the methodology for a specific domain and a specific learning situation, such as biology with primary school learners.

9.10   Linkages between Learning Space and Context

In order to investigate the linkage between learning space and context, we will adopt a methodology which takes into account knowledge engineering, instructional design models and design-based research, among others. The methodology would involve the conception and modelling of the context model of different learning spaces (see Fig. 9.5). The model draws from Bazire and Brézillon (2005), which considers various forms of the environment (physical, virtual, social and cognitive).

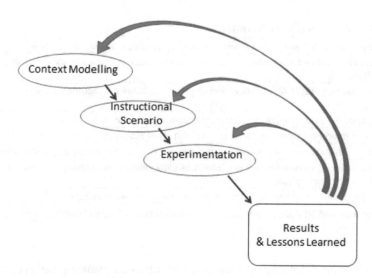

Fig. 9.4 The design-based research process in the TEEC project

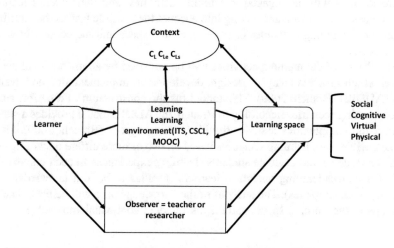

Fig. 9.5 Learning context model. (Adapted from Bazire & Brézillon, 2005)

In this model we take into account the "spatiotemporal location" component of the context, where the "items" represents any learning systems (e.g. in intelligent tutoring, computer-supported collaborative learning systems or massive open online courses).

The second phase of the methodology will involve the construction of an ontology of learning spaces. This ontology will be built from a reflexion on the relationship between learning space and learning situation (see Table 9.1). The ontology

Table 9.1 Examples of learning context

Physical classroom + group[a] of learners + individual learning + teacher as an instructor
Physical classroom + team inside the group + collaborative learning + teacher as an instructor and a facilitator
Virtual classroom with CSCL tools + team or group + collaborative learning + teacher as facilitator
CSCL system + team or group + collaborative learning + teacher as a facilitator
Intelligent tutoring system + one learner + individual learning + teacher = intelligent tutor
MOOC + one learner inside a massive virtual group + individual learning + no tutor/facilitator
Social networking environments + one learner inside a virtual team or group + individual learning + no tutor/facilitator
Social networking environment + team or group + collaborative learning + moderator

[a]A group = around 30 students; a team = 3 to 6 students and individual learning = 1 to 3 students

will inform the development of use case scenarios demonstrating various forms of learning spaces.

The third phase of the methodology would involve running user experiments, where learners will be engaged in learning activities, and their overall learning experiences will be evaluated, taking into account the possible factors they attribute to enhanced learning outcomes or experience and space and the context in which this occurs.

The design of the learning activities will be informed by an instructional design model which involves analysis, design, development, implementation and evaluation (ADDIE) (Branch, 2009). This model fits the construction of a performance-based learning scenario. According to Branch, the ADDIE model provides a way to address the complexities associated with learning spaces (Branch, 2009). Using ADDIE will design learning scenarios based on two or more different learning contexts for students and evaluate students' learning experiences in each context. We will also collect learning analytics, learners' profiles to build a knowledge base (activity trace templates). The analysis of the knowledge base will help validate the ontology of the learning spaces and the discovery of contextual knowledge.

9.11 Discussion

Research shows that the current design of learning spaces in higher education is no longer adequate for innovative pedagogies that take the twenty-first century student learning needs into account (Ramsay et al., 2017). Learning happens in various contexts and different learning spaces. Research relating to the various contexts where learning takes place has become much more critical than before due to the increasing deployment of digital and networking technologies and diverse needs and expectations of students relating to how they access learning resources. For instance, it is critical to the design of context-aware instructional scenario embedded in learning environments (e.g. intelligent tutoring or computer-supported collaborative learning systems or massive open online courses) to provide personalised individual or collaborative learning pathways (Chou, Lai, Chao, Lan, & Chen, 2015; Lavoué, Molinari, Prié, & Khezami, 2015; Van Leeuwen, Janssen, Erkens, & Brekelmans, 2014).

To attract and retain modern students, as well as improve engagement and drive better learning outcomes, institutions of higher education need to explore various contexts and spaces in which students learn, as well as leveraging pedagogical innovations that support blended, virtual and social learning.

The deployment of various educational networked technologies in different learning spaces generates rich data about students and their learning pathways. The growing number of computational and analytical tools that can capture students' interactions (when, what, where and how) provide numerous opportunities to harness such data and optimise learning spaces. Such data can be used to provide students with more accurate and faster feedback (Mayer-Schönberger & Cukier, 2014).

Similarly, teachers can use data obtained from students to reflect on their pedagogical design and optimise the learning environment to meet students' diverse needs and expectations. While the ability to track, harvest and analyse various forms of analytics (social, cognitive, affective and education) can reveal useful insights about learners in interaction with learning environments, methodologies to combine data from learning spaces with data from learner's interaction are lacking (Daniel, 2019; Hood et al., 2015; Hong, Scardamalia, Messina, & Teo, 2015; Kickmeier-Rust & Firtova, 2018). Consequently, there are limited learning context models which include the learning space context.

Also, there are limited instructional design methodologies or instructional models and scenarios which take into account the learning space context. "The organisation of learning space, whether physical or virtual, can have an impact on learning" (Oblinger & Lippincott, 2006). In recent times, learning has become an activity that can occur anywhere and so the idea, that is, learning only happens in physical spaces. In other words, the idea that the students sit and listen to a lecturer for a certain amount of time and then leave is by significance a myth. The transmission form of learning often sustained through physical spaces is becoming an unsustainable model. It does not meet the diverse needs of students and contradicts active learning, an idea that promotes participation and experiential learning.

The constructivist learning approach favours learning that is contextual, active, and social. It may also foster collaboration. In the twenty-first century, learners are highly social and prefer learning spaces that are highly distributed, adaptive and personalised. This is an indication of the importance of the design of learning spaces as they can determine how learning occurs (Temple, 2008).

Universities have made growing investments in redesigning physical learning spaces and equipping them with new digital technologies. These technologies help enhance the physical learning space by providing more diverse activities and influencing the organisation of pedagogy and student learning. Digital learning technologies support active learning activities such as lecture capture, class discussion and provisions of digital lectures (Daniel and Bird in press).

The combination of digital learning technologies and new forms of pedagogy can support social learning spaces. Educators have the opportunity to integrate these new spaces into their students learning environments in order to widen their students' horizons in terms of constructing knowledge. Besides the social learning space, virtual learning spaces provide highly interactive synchronous and asynchronous social networks that are useful for supporting knowledge acquisition.

The nature of these virtual spaces offers the twenty-first century students the ability to be mobile and the capability to multitask. These are some of the traits that are common in current students. The idea of learning spaces, therefore, helps educators envision the various places in which learning can occur. Moreover, it should help them design learning environments that factor in the notion that learning can occur anywhere anytime.

The design of useful learning space requires consideration of the learning content and the activities required for learning. Muñoz-Cristóbal, Gallego-Lema, Arribas-Cubero, Martínez-Monés, and Asensio-Pérez (2017) presented the idea of a learning bucket in virtual learning communities. A learning bucket is a conceptual proposal to help teachers orchestrate learning situations that involve multiple physical and virtual spaces (Munoz-Cristobal et al., 2013). A learning bucket is a configurable container of learning artefacts of multiple types (e.g. 3D models, web pages, artefacts generated with Web 2.0 tools such as Google Docs documents, etc.). The concept of learning bucket aims to introduce flexibility in learning spaces where teachers experiment with various forms of modalities (Muñoz-Cristóbal et al., 2015). With learning buckets, teachers make strategic choices in the design of learning activities and artefacts and choose appropriate pedagogies to support effective learning.

9.12   Conclusion and Open Questions

Learning space refers to an episodic and specific moment where learning occurs. Digital learning technologies have transformed the way students engage and interact off- and online, yet the physical learning spaces in which learning occurs have not changed much. In this chapter, we introduced the concept of learning space in the digital age and the various contexts in which learning occurs. We situated our discussion of learning space about the design of learning space context-aware ontologies, learning scenarios and tools. We aim to construct a transformative theory of context-aware learning spaces such as personal learning networks, virtual learning spaces, social learning spaces and cognitive learning spaces.

The current learning environment has been primarily restricted to the classroom or lecture theatre. Classrooms are learning spaces usually portrayed as students seating in rows and intently listening and taking notes while a teacher or lecturer is standing in front of them delivering knowledge or information. This model of learning environment purports that students' progress towards a degree is determined by the time spent in classrooms and interactions with teachers and other students. However, the model of learning was disrupted in the late 1990s when distributed, and online educational technologies drastically transformed space, time, context and the nature of learning. Critiques of the traditional learning spaces have since emerged. Many of them mainly argue against common learning spaces (classrooms) because they find them inefficient. Institutions of higher education recognise that learning happens in a variety of contexts including informal, formal, non-formal and online.

Although the literature has mostly focused on redesigning physical spaces (Brown & Lippincott, 2003; Johnson et al., 2016), it is essential to look at other dimensions of learning spaces such as social, cognitive, and virtual. We argue that understanding the various forms of learning spaces enable us to better redesign learning spaces to cater for all forms of learning, pedagogues, and students. Further, the design of learning spaces should be in conjunction with the entire institution so as to make them worthwhile and more applicable to their intended audience while providing for the right support that can maintain the design of the learning space since technology learning spaces and support services are mutually dependent (Brown & Lippincott, 2003).

We are aware that providing support to all forms of learning spaces cannot necessarily enable students to transition from one space to another without facing any challenges. It is essential to be aware that students tend to spend more time in informal learning environments rather than formal learning environments (Johnson et al. 2016).

In this chapter, we exposed the foundations for learning spaces and for applying them to context-aware systems for educational networking. We identified several questions that remain open and require further work, among them:

- What are the limits of a personal learning space?
- Can we design it to adapt to collaborative and social learning activities, and what would be the requirements for this development?
- Since CSCL is declined after several levels (individual, interindividual, team, intrateam, interteams, intragroup, intergroup), how can we envisage these multiple dimensions, both conceptually and in terms of implementation?
- Automatic detection and adaptation, the launching of appropriate applications, are significant challenges as well as coping with the learner model and the learner analytics associated. How would this be approached?
- How possible and feasible is it to add more levels of complexity than in the TEEC project, such as having three or more contexts, or different languages?
- How can learning analytics inform and improve the learning design of context-based activities using the DBR methodology?

Our future work involves the modelling of context-aware ontologies and conducting a series of experiments to construct a transformative learning theory that takes into account the various contexts of learning spaces (physical, virtual, social, cognitive and metacognitive).

Acknowledgements We want to thank Roger Nkambou for the CAITS architecture, Christine Richard for the figure of the learning scenario and the whole team of the TEEC project for providing the foundations as well as the experimentations of Context-Aware Collaborative Learning Systems. We also thank the following research funding agencies for supporting our work: ANR and FRQSC.

References

Alterator, S., & Deed, C. (2013). Teacher adaptation to open learning spaces. *Issues in Educational Research, 23*(3), 315.

Anjou, C., Forissier, T., Bourdeau, J., Mazabraud, Y., Nkambou, R., & Fournier, F. (2017, June). Elaborating the context calculator: A design experiment in geothermy. In *International and Interdisciplinary Conference on Modeling and Using Context* (pp. 513–526). Cham, Switzerland: Springer.

Arndt, P. A. (2012). Design of learning spaces: Emotional and cognitive effects of learning environments in relation to child development. *Mind, Brain, and Education, 6*(1), 41–48.

Aronson, E., Blaney, N., Stephan, C., Sikes, J., & Snapp, M. (1978). *The jigsaw classroom*. Beverly Hills, CA: Sage. Google Scholar. Axelrod, S., & Paluska, J.

Bazire, M., & Brézillon, P. (2005, July). Understanding context before using it. In *International and Interdisciplinary Conference on Modeling and Using Context* (pp. 29–40). Berlin, Heidelberg: Springer.

Becker, S. A., Cummins, M., Davis, A., Freeman, A., Glesinger Hall, C., & Ananthanarayanan, V. (2017). *NMC horizon report: 2017 higher education edition*. Austin, Texas: The New Media Consortium. Retrieved October 23, 2019 from https://www.learntechlib.org/p/174879/.

Beemer, J., Spoon, K., He, L., Fan, J., & Levine, R. A. (2018). Ensemble learning for estimating individualized treatment effects in student success studies. *International Journal of Artificial Intelligence in Education, 28*(3), 315–335.

Bennett, S. (2007). First questions for designing higher education learning spaces. *The Journal of Academic Librarianship, 33*(1), 14–26.

Bettini, C., et al. (2010). A survey of context modelling and reasoning techniques. *Pervasive and Mobile Computing, 2010, 6*(2), 161–180.

Biddix, J. P. (2011). "Stepping stones": Career paths to the SSAO for men and women at four-year institutions. *Journal of Student Affairs Research and Practice, 48*(4), 443–461.

Bourdeau, J. (2017, June). The DBR methodology for the study of context in learning. In *International and Interdisciplinary Conference on Modeling and Using Context* (pp. 541–553). Cham, Switzerland: Springer.

Bourdeau, J., Forissier, T., Mazabraud, Y., & Nkambou, R. (2015). Web-based context-aware science learning. In *Proceedings of the 24th International Conference on World Wide Web Companion* (pp. 1415–1418). Republic and Canton of Geneva, Switzerland: Association for Computer Machinery (ACM). https://doi.org/10.1145/2740908.2743048

Branch, R. M. (2009). *Instructional Design: The ADDIE Approach: Proceedings of the Second Sussex Conference, 1977* (Vol. 722). Springer Science & Business Media. New York, NY, USA.

Brézillon, P. (2011). Context and explanation in e-collaborative work. In B. K. Daniel (Ed.), *Handbook of research on methods and techniques for studying virtual communities: Paradigms and phenomena* (pp. 285–302). IGI Global. Cambridge, UK.

Brooks, D. C. (2012). Space and consequences: The impact of different formal learning spaces on instructor and student behaviour. *Journal of Learning Spaces, 1*(2), n2.

Brown, M. B., & Lippincott, J. K. (2003). Learning spaces: More than meets the eye. *Educause Quarterly, 36*(1), 14–17.

Chou, C. Y., Lai, K. R., Chao, P. Y., Lan, C. H., & Chen, T. H. (2015). Negotiation based adaptive learning sequences: Combining adaptivity and adaptability. *Computers & Education, 88*, 215–226.

CIBER Group. (2008). Information behaviour of the researcher of the future (CIBER Briefing paper; 9). London.

Dabbagh, N., & Kitsantas, A. (2012). Personal learning environments, social media, and self-regulated learning: A natural formula for connecting formal and informal learning. *The Internet and Higher Education, 15*(1), 3–8.

Daniel, B. K. (2017). Enterprise lecture capture technologies and value to student learning. *International Journal of Information & Communication Technologies in Education, 6*(2), 23–36. https://doi.org/10.1515/ijicte-2017-0009

Daniel, B. K. (2019). Big data and data science: A critical review of issues for educational research. *British Journal of Educational Technology, 50*(1), 101–113. https://doi.org/10.1111/bjet.12595

Deed, C. (2017). Adapting to the virtual campus and transitions in 'school-less' teacher education. In D. J. Clandinin & J. Husu (Eds.), *Sage international handbook of research in teacher education*. London, UK: Sage.

Dey, A. K., Abowd, G. D., & Salber, D. (2001). A conceptual framework and a toolkit for supporting the rapid prototyping of context-aware applications. *Human-Computer Interaction, 16*(2–4), 97–166.

Ejigu, D., Scuturici, M., & Brunie, L. (2007, March). An ontology-based approach to context modelling and reasoning in pervasive computing. In *Fifth Annual IEEE International Conference on Pervasive Computing and Communications Workshops (PerComW'07)* (pp. 14–19). IEEE.

Erstad, O. (2014). The expanded classroom–Spatial relations in classroom practices using ICT. *Nordic Journal of Digital Literacy, 9*(01), 8–22.

Fletcher, A. K. (2018). Help seeking: Agentic learners initiating feedback. *Educational Review, 70*(4), 389–408.

Forissier, T., et al. (2013). Modeling Context Effects in Science Learning: The CLASH Model. In P. Brézillon, P. Blackburn, & R. Dapoigny (Eds.), *CONTEXT 2013* (pp. 330–335). Springer. Berlin, Heidelberg, Germany.

Garrison, D. R. (2011). *E-learning in the 21st century: A framework for research and practice*. London, UK: Routledge.

Garrison, D. R., Anderson, T., & Archer, W. (2001). Critical thinking, cognitive presence, and computer conferencing in distance education. *American Journal of Distance Education, 15*(1), 7–23.

Griffith, J., Vercellotti, M. L., & Folkers, H. (2019). What is in a question? A comparison of student questions in two learning spaces. *Teaching and Learning in Communication Sciences & Disorders, 3*(1), 7.

Hong, H. Y., Scardamalia, M., Messina, R., & Teo, C. L. (2015). Fostering sustained idea improvement with principle-based knowledge building analytic tools. *Computers & Education, 89*, 91–102. https://www.sciencedirect.com/science/article/pii/S0360131515001207

Hood, N., Littlejohn, A., & Milligan, C. (2015). Context counts: How learners' contexts influence learning in a MOOC. *Computers & Education, 91*, 83–91.

Hung, H. T., & Yuen, S. C. Y. (2010). Educational use of social networking technology in higher education. *Teaching in Higher Education, 15*(6), 703–714.

Jansen, B. A. (2010). Internet filtering 2.0: Checking intellectual freedom and participative practices at the schoolhouse door. *Knowledge Quest, 39*(1), 46–54.

Johnson, L., Adams Becker, S., Cummins, M., Estrada, V., Freeman, A., & Hall, C. (2016). *NMC horizon report: 2016 higher education edition*. Austin, Texas: The New Media Consortium. Retrieved October 23, 2019 from https://www.learntechlib.org/p/171478/.

Kickmeier-Rust, M. D., & Firtova, L. (2018, June). Learning Analytics in the Classroom: Comparing Self-assessment, Teacher Assessment and Tests. In *International Conference in Methodologies and intelligent Systems for Techhnology Enhanced Learning* (pp. 131–138). Springer, Cham.

Kovanović, V., Joksimović, S., Gašević, D., & Siemens, G. (2017). Digital learning design framework for social learning spaces. In *Joint Proceedings of the Workshop on Methodology in Learning Analytics (MLA) and the Workshop on Building the Learning Analytics Curriculum*

(BLAC) co-located with 7th International Learning Analytics and Knowledge Conference (LAK 2017). Central Europe.

Lavoué, É., Molinari, G., Prié, Y., & Khezami, S. (2015). Reflection-in-action markers for reflection-on-action in Computer-Supported Collaborative Learning settings. *Computers & Education, 88*, 129–142.

Lefebvre, H. (1991). *The production of space*. Oxford, UK: Blackwell.

Lin, J. W., & Lin, H. C. K. (2019). User acceptance in a computer-supported collaborative learning (CSCL) environment with social network awareness (SNA) support. *Australasian Journal of Educational Technology, 35*(1), 100–115.

Matthews, K. E., Andrews, V., & Adams, P. (2011). Social learning spaces and student engagement. *Higher Education Research & Development, 30*(2), 105–120.

Mayer-Schönberger, V., & Cukier, K. (2014). *Learning with big data: The future of education*. New York, NY: Houghton Mifflin Harcourt.

McCarthy, J. (1993). Notes on formalizing context. *Proceeding IJCAI'93 Proceedings of the 13th international joint conference on Artificial intelligence* (Vol. 1, pp. 555–560). Chambery, France — August 28 - September 03, 1993.

Milne, A. J. (2006). Designing blended learning space to the student experience. In *Learning spaces*. Washington, DC: Educause.

Miraoui, M. A (2018). *Context-aware smart classroom for enhanced learning environment*. Retrieved on 23 Oct 2019 from https://pdfs.semanticscholar.org/6491/f9ce8a0f3d6b-036fa02322ac0c3b4d3a0406.pdf.

Muñoz-Cristóbal, J. A., Asensio-Pérez, J. I., Martínez-Monés, A., Prieto, L. P., Jorrín-Abellán, I. M., & Dimitriadis, Y. (2015, September). Bucket-Server: A system for including teacher-controlled flexibility in the management of learning artefacts in across-spaces learning situations. *Paper presented at the Tenth European Conference on Technology Enhanced Learning (EC-TEL 2015)*, Toledo, Spain.

Muñoz-Cristóbal, J. A., Gallego-Lema, V., Arribas-Cubero, H. F., Martínez-Monés, A., & Asensio-Pérez, J. I. (2017). Using virtual learning environments in bricolage mode for orchestrating learning situations across physical and virtual spaces. *Computers & Education, 109*, 233–252.

Munoz-Cristobal, J. A., Prieto, L. P., Asensio-Perez, J. I., Jorrín-Abellán, I. M., Martínez-Mones, A., & Dimitriadis, Y. (2013, September). Sharing the burden: Introducing student-centred orchestration in across-spaces learning situations. In *Paper Presented at the 8th European Conference on Technology Enhanced Learning (EC-TEL)*, Paphos, Cyprus.

Oblinger, D. (2004). The next generation of educational engagement. *Journal of Interactive Media in Education, 8*. Retrieved 23 Oct 2019 from https://www-jime.open.ac.uk/articles/10.5334/2004-8-oblinger/.

Oblinger, D. (2005). Leading the transition from classrooms to learning spaces. *Educause Quarterly, 28*(1), 14–18.

Oblinger, D., & Lippincott, J. K. (2006). Learning spaces. Boulder, Colo,.: EDUCAUSE, c2006. 1 v.(various pagings): illustrations. Retrieved on 23 Oct 2019 from http://digitalcommons.brock-port.edu/cgi/viewcontent.cgi?article=1077&context=bookshel.

Osborne, J. (2013). The 21st-century challenge for science education: Assessing scientific reasoning. *Thinking Skills and Creativity, 10*, 265–279.

Prensky, M. (2001). *Digital natives, digital immigrants part 1. On the horizon, 9*(5) (pp. 1–6). UK: Emerald Publishing Limited.

Psyché V., Anjou C., Fenani W., Bourdeau J., Forissier T., & Nkambou R. (2018). Ontology-Based Context Modelling for Designing a Context-Aware Calculator. *Workshop on Context and Culture in Intelligent Tutoring Systems*. ITS 2018.

Ramsay, C. M., Guo, X., & Pursel, B. K. (2017). Leveraging faculty reflective practice to understand active learning spaces: Flashbacks and re-captures. *Journal of Learning Spaces, 6*(3), 42–53.

Schwartz, D. L., Lin, X., Brophy, S., & Bransford, J. D. (1999). Toward the development of flexibly adaptive instructional designs. In *Instructional-design theories and models: A new paradigm of instructional theory* (Vol. 2, pp. 183–213). Mahwah, NJ: Lawrence Erlbaum Associates.

Schwier, R. A., Morrison, D., & Daniel, B. K. (2009). A preliminary investigation of self-directed learning activities in a non-formal blended learning environment. *Online Submission.* Available https://files.eric.ed.gov/fulltext/ED509957.pdf

Scott, K. S., Sorokti, K. H., & Merrell, J. D. (2016). Learning "beyond the classroom" within an enterprise social network system. *The Internet and Higher Education, 29,* 75–90. *Distance Education, 15*(1), 7–23.

Selman, G., Cooke, M., Selman, M., & Dampier, P. (1998). *The foundations of adult education in Canada* (2nd ed.). Toronto, ON: Thompson Educational Publishing.

Selwyn, N. (2012). Making sense of young people, education and digital technology: The role of sociological theory. *Oxford Review of Education, 38*(1), 81–96.

Smith, C. (2017). The influence of hierarchy and layout geometry in the design of learning spaces. *Journal of Learning Spaces, 6*(3), 59–67.

Sottilare, R. A., Baker, R. S., Graesser, A. C., & Lester, J. C. (2018). Special Issue on the Generalized Intelligent Framework for Tutoring (GIFT): Creating a stable and flexible platform for Innovations in AIED research. *International Journal of Artificial Intelligence in Education, 28*(2), 139–151.

Starr-Glass, D. (2018). Building learning spaces. In *Online course management: Concepts, methodologies, tools, and applications (p. 241). Pennsylvania, USA: IGI Global. Hershey.

Stockless, A. (2018). Digital education: learning by opening the walls of the class. *Revue Internationale sur le numérique en éducation et communication, revue-mediations.teluq. ca, 1*(1), P3–P5. Retrieved on 06-02-2019 from https://revue-mediations.teluq.ca/index.php/Distances/article/view/63/31

Temple, P. (2008). Learning spaces in higher education: An under-researched topic. *London Review of Education, 6*(3), 229–241.

Thomas, H. (2010). Learning spaces, learning environments and the dis 'placement' of learning. *British Journal of Educational Technology, 41*(3), 502–511.

Van Leeuwen, A., Janssen, J., Erkens, G., & Brekelmans, M. (2014). Supporting teachers in guiding collaborating students: Effects of learning analytics in CSCL. *Computers & Education, 79,* 28–39.

VanLehn, K. (2016). Regulative loops, step loops and task loops. *International Journal of Artificial Intelligence in Education, 26*(1), 107–112.

Vercellotti, M. L. (2018). Do interactive learning spaces increase student achievement? A comparison of the classroom context. *Active Learning in Higher Education, 19*(3), 197–210.

Wilson, G., & Randall, M. (2012). The implementation and evaluation of a new learning space: A pilot study. *Research in Learning Technology, 20*(2), 1–17.

Woolf, B. P. (2010). *Building intelligent interactive tutors: Student-centered strategies for revolutionizing e-learning.* Morgan Kaufmann. Burlington, MA, USA.

Xi, L., Yuan, Z., Yun Qui, B., & Chiang, F. K. (2017). An investigation of university students' classroom seating choices. *Journal of Learning Spaces, 6*(3), 13–22.

Part V
Study

Chapter 10
Mexican University Ranking Based on Maximal Clique

Edwin Montes-Orozco, Roman Anselmo Mora-Gutiérrez,
Bibiana Obregón-Quintana, Sergio G. de-los-Cobos-Silva,
Eric Alfredo Rincón-García, Pedro Lara-Velázquez,
and Miguel Ángel Gutiérrez-Andrade

Abstract The evaluation of individual universities generates university rankings according to a set of indicators. In general, the classical rankings are based on reputation, academic achievements, research levels, material and financial resources, etc. In this chapter, a new methodology for university rankings based on the maximum clique problem in networks is presented. To study the method, the Higher Mexican Educational System was modeled through the complex networks approach, using the available information about the main academic characteristics, influence, and importance in the social networks and educational networks of each university. The numerical results show that the Mexican universities can be classified into at least three groups, where the best universities have better-qualified teachers, offer quality undergraduate and postgraduate programs, and are essential and influential in the exchange and dissemination of knowledge using the information and communication technologies (ICTs) and social networks (SN).

Acronyms: Due to the massive quantity of acronyms edited in this chapter, the complete relations appears at Table 10.19 of Appendix 10.A.

E. Montes-Orozco (✉)
Posgrado en Ciencias y Tecnologías de la Información, UAM Iztapalapa,
Mexico City, Mexico
e-mail: emontes@xanum.uam.mx

R. A. Mora-Gutiérrez
Departamento de Sistemas, UAM Azcapotzalco, Mexico City, Mexico
e-mail: mgra@azc.uam.mx

B. Obregón-Quintana
Departamento de Matemáticas, Facultad de Ciencias, UNAM, Mexico City, Mexico
e-mail: bobregon@ciencias.unam.mx

S. G. de-los-Cobos-Silva · E. A. Rincón-García · P. Lara-Velázquez ·
M. Á. Gutiérrez-Andrade
Departamento de Ingeniería Eléctrica, UAM Iztapalapa, Mexico City, Mexico
e-mail: cobos@xanum.uam.mx; rincon@xanum.uam.mx; plara@xanum.uam.mx;
gamma@xanum.uam.mx

© Springer Nature Switzerland AG 2020
A. Peña-Ayala (ed.), *Educational Networking*, Lecture Notes in Social
Networks, https://doi.org/10.1007/978-3-030-29973-6_10

Keywords Social networks · Complex system · Educational system · Optimization · Social networking · Educational networking

10.1 Introduction

University rankings are generally performed annually to determine the best universities using a set of indicators (Doğan & Al, 2018). These rankings are classified into two types: specific/local and global (Sánchez Hervás et al., 2017; Liu, 2015; Zheng & Liu, 2015).

Local rankings are generated using certain aspects in which institutions can stand out individually. On the other hand, global rankings take into account several indicators at the same time.

For example, the newspaper *The Times* presents its ranking called "Higher Education Supplement" (Altbach, 2015) using the following ratings: 60% to the quality of the research, 20% to the student/academic ratio, 10% to the international presence, and 10% to the capacity that a graduate gets a job.

On the other hand, the Institute of Higher Education of Jiao Tong University of Shanghai, China, presents its classification known as "Academic Ranking of World Universities" (Chang & Ouyang, 2018), where the following evaluations are used: the number of Nobel Laureates with 10% to professors retired and 20% to active professors, 20% to the number of highly cited researchers, 20% to the number of articles published in the scientific journals *Science* and *Nature*, 20% to the number of academic papers registered in the Science Citation Index (SCI) and the Social Science Citation Index (SSCI), and finally, 10% to the score of all previous indicators divided by the number of full-time academics.

These lists have periodic changes; however, there are no drastic changes in it; that is, a university rarely goes up or down several places. The group of the best or worst universities only changes positions among themselves.

The primary motivation of this work is based on the fact that classic university rankings only use the information on academic activities such as the production of scientific articles, the quality of graduates, etc. On the other hand, our work aims to generate a ranking using the data on the use of Information and Communication Technologies (ICTs) and digital platforms to share and disseminate knowledge, such as:

- e-Learning: It is a natural evolution of distance learning, as it combines and exploits the advantage produced by technological tools, in addition to changes in the socio-cultural context, which performance a new structured education (Klašnja-Milićević, Vesin, & Ivanović, 2018; Laeeq, Memon, & Memon, 2018). Alonso, López, Manrique, and Viñes (2005) present another definition of e-learning, which denotes that e-learning is the use of new multimedia and Internet technologies to improve the quality of learning by facilitating access to resources and services, as well as foreign exchange and collaboration.

- Educational networking: The term is defined as the work of social networks focused on education that potentially includes features such as file sharing, communication, information network technologies, etc. (Nee, 2014).

In recent years, a growing number of institutions are using educational networks to communicate and create a presence in social networks online or on the web.

- Social networking: It is the use of social media sites based on the Internet to connect with friends, family, colleagues, customers, etc. (Pempek, Yermolayeva, & Calvert, 2009). Social networks have become an important base for marketing professionals who seek to attract customers. Also, they may have a social purpose, a commercial purpose, or both through sites such as Facebook, Twitter, LinkedIn, and Instagram, among others.

Because the methodology presented in this work is based on social network analysis, it is necessary to describe the following definitions:

- Complex system: A complex system arises as a consequence of the emerging and random behaviors of the components within it and not due to a predetermined plan. Formally, there is not a universally accepted definition of complex systems. However, their main characteristics are self-organization, emergency, independence, and interdependence (Shore, 2018).
- Complex network: It is a representation of the components of a complex system and their relationships through the topological characteristics of this representation, such as the degree distribution, the clustering coefficient, the shortest path between two nodes, the average distance of the network, etc.; and the dynamics of the interactions between its components over time (Newman, 2010).
- Social network: It is a network that is composed for a set of nodes that generally represent people and/or institutions and a set of links that can represent tastes and preferences in common among the elements of interest (Zachary, 1977).
- Network analysis: It is the study of relationships and flows between actors such as people, groups, organizations, or other entities processing information and/or knowledge (Gretzel, 2001).
- Social networks analysis: It provides a mathematical and visual analysis of complex human systems (Gretzel, 2001). It is worth mentioning that, online education systems and social network services (e.g., Facebook, Twitter, YouTube), are increasingly used by universities (Antonio & Tuffley, 2014), both in official courses and communicate or share information and knowledge with students. However, the scope and benefits are still being studied (Eid & Al-Jabri, 2016; Silva, da Silva, & Araujo, 2017; Thomas, West, & Borup, 2017).

Currently, universities are challenged to offer high-quality education in a sociocultural and technological system; therefore, they should look for education mechanisms that are considering social inclusion (AbuJarour & Krasnova, 2018).

Also, ICTs are changing the perspectives of access to knowledge and interaction. In education, they have incorporated into curricular courses and degrees at all levels.

Buxarrais-Estrada and Ovide (2011) describe that teachers should be prepared for the use of ICTs, both as an object of knowledge and as an educational tool. Also, at present, the educational model at all levels has begun to include these tools.

Based on the above, it can be seen that educational systems have the essential characteristics of complex systems since they are composed of several interdependent or different types of entities (use of ICTs, quality of research, quality of professors, etc.).

On the other hand, as mentioned earlier it is essential to mention that a complex system is often represented as a network, formed by a set of components (nodes) that interact with each other by one or more types of relations (connection), that in most cases is non-linear (non-linear systems represent those whose behavior is not expressible as the sum of the action of their components.) (Latora, Nicosia, & Russo, 2017; Sayama, 2015).

The present work presents a methodology for university rankings using the maximum clique problem. And it is focuses on the studies of the universities belonging to the Mexican Higher Education System (MHES) through the datification of three types of information: The first, using the main academic characteristics taken in some classical rankings; the second, using the information about the uses of ICTs in the universities (e.g., use of ICTs in traditional courses, online courses, etc.); and the third, using the information about the use of social networks for knowledge sharing and cultural diffusion.

Further, the maximum cliques are obtained through the application of an Ant System (AS) algorithm (Dorigo, Maniezzo, & Colorni, 1996) to get the set of elements that share similar characteristics and that help to identify the best and worst aspects of the network.

This work seeks to answer, among others, the following questions: What characteristics can be relevant among universities? Can an university be considered of high academic level if it is influential in social networks? What impact do the social networks of the different universities have on technological development and for the exchange of knowledge? How does the study of information on the use of ICTs for dissemination and sharing of knowledge affect the ranking?

For this, the study plans of severa universities are presented, and some official online and distance programs are considered that are separated from their traditional parts. Also the impact generated by publications in social networks of the different universities is studied, that is, as the social networks help or affect their reputation in society, as well as the importance for their academic development of the use of them to share knowledge.

The present work is organized as follows: Sect. 10.2 presents a literature review about the main topics used in the body of work (e.g., online educational systems, social networks, Maximum Clique Problem (MCP), university systems, rankings, etc.).

Section 10.3 describes the modeling process in different kinds of networks, the development, and tools used for statistical studies and the elaboration of the ranking using the PCM.

Section 10.4 shows the numerical results and their studies for each network model; Sect. 10.5 shows the generalization of results; and finally, in Sect. 10.6, the conclusions about the research developed are presented.

10.2 Background and Literature Review

In order to establish the methods, models, and criteria used in this work, the specific literature of the following topics was reviewed:

- The ranking of universities (Clauset, Arbesman, & Larremore, 2015; De Bacco, Larremore, & Moore, 2018; De Witte & Hudrlikova, 2013; Doğan & Al, 2018; Karimi, Génois, Wagner, Singer, & Strohmaier, 2018; Wur, 2017).
- Characteristics used in the main rankings (Builes, Castaño, & Zuluaga, 2018, Ordorika & Rodríguez Gómez, 2010 and; Rust & Kim, 2016).
- The diffusion of knowledge through social networks (Larner, 2015).
- The effect they have on people and specifically on students (Burov, 2016; Nee, 2014).
- The development of educational networks (Antonio & Tuffley, 2014; Davydova, Dorozhkin, Fedorov, & Konovalova, 2016; Versteijlen et al., 2017).
- Modeling and solution of the optimization problem (Benlic & Hao, 2013; Conte, De Virgilio, Maccioni, Patrignani, & Torlone, 2016; Goeke, Moeini, & Poganiuch, 2017; Nogueira & Pinheiro, 2018).

10.2.1 Ranking of Universities

Concerning university rankings, Doğan and Al (2018) describe that the main university rankings can be done using and quantifying a smaller number of characteristics, and the same results would be obtained. Therefore, it is essential to know what components to take to perform the classification. In recent years, there have been some types based on the statistical analysis; however, none contemplates the use of complex networks and optimization. For example:

- De Witte and Hudrlikova (2013) propose a fully non-parametric methodology for university rankings, obtaining universities in English-speaking countries benefiting from the benevolent classification. On the contrary, they show that the rankings with fixed weighting systems favor major universities and research-oriented centers.
- Daraio, Bonaccorsi, and Simar (2015) propose an alternative approach to create scores QS, which are known as the methodology of Composite I-Distance Indicator (CIDI) (Dobrota et al., 2016) based on the proposal of a correction weights composite indicators based on the methodology of CIDI.
- Zornic et al. (2015) present a methodology that ranks institutions according to the number of articles published, the average score of regular appointments (MNCs) and four indicators based on percentiles using the I-distance method.

- Daraio et al. (2015) present a contribution to the birth of a new generation of rankings. The authors integrate a new type of information and use a new sorting technique implementing confidence limits for "management efficiencies."

Based on these features and methodologies, over recent decades, the development of university rankings has grown in Europe and Latin America. Below are some examples of the main classifications of universities in the world:

- British Newspaper, *The Times,* publishes an annual world ranking of universities, in which various global characteristics and areas are analyzed for the best universities in the world (Ordorika & Rodríguez Gómez, 2010).
- The ranking compiled by Shanghai Jiao Tong University (China) orders the world's top 500 universities based on quality criteria, such as the level of students, teachers, and schools, research activities, publications, etc. (Rust & Kim, 2016).
- *The Economist* newspaper produces annually a world ranking that ranks the best graduate programs offered worldwide under the title Which MBA? (Builes et al., 2018).
- The classification developed by the Scientific Information and Documentation (CINDOC) Laboratory of Internet of the Spanish National Research Council (CSIC) uses the web impact factor of each university. Its database consists of 12,000 universities; the list is arranged according to a dictator who combines the volume of published information, visibility, and impact of these pages according to the number of external links received (Builes et al., 2018).

Based on the above and with the results shown in the work of Clauset et al. (2015), we can see that a greater institutional prestige leads to more excellent production of teachers, a position of better education and a more influential position within the discipline.

Regarding the university rankings in Mexico, Hernández, Leyva, Márquez, and Cerda (2014) developed several studies about the integration of higher education in the country through university rankings. The authors describe that the methodologies used by the world ranking of universities are based mainly on research, teaching, and linking. In addition, they emphasize that national indicators only approximate international indicators, suggesting that Mexican universities are not designed to comply with the standards of the global context.

In addition, Ordorika (2015) mentions that the universities that focus more on research obtain the best places in Mexican and International classifications.

10.2.2 Social Networking, Educational Networking, and e-Learning

The review on this topic was carried out on Social Networking (SN), Educational Networking (EN), and e-learning in the field of ICT in higher education and the influence on learning. Next, some works that focus on these issues are shown.

10.2.2.1 Social Networking

For SN, Sadowski, Pediaditis, and Townsend (2017) study how institutions are incorporating online technologies into management frameworks and the delivery of work and tasks, both through internal learning management systems and social networks. The authors describe that SN is an excellent tool to promote the connection between classmates with other students, but it is necessary to have a balance between learning and digital connection.

On the other hand, Hung and Yuen (2010) present a study of how SN can be used to complement classroom courses in higher education. As in the work of Sadowski et al., 2017, the results showed that most of the students developed strong feelings of social connection and expressed positive feelings about their learning experiences in the classrooms in which SN was used.

Finally, the work of Antonio and Tuffley (2014) describes how the adoption of social network technologies by higher education institutions reflects a significant change in learning paradigms and knowledge exchange.

10.2.2.2 EN and Use of ICT

An essential aspect of EN is the approach in educational games (gamification) because they are becoming increasingly popular. While games explicitly designed for educational purposes have been used for decades, gamification is particularly new and contrasting evidence about its effectiveness was presented in (de-Marcos, Garcia-Lopez, & Garcia-Cabot, 2016).

On the other hand, Bustos and Román (2016) report that ICT should be included in education at all levels to meet the growing demand of students to share and disseminate knowledge. In addition, Vera Noriega, Torres Moran, and Martínez García (2014) assess the necessary ICT skills of higher education teachers in Mexico and emphasize that teachers with moderate to top ICT skills are of great importance for the inclusion of these tools.

With regard to teaching and virtual tutoring, Martínez Clares, Pérez Cusó, and Martínez Juárez (2016) studied the perception of the usefulness that students have for these tools in relation to other forms of university education; the authors discovered that virtual tutoring should be better used as a tool to support the general development of students, provided that the platforms created for this purpose have been optimized and profitable.

Finally, Díaz-García, Cebrián-Cifuentes, and Fuster-Palacios (2016) study the impact of ICT for universities in the field of learning strategies. The results obtained describe that there is a broad relationship between ICT skills and the learning strategies that students put into practice in teaching and learning, so they should be exploited in the field of EN.

10.2.2.3 e-Learning

Regarding e-learning, Avello Martínez and Duart (2016) report that the introduction and intensive use of ICTs in social dynamics and the educational reality much encourage social interaction and the dissemination of knowledge. In recent years, new trends have emerged in e-learning, and mobile learning, the "Flipped Classroom" model, personal learning environments, game-based learning and learning analysis.

In addition, Al-Samarraie, Teng, Alzahrani, and Alalwan (2018) presented a study to determine the key factors that affect the continuous satisfaction of students and instructors with e-learning in the context of higher education. The authors describe that their findings provide new insights into how higher education institutions can promote the use of e-learning to ensure the continuity of e-learning.

On the other hand, Cabrera, Ortega-Tudela, Hita, Ruano, and Colón (2010) study how e-learning has impacted professional training in Spain. The authors detail that the continuous increase in the number of students enrolled in distance and online education and the use by some autonomous communities of learning materials created specifically for this established modality and for the traditional way helps to improve the educational level of the country.

10.2.3   Social Networks in Culture and the Creation of Educational Networks

As for social networks in culture, Kramer, Guillory, and Hancock (2014) study the emotional states and their transmission through an experiment with people who use Facebook. In this analysis, the authors tested whether emotional contagion occurs outside of the personal interaction between individuals, decreasing the amount of sensitive content in the news.

The results show that, when positive expressions were reduced, people produce fewer positive and more negative publications. When the negative feelings were reduced, the opposite pattern occurred.

On the other hand, Vohra and Hallissey (2015) study how society uses and exploits electronic media and devices. The authors note that the majority of the adult population has a smartphone, where only a quarter of its use is invested in making telephone calls, and the rest is spent on the Internet and participation in social networks. Therefore, they emphasize that the use of this technology can revolutionize the field of education.

Regarding the use of ICT to create educational activities (Cortés & Lozano, 2014) indicate that a student not only learns from the teacher and from the textbook, but also from many other actors, such as the media, partners, and society in general, so ICT should be exploited for the dissemination of knowledge.

10.2.4   *Maximum Clique Problem*

Based on the approaches proposed by (Vohra & Hallissey, 2015; Yang & Leskovec, 2015 and Zhu, 2014), we can describe a clique as the analogy of parent-child relationships, where:

- The parents have a marital relationship.
- The children have a relationship of brotherhood.
- Parents and children have a father–son relationship.

Here, we can see that all the components are directly connected to each other. Therefore, a maximum clique in a graph or network is the one that contains the most significant number of nodes that have direct links to each other.

Therefore, we established our study concerning the system of higher education in Mexico is through the MCP to obtain the university groups that share more features with each other.

Formally, the MCP is an optimization problem whose objective is to find the maximum set of elements in a network that have direct connections with each other. In this work, a network R is defined through a set nodes $V = \{v_1, v_2,..., v_n\}$ and a set of ordered pairs $E = \{v_i, v_j\} \subset V \times V$ (connections). If the pair $\{v_i, v_j\} \in E$ exists, implies a connection between the nodes v_i and v_j.

Then, a clique is a set of nodes C, where every pair of nodes belonging to C is connected with an edge in R, that is, C is a complete sub-network. A clique is partial if it is part of another clique. Otherwise, it is maximal. The objective of the MCP is to find a clique with the highest cardinality. Figure 10.1 shows an example in an 8-nodes network.

In Fig. 10.1 (a), it can be seen an example of 8 nodes and 12 connections (*Rp* network); Fig. 10.1 (b) shows the case of a clique formed by the nodes D, F, and G; and Fig. 10.1 (c) contains the example of a clique formed by nodes D, E, G, and H, which is the maximal clique for *Rp*.

It is essential to mention that MCP is a problem of combinatorial optimization that is classified as an NP-complete problem, which is difficult to solve (Ullman, 1975). Due to its complexity, exact methods take a long time to provide a solution, sometimes as much as the age of the universe (De-los-Cobos, 2010); therefore, it is necessary to develop heuristic techniques that solve it to reach a solution close to the optimum in a reasonable time.

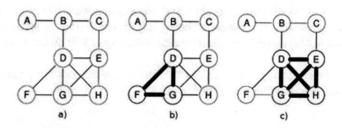

Fig. 10.1 Cliques in an 8-nodes network (*Rp*)

Because it has been shown that the Ant System (AS) algorithm (Dorigo et al., 1996) has good behavior in problems presented as graphs, this technique was chosen to find the MCP.

10.3   Materials and Methods

In this Section, the materials used and the resolution methods for the calculations and problems used to carry out this work are presented. In the first subsection (Sect. 10.3.1), the different data obtained through the various platforms used to get the information for modeling the networks are described and detailed.

On the other hand, Sect. 10.3.2 contains information about the way to model each kind of network used in this work, as well as the methodology for the resolution of the PCM to obtain the university rankings.

10.3.1   Materials

The materials that were used during the development of this work are three databases of different types:

The first type is based on the information available in the EXECUM repository of the National Autonomous University of Mexico "NAUM" (Universidad Nacional Autónoma de México) (ExECUM UNAM, 2017), which contains the information gathered in the project known as ECUM (the comparative study of Mexican universities). ECUM is the collection and analysis of information obtained from official sources such as: Secretariat of Public Education "SPE" (Secretaría de Educación Pública), Mexican National Council of Science and Technology "MNCST" (Consejo Nacional de Ciencia y Tecnología), National Institute of Copyright "NIC" (Instituto Nacional de Derechos de Autor), Scopus, among others).

This project contains the data for the period 2007–2017, and all the information is in the repository available at http://www.execum.unam.mx. On this website, the information can be visualized in three ways through the data explorer: personalized selection, first 20 universities, and selected universities.

For this work, the option of selected universities was chosen, where the information of a set of 60 institutions (public and private) that were incorporated in this Section are presented. The reason for being only 60 universities is because they are representative of each sector of the higher education system in different educational areas.

On the other hand, for the modeling of the clustering networks, the available information was taken for the following items:

- Type of contract for professors: by the hour, part time, and full time.
- The number of students per level: higher university technician, bachelor's degree, specialty, master's degree, and doctorate.
- Level of studies of professors: bachelor, specialty, masters, and doctorate.
- Teachers of the institutions in the National System of Researchers "NSR" (Sistema Nacional de Investigadores) of the MNCST.
- Professors recognized in the National Program for the Improvement of Teachers "NPIT" (Programa de Mejoramiento del Profesorado) or PROMEP.
- Documents, articles, and appointments in ISI and SCOPUS for each institution.
- Participation in documents and articles indexed in the databases of journals indexed by MNCST by each institution.
- Degree programs by the agencies recognized by the Council for Accreditation of Higher Education "CAHE" (Consejo para la Acreditación de la Educación Superior, A.C.).
- Postgraduate programs (specialty, masters, and doctorate) recognized in the National Register of Postgraduate Quality "NRQP"(Padrón Nacional de Posgrados de Calidad).

The second type is based on the information obtained by the XOVI Social Analytics tool (Social Analytics Tool, X. G., n. d.) This tool is used to get figures and key data that allow comparing different profiles on platforms such as Facebook, Twitter, Google+, and YouTube.

With this tool, it is possible to evaluate the performance of specific keywords to get more information about the trends of a particular topic. Also, XOVI has an export platform and report creation, where all the data can be extracted effectively. For this work, we use the information regarding:

- The use of digital platforms for the dissemination and sharing of culture and education for the leading universities in Mexico (the same universities were considered as for the EXECUM database).
- The impact and influence that universities have on social networks.
- The perspective of Latin American society for each of the universities (if there is violence if they have a good reputation, prestige, etc.).

Finally, the third type is based on the information obtained through the site https://public.tableau.com/profile/coordinaci.n.de.universidad.abierta.y.educaci.n.a.distancia#!/vizhome/PoblacinEscolar2014-2015/PoblacinEscolar2014-2015 belonging to the profile of the Open University and Distance Education Coordination of NAUM "OUDEC" (Coordinación de Universidad Abierta y Educacion a Distancia).

In this platform, information about the online and distance programs offered by the NAUM is obtained, where we can get information about the following items:

- The total of men and women of first income.
- The total of men and women of re-entry (students who were in another program either online or traditional).

For the first type of information, the separation of the data was carried out, taking the total values and the disaggregated data. Regarding the complete data, the value of all the individual characteristics of a particular topic is added. For example, for the issue of enrolled students, the sum of the students enrolled at the associate, bachelor's, master's, specialty, and doctorate level is done; meanwhile for the disaggregated data, the value of each is taken into account; following the previous example, in this case, the value of each subtopic (scholar level) influences.

Besides, it is essential to mention that for the second and third type, only the disaggregated data are contemplated (since there is not as much information as in the first type).

10.3.2   Methods

In the current subsection, the modeling processes of the three kinds of networks used in this work are detailed, as well as how to calculate the main metrics and the maximum clique.

It is important to emphasize that the present work was carried out in Mexico; therefore, the academic institutions and organizations mentioned in the rest of the work have a proper name in Spanish; so first the similar name in English is published, following by the corresponding acronym.

10.3.2.1   Modeling of Clustering Networks

Based on the information available in the EXECUM repository, 16 networks were modeled using two statistical tests with a level of significance of 5% applied to the information about the total and disaggregated data of the main characteristics of universities in Mexico:

- The first is a parametric test known as t-student, which is a test in which the statistic used has a t-student distribution if the null hypothesis is correct; in general, it is applied when the population studied follows a normal distribution, but the sample size is too small for the statistic on which the inference is based to be normally distributed, using an estimate of the standard deviation instead of the real value (Fadem, 2008).
- The second is a non-parametric test that uses the Wilcoxon statistic where the sample is not subject to specific requirements that are common to the parametric tests (Badii et al., 2012). Fundamentally, these requirements refer to the distribution that the variable presents in the population. On the other hand, they are especially useful for small sample sizes or, in cases in which the variable that interests us is an ordinal scale (Hodges, 1990).

For this, it is considered that if the characteristics of the j-th and k-th universities are statistically similar for the i-th year, a link between them is added. Next, Fig. 10.2

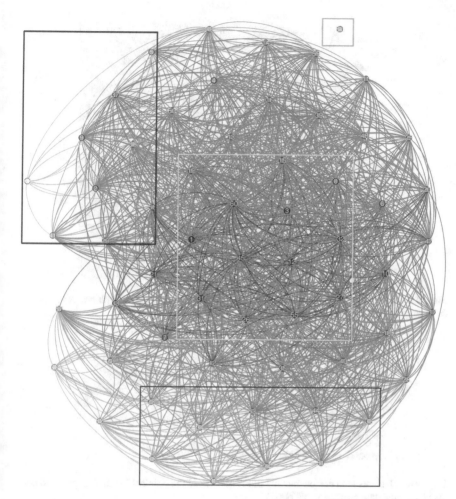

Fig. 10.2 Graph of the network of the year 2015 with non-normalized disaggregated data (considering the NAUM)

shows the example of the network modeled with the data broken down for the year 2015. Based on the information in Fig. 10.2, it can be seen that most of the nodes present a high value in the degree, that is, they are very connected. Also, we can see that the universities that present excellent characteristics according to the classic rankings (red box) are more linked to each other, as well as the universities that obtain the worst scores in the rankings (blue box) (Márquez-Jiménez, 2010).

Based on the above, Fig. 10.3 shows the universities belonging to the group of the best and Fig. 10.4 illustrates the universities belonging to the group of the worst. The elements shown in Fig. 10.3 are those that are positioned in the upper left of Fig. 10.2 and represent the universities that obtain the best classification values in Mexico. For example, they are National Polytechnic Institute "NPI"

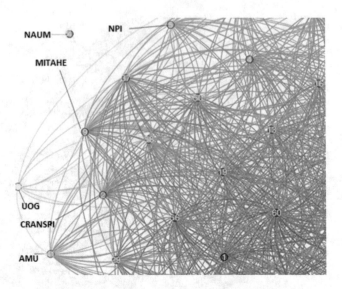

Fig. 10.3 Elements with a high ranking of the network of the year 2015 with non-standardized disaggregated data (considering the NAUM)

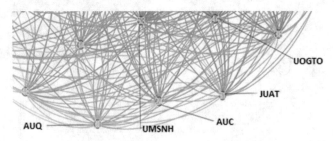

Fig. 10.4 Elements with the low ranking of the network of the year 2015 with disaggregated non-standardized data (considering the NAUM)

(Instituto Politécnico Nacional), Autonomous Metropolitan University "AMU" (Universidad Autónoma Metropolitana), and Technological Institute of Superior Studies of Monterrey "TISSM" (Instituto Tecnológico de Estudios Superiores de Monterrey), among others.

The elements shown in Fig. 10.4 are positioned in the lower right part of Fig. 10.2 and represent the universities that obtain the worst classification values in Mexico, for example, University of Michoacán of San Nicolás de Hidalgo "UMSNH" (Universidad Michoacana de San Nicolás de Hidalgo), Juarez University Autonomous of Tabasco "JUAT" (Universidad Juárez Autónoma de Tabasco), and the University of Guanajuato "UOGTO" (Universidad de Guanajuato), among others.

It is essential to mention that all the networks modeled in this work follow the model of small-world networks proposed by Watts and Strogatz (1998), whose main characteristics are as follows: any pair of nodes can communicate through a relatively small route (global property) and high grouping value of the coefficient for each node (local ownership). Now, to understand better the results shown in the following sections, the identifiers used for each network of this type are:

- Networks 1 and 2 correspond to the year 2013, networks 3 and 4 correspond to the year 2014, networks 5 and 6 correspond to the year 2015, networks 7 and 8 correspond to the year 2016, and networks 9 and 10 correspond to the year 2017.
- The identifiers correspond to disaggregated data and the even identifiers for complete data.
- The sub-index .1 corresponds to the first technique that assumes that the data is normalized, while .2 identifies the second technique that assumes non-normalized data.

10.3.2.2   Modeling of Influence Networks

The information for this kind of networks was obtained through the social analytics tool of XOVI, and the numerical values were taken for each metric studied. Like the clustering networks, the networks were generated by performing a parametric test and a non-parametric analysis with a level of significance of 5%.

For this analysis, there are three kinds of networks: (i) modeled networks with information about the influence of social networks in the different universities that belong to the system; (ii) modeled networks through information about the use of ICTs in traditional courses and exchange of knowledge in the higher education system in Mexico and; (iii) networks based on the perspective of society (impact on social networks).

For each kind of network, there are three sub-types: sub-type 1 is based on the behavior of the keywords presented for each study; sub-type 2 is based on the ranking obtained by the keywords given for each research and; sub-type 3 is based on the historical ranking obtained by the keywords.[1,2,3] In turn, for the studies referring to the three kinds of networks, extraction of the data accumulated by XOVI was realized weekly for 32 days (November 11 to December 14, 2018); therefore, for the sub-types of studies (historical data, keywords, and ranking), the identifiers are handled in the following way:

- The Week from November 11 to 16, 2018 (10.11.16.18).
- The Week from November 17 to 24, 2018 (10.11.24.18).
- The Week from November 25 to December 1, 2018 (10.12.1.18).
- The Week from December 2 to 9, 2018 (10.12.9.18).
- The Week from December 10 to 14, 2018 (10.12.14.18).

---

[1] Table 10.21 shows the identifier for each university for influence networks in social networks.

[2] In the Table 10.22 shows the identifier for ICT use networks in universities.

[3] Table 10.23 shows the identifier of the elements for the perspective networks that the society has for universities.

To graphically view the modeled networks, Fig. 10.5 contains the graph of the network 10.11.16.18 using the parametric test.

In Fig. 10.5, it can be seen that three connected communities were formed; however, none of them becomes an independent connected component. Therefore, to analyze which elements are found in each community, Figs. 10.6 and 10.7 show which the elements are corresponding to each one.

The universities that are in the community shown in Fig. 10.6 are: NAUM, AMU, NPI, Center for Research and Advanced Studies of the National Polytechnic Institute "CRANSPI" (Centro de Investigación y Estudios Avanzados del Instituto Politécnico Nacional), Autonomous University of Nuevo Leon "AUNL" (Universidad Autónoma de Nuevo León), Meritorious Autonomous University of Puebla "MAUP" (Benemérita Universidad Autónoma de Puebla), Autonomous University of Chapingo "CAU" (Universidad Autónoma de Chapingo), etc.

The universities that are located in the community of Fig. 10.7 are: Universidad Veracruzana "VU" (University of Veracruz), Sonora Institute Technological "SIT" (Instituto Tecnológico de Sonora), Autonomous University of Baja California "AUBC" (Universidad Autónoma de Baja California), Autonomous University of Querétaro "AUQ" (Autonomous University of Queretaro), Autonomous University

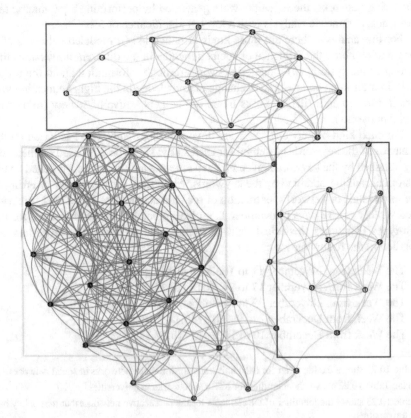

Fig. 10.5 Modeled network with the information of the influence in the social networks of the universities (the year 2018)

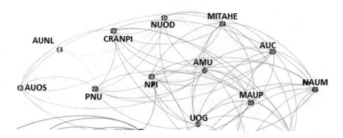

Fig. 10.6 Elements with high value in classic rankings taken from Fig. 10.5

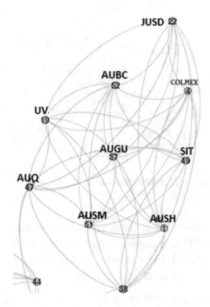

Fig. 10.7 Elements with a low value in classic rankings taken from Fig. 10.5

of the State of Morelos "AUSM" (Universidad Autónoma del Estado de Morelos), and College of Mexico "COLMEX" (El Colegio de México), etc.

In Sect. 10.4, the information and description of each university belonging to each community will be given. Figure 10.8 shows the graph of the network 10.11.24.18 modeled with information about the keywords for the ICTs' influence in the higher education system in Mexico.

In Fig. 10.8, it can be seen that in comparison with the previous network kind (Fig. 10.5), fewer nodes are considered since the other nodes do not provide information or the Social Analytics tool cannot find relevant information about the social networks of some universities.

On the other hand, it can be seen that four connected components are generated: the first and second connected components are made by the universities that obtain the best values in the classic rankings (NAUM, MAU, NPI, CRANSPI, etc.), the third by universities that get intermediate values, and the fourth formed by private universities that receive favorable values.

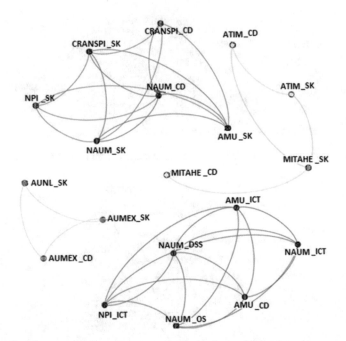

Fig. 10.8 Modeled network with information about the use of ICTs in traditional courses and knowledge sharing (the year 2018)

It is essential to mention that in some cases, there is more than one node to represent the universities; which is presented because several characteristics are known as "keywords" are analyzed. For example, in Fig. 10.8 the NAUM is found five times, where each node represents the following:

- Use of ICTs in traditional courses (ICT).
- Use of ICTs in online and/or distance programs (OS/DSS).
- Use of ICTs to share knowledge (SK).
- Use of ICTs to disseminate the activities of the university (CD).

In some cases, such as Monterrey Institute of Technology and Higher Education "MITAHE" (Instituto Tecnológico y de Estudios Superiores de Monterrey), there are fewer nodes that represent it; this is because there is no information on a particular topic to be evaluated.

Now, Fig. 10.9 shows the graph of the network 10.12.1.18 modeled with the information about the perspective that Mexican and Latin American societies have for the universities of the National higher education system.

In Fig. 10.9, it can be seen that there are fewer nodes concerning the previous networks. Also, as in the case of clustering networks, it can be seen that there are at least two types of universities. In the red box, there are the universities that obtain the best values in the classic rankings; while in the blue box are the elements that obtain lower values (Márquez-Jiménez, 2010).

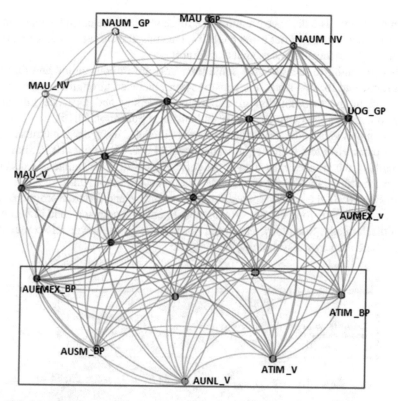

Fig. 10.9 Modeled network with the information of the perspective that the society has for the universities (the year 2018)

It is essential to mention that, as for the previous type (Fig. 10.8), in some cases, there is more than one node representing each university; since when they are built with information about the perspective that society has towards them, there are four types of data:

- Right academic perspective (GP).
- Wrong academic perspective (BP).
- Violence and abuse in the university (V).
- No violence (NV).

In Sect. 10.4.4, the representation of each type of node (GP, BP, V and NV) for universities is described.

10.3.2.3 Modeling of Online & Distance Degree Networks

Due to the importance of NAUM, the current characteristics of ICT in their online and distance programs were studied. The online programs are those in which the consultancies are programmed in the educational institution and can be individual or in the group. The distance programs are those in which the consultancies and all the activities are carried out using the ICTs; the consultancies are group or individual through videoconferences, chats, or forums.

It is essential to mention that some programs are in both online and distance programs, and for this study, the networks were modeled through the information obtained in the Coordination of the Open University and Distance Education (Coordination of the Open University and Distance Education, 2016).

As for previous studies, these networks were modeled using a parametric test and a non-parametric test with a significance level of 5%. Then, Fig. 10.10 shows an example of the graph for the data of the period 2014–2015 for online and distance programs offered by NAUM (for the identifiers see Appendix 10.D).

In Fig. 10.10, there are at least three types of online and distance programs. For example, at the top of the graph are Spanish Language and Literature, Bibliotechnology, Sociology, etc. (blue box), while in the lower left corner (red box), there are Laws, Psychology, Nursing, etc.

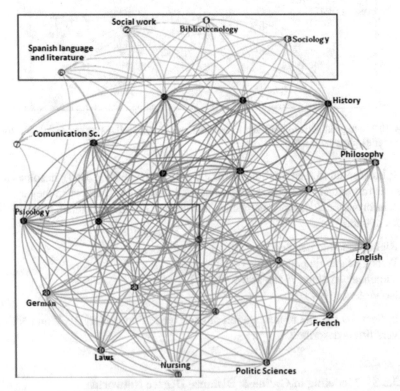

Fig. 10.10 Modeled network through the information of the online and distance programs offered by the NAUM

10.3.2.4   Resolution Method for Maximum Clique Problem

After analyzing the techniques that have been used to solve the MCP, it was noted that swarm intelligence (Wu and Hao 2015a, b) (swarm intelligence systems are typically formed by a population of simple agents that interact locally with each other and with their environment, since they are mostly inspired by nature, especially by specific biological systems.) techniques are better. Therefore, in this work, we used an Ant System (AS) algorithm developed in C language.

AS is an algorithm of interaction among agents that is inspired by a natural metaphor, which is the communication that exists between ants that allows them to find paths from their nest to the sources of food.

The implementation details of the AS algorithm for the optimization problem used for the university ranking are shown below.

(a) Fitness Function. Once the search has started generating initial solutions that are in the feasible region (solutions in which the aggregated nodes have connections to all the other nodes), the fitness function is given by the number of nodes that they are in the solution x (Eq. 10.1)

$$\text{fitness}(x) = \sum_{(i=0)}^{n} c_i$$

$$(10.1)$$

Where, N is the total number of nodes in the original graph and c_i is the cost that the node has in the solution (which will be '1' if it is added and '0' if it is not added).

(b) Representation of the solutions. Each solution is represented by a vector of size n, with $n = |V|$. That is, each element of the vector represents a node $v_i \in V$. The vector at the beginning is with values '0' in each position and when adding a particular node to the solution, it is respective position takes a value of '1'.

If we take as an example the graph seen in Sect. 10.2, the solution that represents the maximum clique in Fig. 10.1 can be seen as follows:

0	0	0	1	1	0	1	1

(c) Initial solution. The process to generate the initial solution is defined as follows:

- To each element in the set of nodes that form the graph, it is given the same probability of being chosen (at the start).
- Once the first node is added to the solution, the neighbors are given the same probability of being chosen (those who are not neighbors of this first aggregate node are discarded).

- As long as there are candidates to add (nodes that are neighbors of all the nodes that are already in the solution that is being built), they are assigned the same probability of being chosen for each one.

Following this process, only the nodes that meet the condition of being neighbors of all the nodes that are already in the solution will be added to the solution. With this, it is ensured that every time a new node is added to the solution, we have a click. Pseudocode 10.1 formalizes the process of generating initial solutions.

Pseudocode 10.1 Generation of initial solutions

```
1: Choose randomly the first node v_i ∈ V
2: C ← v_i
3: Candidates←{v_j/(v_i, v_j) ∈ E}
4: while Candidate! = 0 do
5: Choose a node v_j ∈ Candidates with a probability p(v_j): C ← v_j ∪ C
7: Candidates ← Candidates ∩ {v_k/ (v_j, v_k) ∈ E}
8: end while
9: Return C
```

(d) Definition of the characteristics of the algorithm. This technique requires four control parameters, which were calibrated using the Differential Evolution algorithm *(DE)* (Storn & Price, 1997), where the following values were obtained for each of them:

- Number of iterations: 100.
- The number of ants: 15.
- Evaporation coefficient $(\rho) = 0.25$.
- Coefficient of importance for the pheromone $(\alpha) = 2$.

Also, to attack this problem, the calculation of the selection probability $p(v_j(a + 1))$ of the elements was modified to construct the solution $x(a + 1)$, where k is the ant number and, the iteration number (Eq. 10.2)

$$p\left(v_j\left(a+1\right)\right)=1-\rho\times\frac{\left[\tau_j\left(a+1\right)\right]^{\alpha}}{\sum_{n=0}^{n=\text{candidates}}\left[\tau_{n(a+1)}\right]^{\alpha}} \tag{10.2}$$

Where $v_j (a + 1)$ is the value if the pheromone that has the element j for the iteration $(a + 1)$, α denotes the importance of the pheromone level and $[\tau_l(a + 1)]$ with $l = 1,2,...,n$, is the value of the pheromone.

Of all the nodes. On the other hand, the value of the pheromone is calculated according to Eq. (10.3):

$$\tau_j\left(a+1\right)=\tau_j\left(a\right)+\Delta \tag{10.3}$$

Where each element j is part of the best solution of the current iteration (a); and Δ is defined as the value of the *fitness function* for the best solution found in iteration a. With this, only the pheromone of the nodes that have been part of the best solutions found in each iteration will be strengthened, and the ants will be more likely to add those nodes.

(e) Stop criteria. When the best solution does not change after 20 iterations, or when the total number of iterations is reached.

For more information about the technique, see the work of Dorigo et al. (1996).

10.4   Results

In this Section, the numerical results of the primary metrics of complex networks for the three kinds of network models: clustering based on the main characteristics, influence on social networks, and online and distance programs offered by the NAUM (Sect. 10.4.1) and the results obtained by the AS algorithm when solving the MCP in each of the modeled networks (Sects. 10.4.2, 10.4.3, 10.4.4, and 10.4.5) are presented.

10.4.1   Main Structural Metrics

In this subsection, we show the numerical results for the primary metrics of complex networks, whose concepts are described below:

- Average degree (Erdös & Rényi, 1959). The degree of a node is the number of links connected to it. Then, the average degree is calculated with the degrees of all the nodes that belong to the network.
- The diameter of the network (Albert & Barabási, 2000). It is given by the longest route of all the shortest routes between any pair of nodes.
- The density of the network (Gockel & Werth, 2011). The density describes the part of the potential links in a network of all the real links. A potential link is one that could exist between two nodes. In short, the density measures the degree of connectivity of the network at the global level.
- Modularity (Newman, 2006). Measures the strength of the division of a network into modules or communities. Networks with high modularity have dense links between the nodes within the modules but scattered links between the nodes in different modules. However, it has been shown that modularity cannot detect small communities.
- Clustering coefficient (Watts & Strogatz, 1998). It quantifies how much a node is interconnected with its neighbors.
- Average route length (Albert & Barabási, 2002). Defines the average number of steps that must be traveled through the shortest path for all possible pairs of nodes.

10.4.2 Clustering Networks

Based on the identifiers shown in the previous subsection, the comparisons of the primary metrics for the modeled networks through the data about the universities belonging to the higher education system in Mexico are performed. Because networks 1, 3 and 5, are generated with disaggregated data and networks 2, 4 and 6, with the complete data, the comparison was made for the modeled networks considering and not considering NAUM.

Table 10.1 contains the numeric values of the metrics from the modeled networks with disaggregated data, and Table 10.2 shows the results of the metrics for the modeled networks with complete data.

Table 10.1 Results of the primary metrics from the networks modeled with disaggregated data (with and without NAUM)

ID	A. Degree	W. degree	Diameter	Density	Mod.	Clustering C.	Route L.
10.1.1 with NAUM	37.119	74.238	4	0.64	0.214	0.851	1.41
10.1.1 without NAUM	35.492	70.984	4	0.612	0.201	0.838	1.484
10.1.2 with NAUM	35.695	71.39	3	0.615	0.247	0.852	1.444
10.1.2 without NAUM	35.017	70.034	5	0.604	0.227	0.84	1.508
10.3.1 with NAUM	37.567	75.134	5	0.637	0.202	0.854	1.445
10.3.1 without NAUM	34.068	68.136	5	0.587	0.21	0.841	1.546
10.3.2 with NAUM	37	74	5	0.627	0.203	0.848	1.462
10.3.2 without NAUM	33.356	66.712	4	0.575	0.241	0.837	1.546
10.5.1 with NAUM	41.119	82.238	3	0.709	0.171	0.866	1.31
10.5.1 without NAUM	33.356	66.712	4	0.575	0.241	0.837	1.546
10.5.2 with NAUM	41.22	82.44	3	0.711	0.168	0.869	1.308
10.5.2 without NAUM	33.186	66.372	4	0.572	0.27	0.839	1.54
10.7.1 with NAUM	34.3	68.6	4	0.581	0.256	0.844	1.521
10.7.1 without NAUM	35.898	71.797	4	0.619	0.222	0.849	1.448
10.7.2 with NAUM	38.3	76.6	3	0.649	0.226	0.859	1.396
10.7.2 without NAUM	36	72	4	0.621	0.219	0.848	1.448
10.9.1 with NAUM	34.867	69.733	4	0.591	0.255	0.853	1.503
10.9.1 without NAUM	34.407	68.814	5	0.593	0.238	0.862	1.513
10.9.2 with NAUM	35	70	4	0.593	0.244	0.861	1.504
10.9.2 without NAUM	35.051	70.102	5	0.604	0.229	0.871	1.493

Table 10.2 Results of the main metrics from networks modeled with complete data (with and without NAUM)

ID	A. Degree	W. degree	Diameter	Density	Mod.	Clustering C.	Route L.
10.2.1 with NAUM	52.933	105.867	3	0.897	0.046	0.941	1.104
10.2.1 without NAUM	52	104	2	0.897	0.048	0.939	1.103
10.2.2 with NAUM	46.867	93.733	3	0.794	0.094	0.896	1.218
10.2.2 without NAUM	48.203	96.407	3	0.831	0.076	0.915	1.172
10.4.1 with NAUM	51.633	103.267	2	0.875	0.062	0.928	1.125
10.4.1 without NAUM	50.847	101.695	3	0.877	0.053	0.935	1.126
10.4.2 with NAUM	52.467	104.933	3	0.889	0.053	0.936	1.112
10.4.2 without NAUM	50.136	100.271	3	0.864	0.067	0.923	1.137
10.6.1 with NAUM	46.9333	93.867	2	0.795	0.118	0.897	1.205
10.6.1 without NAUM	42.441	84.881	3	0.732	0.129	0.888	1.303
10.6.2 with NAUM	53.9	107.8	2	0.914	0.037	0.945	1.086
10.6.2 without NAUM	39.932	79.864	3	0.688	0.144	0.86	1.342
10.8.1 with NAUM	47.7	95.4	3	0.808	0.88	0.911	1.2
10.8.1 without NAUM	48.305	96.61	3	0.833	0.078	0.919	1.171
10.8.2 with NAUM	47.56	95.133	3	0.806	0.09	0.911	1.199
10.8.2 without NAUM	51.322	102.644	3	0.885	0.043	0.94	1.117
10.10.1 with NAUM	50.4	100.8	3	0.854	0.072	0.926	1.147
10.10.1 without NAUM	49.356	98.712	2	0.851	0.077	0.916	1.149
10.10.2 with NAUM	50.33	100.667	3	0.853	0.074	0.926	1.146
10.10.2 without NAUM	51.559	103.119	2	0.889	0.053	0.934	1.111

Based on the information shown in Tables 10.1 and 10.2, the numerical values for the average degree (AD) in the modeled networks with disaggregated data are higher in the networks that contemplate to the NAUM, for example: for the network 10.1.1 with NAUM we have value of AD = 37.119, while for the network 10.1.1 without NAUM we have a value of AD = 35.492.

The same fact is repeated for the modeled networks with complete data, as observed in the network 10.4.1 with NAUM, we have a value of AD = 51.633, while for the network 10.4.1 without NAUM we have a value of AD = 50.847.

Also, in the modeled networks with complete data, Clustering Coefficients (CC) values are higher, and Modularity Coefficients (MC) are lower respect to CC and

MC of the networks modeled with disaggregated data. For example, networks 10.2.1 with NAUM and 10.8.1 without NAUM have values of CC = 0.941 and 0.919 and Modularity (MC) of 0.046 and 0.078, respectively; networks 10.10.1 with NAUM and 10.5.2 without NAUM have the values of CC = 0.926 and 0.837 and MC = 0.072 and 0.241 respectively.

Based on the results shown for AD, CC, and MC for all the networks (with and without NAUM using disaggregated and complete data), we can say that the modeled networks with the complete data are more connected than the modeled networks with disaggregated data. The above gives a guideline to say that the NAUM has a vital role in the links of the network, since by eliminating it, the resulting clustering changes.

An important reason for this behavior could be that most universities take NAUM as a model for their development, following their behavior. On the other hand, as mentioned above, NAUM always obtains the best values in each characteristic, so that its elimination entails modifying the maximum evaluation parameters for clustering the rest of the universities.

10.4.3   Influence in Social Networks of the Universities in Mexico

Based on the identifiers shown in the previous subsection, a comparison of the structural characteristics for the modeled networks is made using the information on the influence on social networks of the universities belonging to the higher education system in Mexico obtained through Social Analytics.

Table 10.3 shows numerical results for the modeled networks with the historical data about the influence of the ICTs in the universities for November 16 and 24 and December 1, 9, and 14, 2018.

Table 10.4 shows the results of the main metrics for the modeled networks with the data about the keywords used for the influence of the ICTs in the universities for November 16 and 24 and, December 1, 9 and 14, 2018.

Table 10.3 Results for the primary metrics of the technological influence networks of the universities in Mexico (historical data)

Historical data							
ID	Degree	W. degree	Diameter	Density	Mod.	Clustering C.	Route L.
10.11.16.18	4.7	9.4	7	0.247	0.499	0.771	2.610
10.12.1.18	4.9	9.8	7	0.257	0.427	0.694	2.893
10.12.9.18	5.2	10.4	7	0.273	0.375	0.734	2.825
10.12.14.18	4.8	9.6	7	0.252	0.435	0.7	2.597

Table 10.4 Results for the main metrics for the technological influence networks of the universities in Mexico (keywords)

Keywords ID	Degree	W. degree	Diameter	Density	Mod.	Clustering C.	Route L.
10.11.16.18	3.6	7.2	2	0.189	0.623	0.894	1.133
10.11.24.18	3.6	7.2	2	0.189	0.623	0.894	1.133
10.12.1.18	3.9	7.8	1	0.205	0.641	0.950	1.052
10.12.9.18	3.7	7.4	2	0.195	0.620	0.915	1.106
10.12.14.18	3.4	6.8	3	0.179	0.646	0.890	1.236

Table 10.5 Results for the main metrics for the technological influence networks of the universities in Mexico (ranking)

Ranking ID	Degree	W. degree	Diameter	Density	Mod.	Clustering C.	Route L.
10.11.16.18	8.8	17.6	4	0.463	0.411	0.902	1.857
10.11.24.18	9.7	19.4	4	0.511	0.319	0.881	1.763
10.12.1.18	9.1	18.2	4	0.479	0.385	0.867	1.742
10.12.9.18	9.7	19.4	3	0.511	0.343	0.851	1.668
10.12.14.18	9.4	18.8	3	0.495	0.369	0.873	1.7

Table 10.5 shows the results of the main metrics for the modeled networks through the ranking formed with the historical data of the influence of the ICTs in the universities for November 16 and 24 and December 1, 9, and 14, 2018.

Based on the results shown in Tables 10.3, 10.4, and 10.5, in the modeled networks with the historical data and with the keywords used, the values of AD are lower compared with the values of the modeled networks with the data obtained of the ranking. For example, the networks with identifier 10.12.9.18 have the following values are available for AD: Historical = 5.2, keywords = 3.7, and ranking = 9.7.

On the other hand, the values of CC for the modeled networks with the historical data are smaller compared to the other two kinds of networks. For example, the networks with identifier 10.11.16.18 have the next values: Historical = 0.771, keywords = 0.0.894, and ranking = 0.902.

Furthermore, based on the values obtained for the average length route (ALR) and the diameter (DC), the values are larger for the modeled networks with the historical data.

For example, the networks with identifier 10.12.1.18 have the following values for the DC: Historical = 7, keywords = 1 and, ranking = 4; while for ALR the following values are obtained: Historical = 2.893, keywords = 1.052, and ranking = 1.742.

This behavior is because the modeled networks with keywords and ranking data are almost complete (for the vast majority of the nodes there are direct links between them).

Moreover, it can be seen that the density for the networks modeled with the keywords and historical data has a low value (around of 0.25%); meanwhile, the density for the networks modeled with the ranking data has a value around of 0.5%.

Now, the results of the main metrics for the modeled networks with the information of the influence in social networks of the universities in Mexico are shown. Table 10.6 shows the numerical results of the main metrics for the modeled networks with historical data.

Table 10.7 shows the results of the main metrics for the modeled networks with the data obtained from the keywords used to study the influence on social networks of the universities in Mexico.

Table 10.8 shows the numerical results of the main metrics for the modeled networks within the ranking of historical data of the influence in social networks of the universities in Mexico.

Based on the results of the modeled networks considering historical and keywords data shown in Tables 10.6, 10.7, and 10.8, their AD values are lower in

Table 10.6 Results of the main metrics for the networks modeled with the information about the influence in social networks of the universities in Mexico (historical)

Historical							
ID	Degree	W. Degree	Diameter	Density	Mod.	Clustering C.	Route L.
10.11.16.18	13.808	27.615	8.00	0.271	0.487	0.822	3.065
10.11.24.18	14.231	28.462	6.00	0.279	0.480	0.858	2.166
10.12.1.18	13.077	26.154	9.00	0.256	0.515	0.835	3.344
10.12.9.18	12.500	25.000	9.00	0.245	0.548	0.840	3.408
10.12.14.18	13.654	27.308	9.00	0.268	0.440	0.827	3.381

Table 10.7 Results of the main metrics for influence networks in social networks of the universities in Mexico (keywords)

Keywords							
ID	Degree	W. Degree	Diameter	Density	Mod.	Clustering C.	Route L.
10.11.16.18	15.962	31.923	4.00	0.313	0.506	0.947	1.297
10.11.24.18	15.385	30.769	6.00	0.302	0.513	0.928	1.650
10.12.1.18	16.000	32.000	3.00	0.314	0.514	0.956	1.253
10.12.14.18	15.231	31.846	3.00	0.312	0.518	0.969	1.301

Table 10.8 Results of the main metrics for influence networks in social networks of the universities in Mexico (ranking)

Ranking							
ID	Degree	W. Degree	Diameter	Density	Mod.	Clustering C.	Route L.
10.11.16.18	43.154	86.308	2	0.846	0.031	0.932	1.154
10.11.24.18	41.962	83.923	2	0.823	0.041	0.918	1.177
10.12.1.18	45.846	91.692	2	0.899	0.010	0.955	1.101
10.12.9.18	46.423	92.846	2	0.910	0.007	0.961	1.090
10.12.14.18	46.538	93.077	2	0.913	0.007	0.967	1.087

comparison to the modeled networks with the ranking data. For example, the networks with ID 10.12.14.18 have the following values: Historical = 13.654, keywords = 15.923, and ranking = 46.538.

On the other hand, all networks have a high CC (above 0.8), being even higher for networks modeled with keywords and classification data. In turn, they have lower ALR values than networks modeled with historical data.

The above is observed in the networks with identifier 10.11.16.18, which have the following values for CC: Historical = 0.822, keywords = 0.947, and ranking = 0.932 and for ALR have the next values: Historical = 3.065, keywords = 1.297, and ranking = 1.154.

Additionally, it is important to mention that there are links between the vast majority of the nodes in the modeled networks with the keywords and the ranking data. Also, the modeled networks with the ranking data are denser and have lower modularity compared with the other types.

For example, the networks with the identifier 10.12.1.18 have the following values for density (DN): Historical = 0.256, keywords = 0.314, and ranking = 0.899 and for modularity (MD): Historical = 0.515, keywords = 0.956, and ranking = 0.010.

Next, numerical results of the main metrics for the modeled networks with the data obtained for the perspective that the society has towards the universities are presented. Table 10.9 shows the results of the main metrics for the modeled networks with historical data.

Table 10.10 shows the results of the main metrics for the modeled networks through the data obtained about the keywords used for the studies from the perspective of society towards the universities.

Table 10.11 shows the results of the main metrics for the modeled networks through the historical ranking of the data obtained from the perspective of the Mexican and Latin American society on the universities.

Based on the information shown in Tables 10.9, 10.10, and 10.11, the values for DC and ALR for the modeled networks with the historical data are greater in comparison to the values of the modeled networks with the keywords and ranking data; this is a consequence of having a lower clustering coefficient value.

For example, the networks with ID 10.11.24.18 have the following DC values: Historical = 9, keywords = 2, and ranking = 2 and for ALR: Historical = 3.411, keywords = 1.688, and ranking = 1.277.

Regarding the density, the modeled networks with ranking data are denser concerning the other two kinds of networks. This case is observed in networks with identifier 10.12.9.18 that have the following values: Historical = 0.221, keywords = 0.303, and ranking = 0.771; that is, there are more direct links for the nodes corresponding to the modeled networks through the ranking data.

From the results shown for the three types of studies, it can be concluded that the modeled networks through the ranking are denser, have a higher clustering coefficient, as well as a smaller diameter and route length compared to the modeled networks with the historical and keywords data.

Table 10.9 Results of the main metrics for networks from the perspective that society has towards universities (historical)

Historical							
ID	Degree	W. degree	Diameter	Density	Mod.	Clustering C.	Route L.
10.11.16.18	7.182	14.364	2	0.342	NA	0.861	1.667
10.11.24.18	3.545	7.091	4	0.169	0.62	0.747	2.305
10.12.1.18	4.636	9.273	9	0.221	0.59	0.759	3.411
10.12.9.18	4.636	9.273	9	0.221	0.59	0.759	3.411

Table 10.10 Results of the main metrics for modeled networks with the information of the perspective that the society has towards universities (keywords)

Keywords							
ID	Degree	W. degree	Diameter	Density	Mod.	Clustering C.	Route L.
10.11.24.18	6.273	12.545	2	0.299	NA	0.928	1.706
10.12.1.18	6.636	13.273	2	0.316	NA	0.948	1.688
10.12.9.18	6.364	12.727	2	0.303	NA	0.936	1.701
10.12.14.18	6.455	12.909	2	0.307	NA	0.892	1.697

Table 10.11 Results of the main metrics for the modeled networks with the information of the perspective that the society has towards universities (ranking)

Ranking							
ID	Degree	W. degree	Diameter	Density	Mod.	Clustering C.	Route L.
10.11.16.18	19.364	38.727	2	0.922	NA	0.944	1.087
10.11.24.18	15.727	31.455	2	0.749	NA	0.873	1.260
10.12.1.18	15.364	30.727	2	0.732	NA	0.870	1.277
10.12.9.18	16.182	32.364	2	0.771	NA	0.887	1.238
10.12.14.18	14.636	29.273	2	0.697	NA	0.871	1.312

10.4.4   Online and Distance Programs of NAUM

In this subsection, the studies of the main metrics for the modeled networks through the information obtained about the online and distance programs offered by the NAUM are presented. Table 10.12 shows that the networks for the 2013–2014 and 2014–2015 periods have a higher average grade than the network that represents the years 2012–2013. However, it is important to mention that the network for the period 2012–2013 is a complete graph (any pair of nodes is directly connected).

In addition, it is observed that for all networks, the MC is almost null (values close to zero) and has a higher DN value (values: 0.926 for the period 2013–2014 and 0.8 for the period 2014–2015); therefore, the CC is high (0.96 for the period 2013–2014 and 0.93 for the period 2014–2015).

Table 10.12 Results of the main metrics for NAUM online and distance programs networks

ID	Degree	W. degree	Diameter	Density	Mod.	Clustering C.	Route l.
2012–2013	10	20	1	1	0	1	1
2013–2014	23.153	46.307	2	0.926	0.003	0.96	13
2014–2015	19.2	38.4	2	0.8	0.052	0.93	12

10.4.5 Results for MCP Considering the NAUM

Next, the results obtained using the AS algorithm to solve the instance (specific case of a problem obtained when specifying particular values for all the parameters of the same) of the MCP in the networks modeled considering the NAUM are shown.

Table 10.13 contains the information about the cardinality and the numerical identifiers of the elements that belong to the maximum clique for each network (for more information about the identifiers, see Appendix 10.B).

Based on the information of Table 10.13, for the modeled networks with total data, more elements belong to the clique than the modeled networks with disaggregated data. For example, networks belonging to the year 2013 (10.1.1, 10.1.2 (disaggregated data) and, 10.2.1 and 10.2.2 (total data) have the following results: 10.1.1 = 31, 10.1.2 = 29, 10.2.1 = 51, and 10.2.2 = 44.

Now, to see graphically the elements belonging to the original network and, in turn, the elements that are inside and outside of the clique, Figs. 10.11 and 10.12, which belong to the network 10.1.1 with NAUM, are shown.

In the graph of Fig. 10.11, all the nodes and their original links are shown; meanwhile, in Fig. 10.12, the nodes belonging to the maximum clique and the nodes that do not belong to the maximum clique (in isolation) are shown. Also, it is observed that most of the nodes are outside the clique, so it is important to study the special characteristics of each node.

To better study the network shown in Figs. 10.11 and 10.12, Fig. 10.13 shows the elements that are outside the clique.

In the set of nodes that are outside the clique, there are two types of universities. In the red box, are the universities that obtain the best values in the classic rankings; while in the blue box, are the universities that obtain low values.

Next, Fig. 10.14 shows the complete graph and Fig. 10.15 shows the elements that are outside the maximum clique found for the network 10.4.2 with NAUM.

In Fig. 10.15, it is observed that the elements belonging to the clique are more in comparison to the network of Fig. 10.12; therefore, it is important to study the special characteristics of each node.

As in the previous case, Fig. 10.16 contains the elements that are outside the clique for the network 10.4.2 with NAUM.

In the set of nodes that are outside the clique, there are two types of universities. In the red box are the universities that obtain the best values in the classic rankings, while in the blue box are the universities that obtain low values.

Table 10.13 Maximum clique results (Networks with NAUM)

Network	Clique size	Clique elements
10.1.1 with NAUM	31	1 3 4 6 7 8 9 10 11 13 15 16 18 19 21 25 26 27 29 31 35 37 39 45 46 47 49 53 55 56 60
10.1.2 with NAUM	29	1 6 8 9 10 11 13 14 15 16 18 19 20 21 25 27 29 31 37 39 45 46 47 49 53 55 56 59 60
10.2.1 with NAUM	51	1 3 4 6 7 8 9 10 11 12 13 14 15 16 17 18 19 20 21 23 25 26 27 28 29 31 32 33 34 36 37 39 40 41 42 44 45 46 47 48 49 50 52 53 54 55 56 57 58 59 60
10.2.2 with NAUM	44	1 3 4 6 7 8 9 10 11 12 13 14 15 16 17 18 19 20 21 25 26 27 28 29 31 32 34 35 36 39 40 41 42 43 44 45 47 50 52 53 55 56 57 60
10.3.1 with NAUM	28	1 2 4 6 7 8 9 11 13 14 15 19 20 25 26 29 31 32 37 39 46 47 49 50 56 58 59 60
10.3.2 with NAUM	29	1 3 4 6 7 8 9 10 11 13 14 15 19 20 25 26 31 32 37 39 45 46 47 49 55 56 58 59 60
10.4.1 with NAUM	47	1 3 4 6 7 8 9 10 11 12 13 14 15 16 18 19 20 21 25 26 27 28 29 31 32 34 36 37 38 39 41 42 44 45 46 47 49 50 52 53 54 55 56 57 58 59 60
10.4.2 with NAUM	49	1 3 4 6 7 8 9 10 11 12 13 14 15 16 17 18 19 20 21 23 25 26 27 28 29 31 32 34 35 36 37 39 40 42 44 45 46 47 48 49 50 52 53 55 56 57 58 59 60
10.5.1 with NAUM	32	1 3 4 7 8 9 10 12 13 16 17 19 20 21 25 26 27 31 32 33 34 35 39 40 42 45 46 47 53 55 56 60
10.5.2 with NAUM	33	1 2 3 4 6 8 9 10 11 12 13 14 15 16 19 20 21 25 26 27 29 31 32 37 39 45 46 47 50 55 56 59 60
10.6.1 with NAUM	41	1 3 4 6 7 8 9 10 12 13 14 15 16 18 19 20 21 23 25 26 27 28 29 31 32 34 36 37 39 41 43 44 45 47 50 53 55 56 57 58 60
10.6.2 with NAUM	49	1 3 4 6 7 8 9 10 11 12 13 14 15 16 18 19 20 21 23 24 25 26 27 28 29 31 32 34 36 37 39 41 44 45 46 47 48 49 50 51 52 53 54 55 56 57 58 59 60
10.7.1 with NAUM	29	2 3 5 6 7 8 9 11 13 14 15 16 17 20 25 26 28 29 32 39 42 46 47 50 54 56 57 58 59
10.7.2 with NAUM	32	2 3 5 6 7 8 9 10 11 13 14 15 16 17 19 20 26 27 28 29 32 42 43 46 47 49 50 54 56 57 58 59
10.8.1 with NAUM	45	1 2 3 4 5 6 7 8 9 10 12 13 14 15 16 18 19 20 25 27 28 29 31 32 36 39 40 41 42 43 44 45 46 47 48 49 50 53 54 55 56 57 58 59 60
10.8.2 with NAUM	45	1 2 3 4 6 7 8 9 10 12 13 14 15 16 18 19 20 25 26 27 28 29 31 32 33 36 39 40 41 43 44 45 46 47 48 49 50 53 54 55 56 57 58 59 60
10.9.1 with NAUM	31	1 2 4 5 6 7 8 9 10 13 14 15 16 19 20 26 27 28 29 32 37 39 42 46 47 49 50 56 58 59 60
10.9.2 with NAUM	33	1 2 3 4 5 6 7 8 9 10 11 13 14 15 16 19 20 25 26 27 29 32 37 39 42 46 47 49 50 56 58 59 60
10.10.1 with NAUM	47	1 2 3 4 5 6 7 8 9 10 11 12 13 14 15 16 17 18 19 20 21 25 26 27 28 29 32 33 34 38 39 40 41 42 45 46 47 48 49 50 54 55 56 57 58 59 60
10.10.2 with NAUM	47	1 2 3 4 5 6 7 8 9 10 11 12 13 14 15 16 17 18 19 20 25 26 27 28 29 32 33 34 36 39 40 41 42 45 46 47 48 49 50 52 54 55 56 57 58 59 60

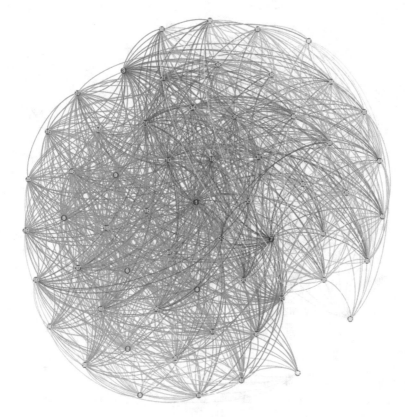

Fig. 10.11 Complete graph of the network 10.1.1 with NAUM

Based on the information shown in this subsection, for both kinds of networks (with and without NAUM) modeled with the disaggregated and total data, the elements that are outside the clique are divided into at least two sets: in the first one, there are the universities that obtain the best values in the classic rankings, while in the second, we can find the universities that obtain the lowest values.

10.4.6   Results for MCP without Considering the NAUM

Next, Table 10.14 contains information about the cardinality and the numerical identifiers of the elements that belong to the maximum clique for each network obtained using the AS algorithm to solve the MCP (for more information about the identifiers, see Appendix 10.B).

Based on the information of Table 10.14, for the modeled networks with the complete data, more elements are belonging to the clique than the modeled networks with disaggregated data. For example, networks belonging to the year 2017 (9.1, 10.9.2 (disaggregated data) and 10.10.1 and 10.10.2 (total data) have the following results: 10.9.1 = 33, 10.9.2 = 34, 10.10.1 = 42, and 10.10.2 = 47.

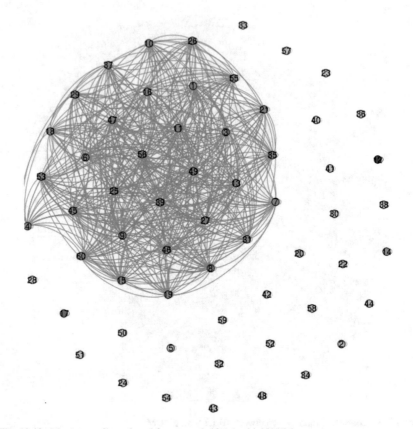

Fig. 10.12 Maximum clique found for network 10.1.1 with NAUM

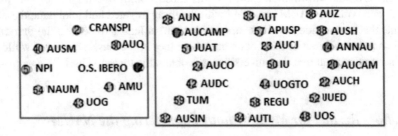

Fig. 10.13 Elements outside the maximum clique found for network 10.1.1 with NAUM

In order to graphically see the elements that belong to the original network and in turn the elements that belong to the clique, Fig. 10.17 shows the elements that belong to network 10.1.1 without NAUM.

On the other hand, Fig. 10.18 shows the nodes that belong and the nodes that do not belong to the maximum clique (in isolation). Also, it can be easily seen that most of the nodes are outside the clique, so it is essential to study the special characteristics of each node.

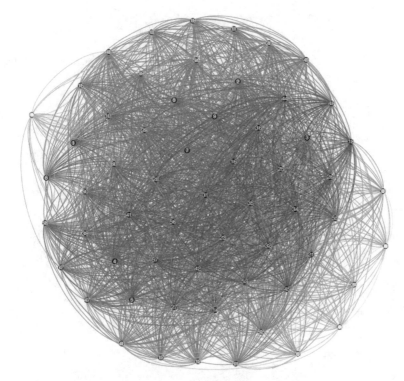

Fig. 10.14 Complete graph for network 10.4.2 with NAUM

Further, to better perform studies on the elements that are outside the clique of the network 10.1.1 without NAUM is shown in Fig. 10.19.

In the set of nodes that are outside the clique, there are two types of universities. In the red box are the universities that obtain the best values in the classic rankings, while in the blue box are the universities that obtain worst values.

Figure 10.20 contains the elements that are outside the maximum clique found for the network 10.4.2 without NAUM. Figure 10.20 shows all the nodes and their original links; while Fig. 10.21 presents the nodes belonging to the maximum clique and the nodes that do not belong to the clique (in isolation). Also, it is essential to mention that the elements belonging to the clique are more in comparison to the network of Fig. 10.18; therefore, it is fundamental to study the particular characteristics of each node. On the other hand, Fig. 10.22 shows the elements that are outside (left) and inside (right) the clique for the network 10.4.2 without NAUM.

Based on the information presented in this subsection, the NAUM has a vital role in connection with other universities, since the elimination of this node causes the clustering changes. As mentioned, this behavior occurs because NAUM is the university that obtains the best values in most of the characteristics used for the modeling of networks.

On the other hand, since the networks are generated from the clustering of the universities, it is observed that the nodes belonging to the cliques of each network are those that are at the middle of the classical rankings, that is, they do not have a high ranking value or low ranking value.

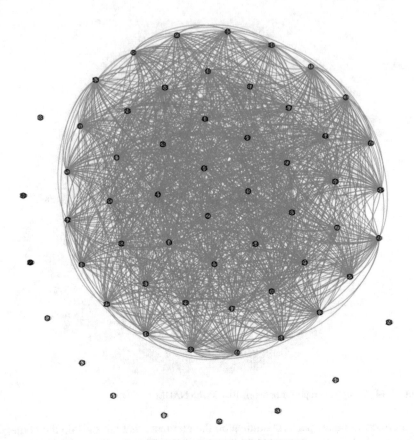

Fig. 10.15 Maximum clique found for network 10.4.2 with NAUM

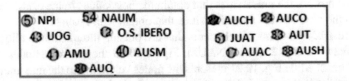

Fig. 10.16 Elements that are outside the maximum clique found for network 10.4.2 with NAUM

To verify the above, the nodes that are inside and outside the maximum clique for each of the networks are studied. Next, the main characteristics that share or distinguish the universities that are inside and outside of the maximum clique are shown.

For inside the clique, the following actions are achieved according to the following relation:

- Most of their teachers work by the hour.
- They offer fewer postgraduate programs.
- They offer a small number of degree programs.

Table 10.14 Maximum clique results without NAUM

Network	Clique Size	Clique elements
10.1.1 without NAUM	30	1 3 7 8 10 12 13 15 16 18 19 21 25 26 27 30 31 32 33 34 35 39 40 42 44 45 46 47 53 54
10.1.2 without NAUM	28	1 3 6 7 8 10 12 13 14 15 16 18 21 25 26 27 28 31 32 35 39 44 45 46 47 53 54 55
10.2.1 without NAUM	50	3 4 6 7 8 9 10 11 12 13 14 15 16 17 18 19 20 21 22 24 25 26 27 28 29 32 33 34 35 36 37 38 39 40 42 44 45 46 47 48 49 50 51 52 53 54 55 56 57 58
10.2.2 without NAUM	46	3 4 6 7 8 9 10 11 12 13 14 15 16 17 18 19 20 21 22 25 26 27 28 29 32 33 34 35 36 37 38 39 42 44 45 46 47 49 50 52 53 54 55 56 57 58
10.3.1 without NAUM	28	1 2 3 6 7 8 10 12 14 15 16 18 20 22 26 27 28 33 35 36 40 42 44 45 51 52 54 56
10.3.2 without NAUM	27	3 7 8 10 12 17 18 21 22 24 26 27 28 30 33 34 36 40 42 44 45 48 51 52 53 54 56
10.4.1 without NAUM	49	1 3 4 6 7 8 9 10 11 12 13 14 15 16 17 18 19 20 21 22 23 24 25 26 27 28 29 32 34 35 36 37 38 39 40 42 44 45 46 47 49 50 52 53 54 55 56 57 58
10.4.2 without NAUM	46	1 3 4 6 7 8 10 11 12 13 14 15 16 17 18 19 20 21 22 23 24 25 26 27 28 29 32 35 36 37 38 39 40 42 45 46 47 48 49 50 53 54 55 56 57 58
10.5.1 without NAUM	26	3 7 8 10 12 17 18 19 21 22 24 26 27 28 32 33 34 38 40 44 45 51 52 54 55 56
10.5.2 without NAUM	24	3 7 8 10 12 16 18 19 21 22 24 26 27 28 32 33 35 40 42 44 45 51 54 55
10.6.1 without NAUM	40	1 3 4 6 7 8 9 10 11 12 13 14 15 16 17 18 19 20 25 26 27 28 29 35 37 38 39 42 43 45 46 47 49 50 54 55 56 57 58 59
10.6.2 without NAUM	37	1 3 4 5 6 7 8 10 12 13 14 15 16 17 18 25 26 27 28 29 32 34 35 36 39 40 42 44 45 47 48 50 52 53 54 55 56
10.7.1 without NAUM	29	1 3 5 6 7 9 11 14 15 16 17 19 20 25 26 27 28 29 32 37 46 47 49 50 54 55 57 58 59
10.7.2 without NAUM	31	1 2 3 5 6 9 11 14 15 16 17 19 20 25 26 27 28 29 32 37 39 46 47 49 50 54 55 56 57 58 59
10.8.1 without NAUM	44	1 2 3 4 6 7 8 9 10 12 13 14 15 16 18 19 20 25 26 27 28 29 31 32 34 36 39 40 41 42 43 44 45 46 47 49 50 53 54 55 56 57 58 59
10.8.2 without NAUM	48	1 2 3 4 6 7 8 9 10 12 13 14 15 16 17 18 19 20 25 26 27 29 30 31 32 33 34 36 38 39 40 41 42 43 44 45 46 47 48 49 50 53 54 55 56 57 58 59
10.9.1 without NAUM	33	1 2 3 6 7 8 9 10 11 13 14 15 16 17 19 20 26 27 28 32 35 37 39 41 46 47 49 50 54 55 57 58 59
10.9.2 without NAUM	34	1 2 3 6 8 9 10 11 13 14 15 16 17 19 20 25 26 27 28 29 32 35 37 39 41 46 47 49 50 54 55 57 58 59
10.10.1 without NAUM	42	1 3 4 6 7 8 10 11 12 14 15 16 17 18 19 20 21 26 27 29 31 32 33 34 35 36 38 40 41 42 44 45 46 47 48 50 51 54 55 56 57 58
10.10.2 without NAUM	47	1 2 3 4 5 6 7 8 10 11 12 13 14 15 16 17 18 19 20 21 25 26 27 29 31 32 33 34 35 36 37 39 40 41 42 44 45 46 47 48 49 50 54 55 56 57 58

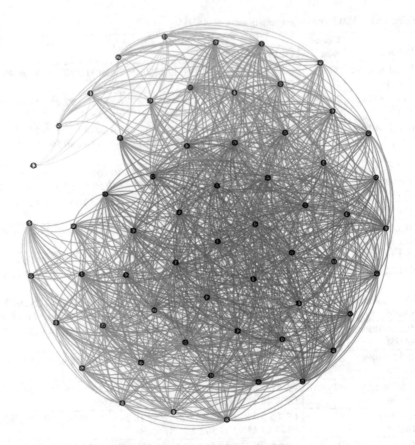

Fig. 10.17 Complete graph for the network 10.1.1 without NAUM

- Not many of its professors belong to the NSR of MNCST.
- Few of its professors have the support of the Program for Professional Development Teacher, For The Superior Type "PPDTST" (Programa para el Desarrollo Profesional Docente, para el Tipo Superior).
- The production of articles and documents in International Scientific Indexing (ISI) is low, as well as their citations.
- The production of articles and documents in the SCOPUS database is low, as are its citations.
- They have a few academic journals.
- Of its undergraduate programs, there are few accredited by the CAHE.
- Have few graduate programs in NRQP of MNCST.
- Most of its professors have only undergraduate studies, and only a handful have a doctoral degree.

Concerning to the outside the clique, a series of actions are performed based on the following sequence:

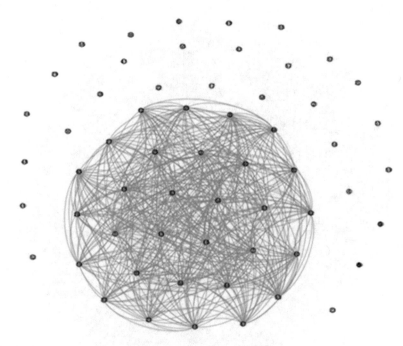

Fig. 10.18 Maximum clique found for the network 10.1.1 without NAUM

Fig. 10.19 Elements that are outside the maximum clique found for the network 10.1.1 without NAUM

- In some universities, almost all of their teachers work full-time, while in others, nearly all of their teachers work by the hour.
- Some universities offer many graduate programs, while others offer fewer or no graduate degrees.
- Some universities offer many undergraduate programs, while others offer less.
- In some universities, the majority of teachers belong to the NSR and have the support of the Program for Professional Development Teacher, for the Superior Type "PPDTST" (Programa para el Desarrollo Profesional Docente, para el Tipo Superior PRODEP), while in a few other professors belong to the NSR and/or have the support of the PPDTST.

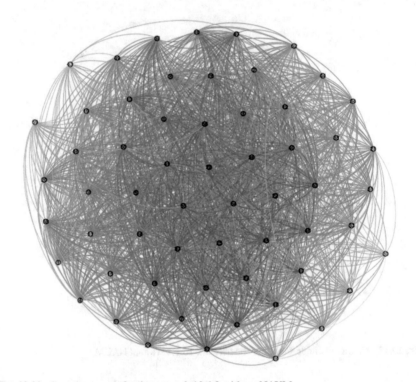

Fig. 10.20 Complete graph for the network 10.4.2 without NAUM

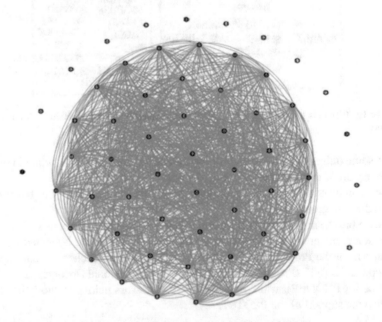

Fig. 10.21 Maximum clique found for the network 10.4.2 without NAUM

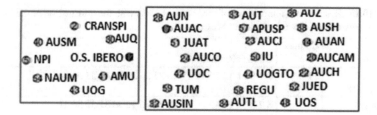

Fig. 10.22 Elements that are outside the maximum clique found for the network 10.4.2 without NAUM

- Some universities have high production of articles and documents in indexes like ISI and SCOPUS, while others have little production.
- Some universities have many academic journals, while others have few.
- Some universities have few graduate programs in NRQP, and in others, most of their graduate programs are in NRQP.

Given the previous lists, the universities that are in the clique share some traits, which do not allow them to have a high ranking. But, studying the nodes that are outside the clique, it can be observed that there are universities that share lower values, in some of these characteristics; contrary to the universities that obtain better ranking, which present higher values in most of these characteristics.

It should be noted that both Postgraduate College "PC" (Colegio de Posgraduados) and CRANSPI are within the clique, although these have a high ranking in some of the features shown above. This is due to both being centers that only offer postgraduate studies, so there are characteristics where they get low ranking.

On the other hand, based on the traits shared by the universities that are outside the clique, there are two groups of universities, with an opposite ranking. Also, the universities that are in the clique have characteristics that could be connected to universities with a high ranking, or with universities with a low ranking, or with both, that is, within the network, these nodes behave as interconnection nodes.

10.4.7   Results for MCP for the Modeled Networks with the Influence in Social Networks of the Universities in Mexico

In this subsection, the results obtained by AS for the MCP in the modeled networks with the information about the influence in social networks of the universities in Mexico is shown. Next, the numerical results are exhibited in Table 10.15.

Based on the results of Table 10.15, for the modeled networks with the ranking data, more elements belong to the clique than the modeled networks with the keywords and historical data. For example, the networks with identifier 10.11.16.18,

Table 10.15 Results maximum clique of modeled networks with the information of the influence in social networks of the universities in Mexico

Network	Clique Size	Clique elements
Historical 10.11.16.18	17	3 7 8 11 12 15 16 17 18 25 26 29 30 31 33 36 41
Historical 10.11.24.18	17	3 7 8 11 12 15 16 17 18 25 26 29 30 31 33 36 41
Historical 10.12.1.18	16	3 7 8 11 12 15 16 17 18 25 26 29 30 31 33 36
Historical 10.12.9.18	16	7 8 11 12 15 16 17 18 23 25 26 29 30 31 33 36
Historical 10.12.14.18	15	2 3 7 8 12 16 18 25 29 30 31 33 36 41 43
Keywords 10.11.16.18	21	1 3 12 15 16 17 20 22 23 24 25 28 30 31 32 34 35 37 40 43 44
Keywords 10.11.24.18	21	1 3 12 15 16 17 20 22 23 24 25 28 30 31 32 34 35 37 40 43 44
Keywords 10.12.1.18	21	1 3 12 15 16 17 20 22 23 24 25 28 30 31 32 34 35 37 40 43 44
Keywords 10.12.9.18	52	1 2 3 4 5 6 7 8 9 10 11 12 13 14 15 16 17 18 19 20 21 22 23 24 25 26 27 28 29 30 31 32 33 34 35 36 37 38 39 40 41 42 43 44 45 47 48 49 50 51 52
Keywords 10.12.14.18	21	1 3 12 15 16 17 20 22 23 24 25 28 30 31 32 34 35 37 40 43 44
Ranking 10.11.16.18	40	3 4 5 7 8 9 11 12 13 15 16 17 20 21 22 23 24 26 27 29 30 32 33 34 35 38 39 40 41 42 43 44 45 47 48 49 50 51 52
Ranking 10.11.24.18	38	1 2 3 4 5 7 8 11 12 13 15 16 17 18 19 20 22 24 26 27 28 31 32 33 36 37 40 41 43 44 45 47 48 49 50 51 52
Ranking 10.12.14.18	48	1 2 3 4 5 7 8 9 10 11 12 13 14 15 16 18 19 20 21 22 23 24 25 26 27 28 30 31 32 34 35 36 37 38 39 40 41 42 43 44 45 47 48 49 50 51 52
Ranking 10.12.9.18	47	1 2 3 4 5 7 8 9 10 11 12 14 15 16 18 19 20 21 22 23 24 25 26 27 28 30 31 33 34 35 36 37 38 39 40 41 42 43 44 45 47 48 49 50 51 52
Ranking 10.12.1.18	45	1 2 3 4 5 7 8 9 10 11 12 14 15 17 18 19 20 21 22 23 24 25 26 27 29 32 33 34 35 36 38 39 40 41 42 43 44 45 47 48 49 50 51 52

have the next values: Historical = 17, keywords = 21, and ranking = 40. To improve the study of the results obtained, in Fig. 10.19, the elements found within the maximum clique found for the modeled network with historical data with identifier 10.12.14.18 are shown.

In Fig. 10.23, if the connections between the elements belonging to the clique and the parts that are outside it are eliminated, three connected components are generated, where, in the component with green elements, are universities that obtain high value in the classical rankings; in the component with blue elements, are the universities that obtain the intermediate values; while in the component with red elements, are the universities that obtain the lowest values. Now, Fig. 10.24 that

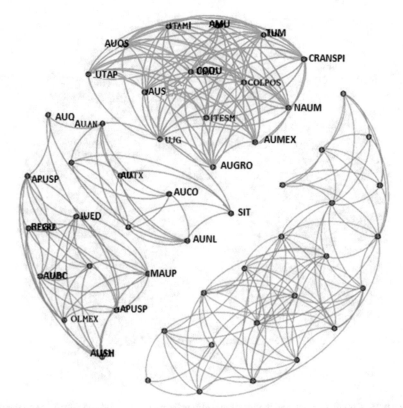

Fig. 10.23 Complete graph and the maximum clique found for the modeled network with historical data of the influence on social networks of the universities in Mexico (identifier 10.12.14.18)

contains the graph of the elements that are inside the maximum click found for the network modeled from the information of the keywords with the identifier 10.11.24.18 is shown.

In Fig. 10.24, it can be seen that at least four connected components are generated, where, in the component with blue elements are the universities that obtain high value in classical rankings (e.g., NAUM, AMU, NPI); in the green, dark green and orange components are the universities that obtain intermediate value; while, in the connected component with pink elements are the universities with low ranking value (University of Yucatan "UY" (Universidad Autónoma de Yucatán), University of Quintana Roo "UOQROO" (Universidad Autónoma de Quintana Roo), University Autonomous of Baja California "AUBC" (Universidad Autónoma de Baja California), etc.).

Finally, Fig. 10.25, contains the graph of the elements that are within the maximum clique found for the modeled network with the information of the ranking with identifier 10.11.24.18 is shown.

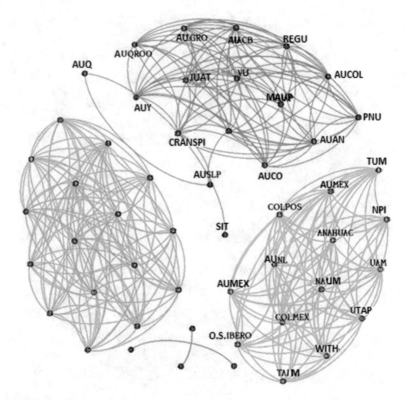

Fig. 10.24 Complete graph and the maximum clique found for the modeled network with the data of the keywords used in the studies of the influence on social networks of the universities in Mexico (identifier 10.11.24.18)

In Fig. 10.25, it can be seen that two connected components are generated, wherein the smallest connected component (red elements), are the universities that are outside the clique, which present high and low values in the classic rankings; while in the component of green elements are the universities that obtain an intermediate value.

Based on the information of Table 10.15 and in Figs. 10.23, 10.24, and 10.25, the universities that obtain an intermediate ranking in the clustering networks are those that are inside the clique as they are: the Autonomous Popular University of the State of Puebla "APUSP"(Universidad Popular Autónoma del Estado de Puebla), Meritorious Autonomous University of Puebla "MAUP" (Benemérita Universidad Autónoma de Puebla), COLPOS, etc.

As for the clustering networks, two remaining groups are generated outside the clique, where, in the first group there are some universities that obtain better values in the original ranking (NAUM, NPI, MAU, MITAHE), the Autonomous University of the State of Mexico "AUSMEX" (Universidad Autónoma del Estado de México), University of Guadalajara "UOG" (Universidad de Guadalajara), etc.). While in the

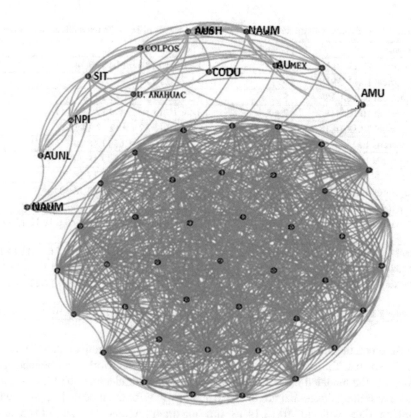

Fig. 10.25 Maximum clique found for the modeled network with the information about the ranking of influence in the social networks of the Mexican universities

second group some universities obtain the worst values in the original ranking Pedagogical National University "PNU" (Universidad Pedagógica Nacional), JUAT, Autonomous University of Guerrero "AUGU" (Universidad Autónoma de Guerrero), etc.).

This fact implies that the universities that obtain the best-ranking values are those that have the best infrastructure, the highest reputation and therefore, highest demand; so social network users focus more on these.

There are cases, such as the Technological University of Mexico "TUM" (Universidad Tecnológica de México), which despite not being a university that has good value in the classical ranking, in this ranking is one of the best, this is probable because it is a private university and has high investment in advertising.

Regarding universities that obtain low ranking values, the fact that those universities with little publicity, low prestige, and weak demand coincides, so they are not impressive for social network users.

In Table 10.16, the results of the two largest cliques are shown for the networks obtained through data about the use of ICTs from universities in Mexico.

Table 10.16 Results maximum clique of networks of use of ICTs in the universities in México

Network	Clique Size	Clique elements
Historical 10.11.16.18	6 y 5	4 5 6 7 10 18 y 9 11 12 13 14
Historical 10.11.24.18	6	4 6 7 10 16 18
Historical 10.12.1.18	5	5 6 7 10 18 y 9 11 12 13 14
Historical 10.12.9.18	7 y 5	3 4 6 10 15 16 18 y 9 12 13 14 17
Historical 10.12.14.18	6 y 5	3 4 6 10 16 18 y 9 12 13 14 17
Keywords 10.11.16.18	6 y 5	1 6 7 9 11 12 y 4 15 17 18 20
Keywords 10.11.24.18	6 y 5	1 6 7 9 11 12 y 4 15 17 18 20
Keywords 10.12.1.18	6	1 6 7 9 11 12 y 4 14 15 17 18 20
Keywords 10.12.9.18	6	1 6 7 9 11 12 y 4 14 15 17 18 20
Keywords 10.12.14.18	5	4 15 17 18 20 y 1 6 9 11 12
Ranking 10.11.16.18	11 y 7	2 3 4 7 8 11 13 14 16 19 20 y 5 6 10 12 15 17 18
Ranking 10.11.24.18	12 y 6	1 2 3 4 7 9 13 14 16 17 19 20 y 5 6 10 11 15 18
Ranking 10.12.14.18	11 y 6	1 3 8 9 10 12 13 16 17 18 19 y 4 5 12 14 15 19
Ranking 10.12.9.18	11 y 7	1 3 8 9 10 12 13 16 17 18 19 y 4 5 6 11 14 15 20
Ranking 10.12.1.18	10 y 8	1 2 3 4 8 9 11 13 17 18 y 5 6 7 10 15 16 19 20

Based on the information shown in Table 10.16, the maximum cliques found are small, so that, the second maximum cliques for each network are obtained. For example, the modeled network with historical data with identifier 10.11.16.18, has two maximum cliques that are of size 6 and 5; while for the modeled network with ranking data with id 10.12.14.18 has maximum cliques of size 11 and 6, respectively.

To better study the elements found in these cliques, Fig. 10.26 presents the modeled network from historical data with identifier 10.12.9.18.

The behavior of the elements found in Fig. 10.26 are as follows: In the purple component, are the universities that use ICTs to share knowledge; in the orange component, are universities that use ICTs for cultural dissemination and have distance and/or online programs; and in the green part are the universities that use ICTs within their traditional educational programs.

On the other hand, Fig. 10.27 shows the graph of the cliques found for the modeled network with the keywords information about the use of ICTs in universities in Mexico (identifier 10.11.24.18).

In the graph of Fig. 10.27, four connected components are generated: In the connected component with green elements, are the universities that make use of ICTs to share knowledge and/or they offer online or distance programs; in the component with purple elements, are the universities that use the ICTs for cultural dissemination and; in the connected component of orange elements, are the universities that have ICTs integrated inside their traditional system.

Further, Fig. 10.28 that contains the graph of the cliques found for the modeled network with the data of the ranking about the use of ICTs in universities in Mexico (identifier 10.12.1.18) is shown.

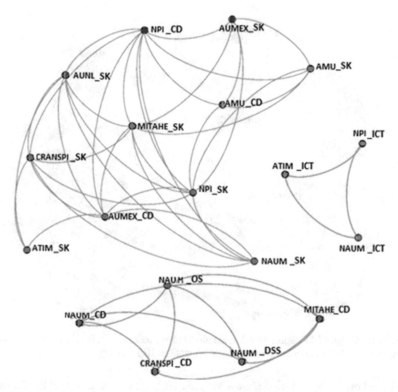

Fig. 10.26 Graph of the cliques found in the modeled network with historical information about the use of ICTs in the universities in Mexico (identifier 10.12.9.18)

In Fig. 10.28, two connected components are generated: The first contains all universities that use ICTs to share knowledge and in its traditional, distance or online programs (purple elements); while the second shows the universities that make use of ICTs for cultural dissemination.

Based on the information of Table 10.16 and Figs. 10.26, 10.27, and 10.28, it can be seen that only the information is modeled for the universities that are leaders in the previous classifications and also for the universities that despite the not being leaders, they have resources to make greater use of ICT (since they are private universities).

At the same time, it can be seen that several components are generated where the universities that have more resources are located in one component (e.g., Autonomous Technological Institute of Mexico "ATIM" (Instituto Tecnológico Autónomo de México), MITAHE, etc.) and in another component, those universities that obtain values high in the previous rankings (NAUM, NPI, AMU, etc.).

Now, regarding the perspective that the society has towards the various universities, in Table 10.17, the results for the maximum cliques belonging to the modeled networks from the information obtained for this topic are shown.

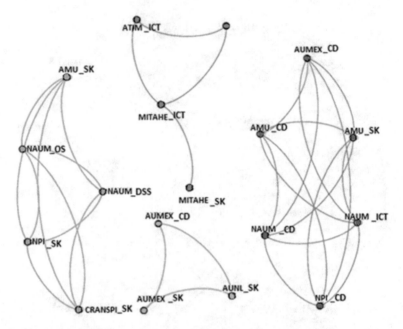

Fig. 10.27 Graph of the cliques found in the modeled network through the keywords information about the use of ICTs in universities in Mexico (identifier 10.11.24.18)

Based on the information shown in Table 10.17, it can be seen that for the modeled networks with the ranking data, there are maximum cliques of large size with respect to the modeled networks with the historical and keywords data. Therefore, the two maximum cliques were obtained for both kinds of networks.

Figure 10.29 contains the graph of the cliques obtained for the modeled network from the historical data about the social perspective for the universities (identifier 10.12.1.18) is shown.

For the graph in Fig. 10.29, there are four components connected: the green and violet components contain the universities that have high prestige and/or are not considered violent; the blue component contains the universities that have low prestige and; the orange component shows the universities that the society perceives as violent (gender violence, discrimination, etc.).

Figure 10.30 shows the graph of the cliques obtained for the modeled network from the keywords data about the social perspective for universities (identifier 10.11.24.18).

As for the graph in Fig. 10.29, there are four connected components: the green and blue components contain the universities that, thanks to advertising on social networks, society has a good impression of; the purple component contains the universities that have a good impression thanks to their academic activities; and the orange component contains the universities that have presented violent acts and/or have taken measures to eradicate them.

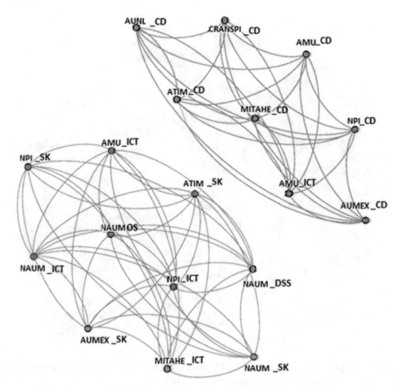

Fig. 10.28 Graph of the cliques found in the modeled network with the data of the ranking about the use of ICTs in the universities in Mexico (identifier 10.12.1.18)

Table 10.17 Results maximum clique of networks modeled with the information of the perspective that the society has for the universities

Network	Clique Size	Clique elements
Historical 10.11.16.18	5	10 11 15 16 19 y 8 9 12 14 17
Historical 10.11.24.18	5 y 4	10 11 15 16 19 y 1 2 3 21 y 7 9 14 17
Historical 10.12.1.18	6 y 5	4 7 8 9 14 22 y 10 11 15 16 17
Historical 10.12.9.18	7 y 5	10 11 15 16 17 19 y 4 7 8 9 14 22
Keywords 10.11.24.18	7 y 5	3 5 9 11 12 17 18 y 6 8 10 14 20
Keywords 10.12.1.18	7 y 5	3 5 9 11 12 17 18 y 1 2 7 13 16
Keywords 10.12.9.18	7 y 5	3 5 9 11 12 17 18 y 6 8 10 14 20 y 1 2 7 13 16
Keywords 10.12.14.18	7 y 5	3 5 9 11 12 17 18 y 6 8 10 14 20
Ranking 10.11.16.18	16	2 4 5 6 7 8 10 11 12 13 14 15 16 17 19 20
Ranking 10.11.24.18	11	1 3 6 7 8 9 11 13 17 19 20
Ranking 10.12.1.18	11	4 6 7 8 9 11 12 15 18 19 20
Ranking 10.12.9.18	12	1 3 4 5 8 10 11 12 15 18 19 20
Ranking 10.12.1.18	12	1 4 5 6 7 9 10 12 15 16 19 20

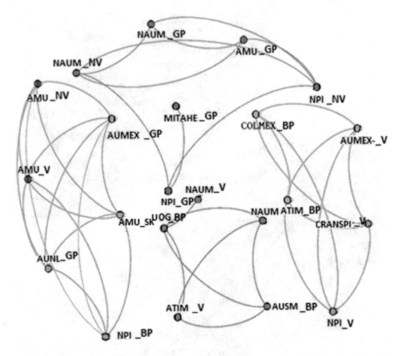

Fig. 10.29 Graph of the cliques obtained for the modeled network with historical data about the social perspective for universities (identifier 10.12.1.18)

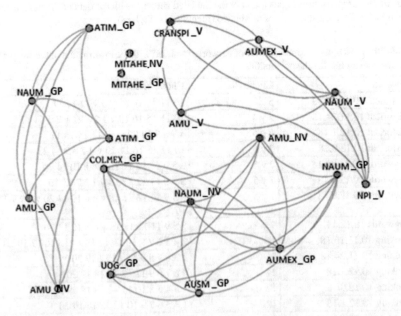

Fig. 10.30 Graph of the cliques obtained for the modeled network from the keywords with identifier 10.11.24.18, on the data of the social perspective for the universities

Finally, Fig. 10.31 shows the cliques obtained for the network modeled from the data of the ranking with the identifier 10.12.14.18 on the data of the social perspective for the universities.

In Fig. 10.31, there are two connected components: the green component contains the universities that have high academic prestige for society; while the pink component is formed by universities that have registered violent acts and/or have low academic prestige.

Therefore, based on the information shown in Table 10.16 and Fig. 10.29, 10.30, and 10.31, it can be seen that the XOVI tool only obtains information about universities that have a high value in the classical classifications, that the others are not relevant in social networks.

In general, the importance of the academic activities realized by each university directly influences the opinion that society has towards it. For example, in research, NAUM, NPI, and AMU are leaders, while in advertising and diffusion of the academic offer, ATIM and MITAHE are leaders; what helps these universities to be considered among the most prestigious universities in the country.

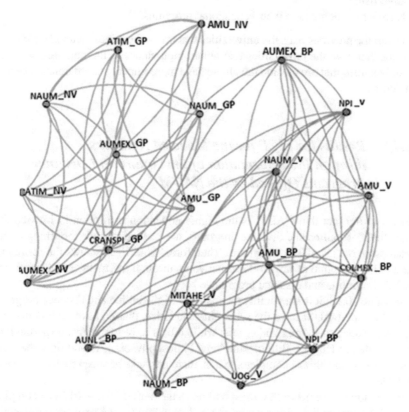

Fig. 10.31 Graph of the cliques obtained for the modeled network with the ranking data about the social perspective for the universities (identifier 10.12.14.18)

To be able to verify the above, the nodes that are inside and outside the maximum clique for each network, are studied. The following are the main characteristics that share or distinguish the universities that are inside and outside of it.

In regards with inside the clique, the next sequence of activities is fulfilled:

- Present intermediates values in university rankings.
- Have a presence in social networks, they use ICTs to disseminate knowledge and culture, but they do not manage to be impressive for society.

Concerning to outside the clique, a series of activities are accomplished according to the following order:

- Present high values in university rankings.
- Have presence and influence in social networks.
- Have the necessary resources to use ICTs in the dissemination of knowledge and culture.
- Have the resources to give a good impression to society through the use of advertising.
- Have a high demand in their educational programs.

Given the previous lists, the universities that are in the clique share some characteristics that have the universities that obtain high values in the ranking. However, the universities that are outside the clique have more impact and influence on social networks.

10.4.8 Results for MCP for the Modeled Networks through the Information of the Online and Distance Programs Offered by the NAUM

Thanks to the fact that NAUM is the higher education institution in Mexico that offers the most online and distance programs, we decided to study the strengths and weaknesses of it is offered programs. Therefore, Table 10.18 shows the results for the maximum clique of the modeled networks through the main characteristics of the NAUM online and distance programs.

In Table 10.18, it is shown that the majority of the online and distance programs offered by the NAUM have the same characteristics, that is, have high demand, a high percentage of ingress and egress; while others have low demand (particularly those related to History, Bibliotechnology, Philosophy, etc.). In addition, the modeled networks are almost complete, that is, there are links between the vast majority of the nodes.

For example, the network corresponding to the period 2012–2013 is a complete network (there are links between all the nodes), while for the network corresponding to the period 2013–2014, only two elements are outside the clique.

Table 10.18 Results of a maximum clique of networks modeled with the data about the online and distance programs offered by the NAUM

Network	Clique Size	Clique elements
Clique 2012–2013	11	1 2 3 4 5 6 7 8 9 10 11
Clique 2013–2014	24	1 2 3 4 5 7 9 10 11 12 13 14 15 16 17 18 19 20 21 22 23 24 25 26
Clique 2014–2015	20	1 3 4 5 8 9 10 12 13 14 15 16 17 19 20 21 22 23 24 25

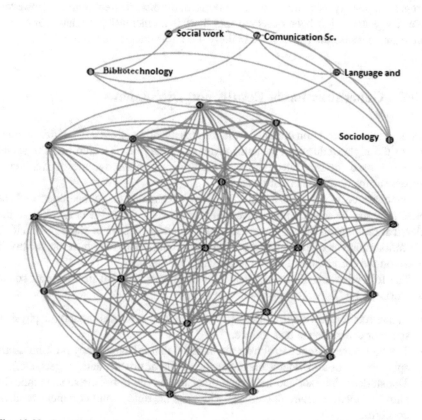

Fig. 10.32 Complete graph and the maximum clique found for the modeled network from the information about NAUM online and distance programs

Because only the network corresponding to the period 2014–2015 can provide certain information, Fig. 10.32 shows the modeled network by the elements that are inside and outside the clique, where it is observed that there are five elements that do not they are inside the clique.

The difference between these programs and others is in demand they present since the demand for these five programs is low.

In the specific case of the network corresponding to the period 2014–2015, the online or distance programs that are found within the clique have high demand, high ingress, re-ingress, and egress. Therefore, they share characteristics that unite them in the representation of the networked system (e.g., Law, Psychology, Administration, etc.).

Meanwhile, other programs that are outside the clique (Social Work, Political Science) have low demand, low ingress and therefore low egress. It should be mentioned that, although NAUM strives to offer a wide range of online or distance programs, society is more inclined towards traditional degree courses. However, over the years, it has been observed that there is an increasing demand for online programs because the use of ICTs in daily life helps interest in them.

10.5 Generalization of Results and Discussion

Based on the information obtained from the classification of the universities belonging to the higher education system in Mexico, it can be observed that the social networks used for the transmission and sharing of knowledge are of significant impact for the development of these.

For example, in networks modeled with the information about the impact and use of ICTs, it can be seen that the universities that always obtain the best values in the classical rankings are those that make use of social networks to share information educational and/or cultural. On the other hand, the impact they have on society is based on the correct use of online information media.

The following are the main advantages of universities that make use of social networks and ICTs for the dissemination of knowledge:

- Create relationship environments for students who cannot approach the physical space or cannot relate face to face.
- Teachers have better control over the material that is shared (they prohibit users, approve photos and videos, make the site public or private, add gadgets, etc.).
- The students interact and help each other to be able to use these sites quickly, allowing them to share documents, videos, talk, reflect, and comment on their concerns.
- Allow an easy peer review.
- Educational networks such as "Moodle" (Moodle, 2018) can be used to educate students about the importance of online platforms for academic development, to migrate to social networking platforms such as Facebook or Twitter.
- Organizations can create communities of students that last beyond a course, a grade, or even beyond graduation.

However, the current society in Mexico does not give fair use to social networks, because in a wrong way, it has been learned that social networks serve only to entertain or play and not conceive as tools that can be of help for professionalism and educational growth at all levels.

In addition, because education in Mexico has always been linked to the use of classrooms and physical spaces to share knowledge, it is essential to motivate the use of ICTs and digital platforms such as Facebook or Twitter for school use from basic levels such as primary or secondary so that students and teachers are increasingly aware of the educational purpose of these tools.

On other hand, the use of ICTs in education has allowed the university community in Mexico to be decentralized, since the offer of distance or online educational programs in states such as Oaxaca, Chiapas or Hidalgo, the demand for different degrees' increases year after year, which increases the participation, and development of communities that do not have physical spaces for higher education.

Also, the use of ICT in Mexican universities has begun to be necessary for the population with a need for professional education that, due to geographical elements or mobility difficulties, cannot access a traditional study system.

It is worth mentioning that most of the undergraduate degrees that have been implemented adequately and that take advantage of most online resources correspond to the socio-administrative, health and education areas, and the degree of nursing degree is obtaining the best results.

Based on the above, we can affirm that the motivational questions for this investigation are answered:

- It is observed that, the characteristics that potentialize the universities are: the number of professors with doctorate and belonging to the NSR, the number of publications in ISI and SCOPUS, that make use and exploitation of the ICTs for the sharing and dissemination of knowledge, make use of social networks, and have excellent references in these.
- The study shown in this paper shows that some universities that are not considered as the best in the classic rankings can have a good reputation in social networks thanks to their publications (usually marketing). However, it does not guarantee that they are an excellent academic option.
- Also, the results show that the impact of social networks for the dissemination and sharing of knowledge influences the potential towards proper functioning of universities; since the universities that are considered the best in the classic rankings make greater use of social networks and ICT in their traditional courses and/or online.
- Finally, the results show that the use of social networks and the perspective that society has towards each university does not have a significant impact, since you only have information about the best universities (public and private).

10.6   Conclusions

In this work, studies about the primary metrics of complex networks and communities that are generated, after they were shown, which was modeled through: the main characteristics of universities in Mexico, its influence for society, its impact and importance in social networks, the use of ICTs for the sharing of knowledge, and for the online and distance programs offered by the leading university in the country (NAUM).

Based on the information from the specific literature, we determined to use the Maximum Clique Problem (MCP), with the objective of rank the universities in Mexico using an Ant System (AS) algorithm. It should be mentioned that the MCP gets communities, where the clustering is based on the principle of homogeneity, that is, the elements belonging to the maximum clique share the same characteristics.

For the modeled networks through the clustering of the universities, we have found that those modeled taking disaggregated data are less dense (DN) and have greater modularity (MD) concerning the modeled networks with the complete data. For modeled networks with information about the influence, use, and importance of social networks for universities, in some cases some universities do not import information for the system, so they do not influence their behavior, also, present greater modularity and are less dense in comparison to the clustering networks.

On the other hand, for the networks modeled with the information of the online and distance programs offered by NAUM, the only relevant information is provided for the period 2014–2015 and. Also, the numerical results of the primary metrics show that the networks follow the topology of the small world since they have a high Clustering Coefficient (CC) and a short average route length.

The numerical results of the Maximum Clique Problem (MCP) for the formation of the ranking show that, for the clustering networks of universities, at least three connected components are generated. The separation of efficient, moderately efficient, and deficient universities can be seen.

This phenomenon is observed in the influence networks in social networks, where there are at least three connected components that correspond to efficient, moderately efficient, and deficient universities.

It is worth mentioning that private universities tend to be part of the dynamic group; which, is due to a substantial investment of these in the management of web resources concerning public universities.

Regarding the online and distance programs offered by NAUM, there are only two connected components generated; where, in the first one, there are all the programs that present high demand and in turn high percentage of egress; while in the second, there are some programs with low demand such as Social Work, Sociology, Bibliotechnology, Philosophy, etc.

Currently, we are working and analyzing a specific kind of multilayer networks which is denoted as multiplex networks, either by a period and/or the different types of information contained in social networks and rankings. These studies can be used to answer the following questions: Is there a difference between the studies by year (monolayer) and joint studies (multiplex)? Is it possible to generalize the characteristics of monolayer networks to multiplex networks? Therefore, we are working about two approaches: the modeling of multiplex networks using supra-adjacency matrices and the MCP considering multiplex connections. Also, we are working in the possible generalization of the primary metrics in monolayer networks for multiplex networks.

Acknowledgments   We thank CONACYT-SNI, UAM, and UNAM for their support. We would like to thank T.A. Chicalote-Jiménez and XOVI for their valuable support in the development of our work, and helping us obtain fundamental data. This *work* was partially supported by PAPIIT-UNAM (IA303418).

Appendix 10.A Acronyms

As mentioned in the body of this chapter, Table 10.19 shows the relation of all the acronyms used and their meaning in Spanish.

Table 10.19   Table of acronyms/abbreviations

ACRONIM	Mean	Spanish (if it is indispensable)
AD	Average Degree	
ALR	Average Length of Route	
AMU	Autonomous Metropolitan University	Universidad Autónoma Metropolitana (UAM)
ANAAU	Antonio Narro Agrarian Autonomous University	Universidad Autónoma Agraria Antonio Narro (UAAAN)
APUSP	Autonomous Popular University of the State of Puebla	Universidad Popular Autónoma del Estado de Puebla (UPAEP)
AS	Ant System Algorithm	
ATIM	Autonomous Technological Institute of Mexico	Instituto Tecnológico Autónomo de México (ITAM)
AUAC	Autonomous University of Aguascalientes	Universidad Autónoma de Aguascalientes (UAAGS)
AUBC	Autonomous University of Baja California	Universidad Autónoma de Baja California (UABC)
AUBCS	Autonomous University of Baja California Sur	Universidad Autónoma de Baja California Sur (UABCS)

(continued)

Table 10.19 (continued)

ACRONIM	Mean	Spanish (if it is indispensable)
AUCAMP	Autonomous University of Campeche	Universidad Autónoma de Campeche (UACAMP)
AUCH	Autonomous University of Chihuahua	Universidad Autónoma de Chihuahua (UACH)
AUCHI	Autonomous University of Chiapas	Universidad Autónoma de Chiapas (UACHI)
AUCJ	Autonomous University of Ciudad Juarez	Universidad Autónoma de Ciudad Juárez (UACJ)
AUCO	Autonomous University of Coahuila	Universidad Autónoma de Coahuila (UACO)
AUDC	Autonomous University of del Carmen	Universidad Autónoma del Carmen (UADC)
AUG	Autonomous University of Guadalajara	Universidad Autónoma de Guadalajara (UAG)
AUGU	Autonomous University of Guerrero	Universidad Autónoma de Guerrero (UAGUERRERO)
AUMC	Autonomous University of Mexico City	Universidad Autónoma de la Ciudad de México (UACM)
AUMEX	Autonomous University of State of Mexico	Universidad Autónoma del Estado de México (UAEM)
AUN	Autonomous University of Nayarit	Universidad Autónoma de Nayarit (UAN)
AUNL	Autonomous University of Nuevo Leon	Universidad Autónoma de Nuevo León (UANL)
AUOS	Anahuac University Online System	Sistema Universidad Anahuac (S.L. Anáhuac)
AUQ	Autonomous University of Queretaro	Universidad Autónoma de Querétaro (UAQ)
AUSH	Autonomous University of the State of Hidalgo	Universidad Autónoma del Estado de Hidalgo (UAEH)
AUSIN	Autonomous University of Sinaloa	Universidad Autónoma de Sinaloa (UASIN)
AUSLP	Autonomous University of San Luis Potosi	Universidad Autónoma de San Luis Potosí (UASLP)
AUSM	Autonomous University of the State of Morelos	Universidad Autónoma del Estado de Morelos (UAEM)
AUT	Autonomous University of Tamaulipas	Universidad Autónoma de Tamaulipas (UAT)
AUTL	Autonomous University of Tlaxcala	Universidad Autónoma de Tlaxcala (UATL)
AUY	Autonomous University of Yucatán	Universidad Autónoma de Yucatán (UAY)
AUZ	Autonomous University of Zacatecas	Universidad Autónoma de Zacatecas (UAZ)

(continued)

Table 10.19 (continued)

ACRONIM	Mean	Spanish (if it is indispensable)
BJAUO	Benito Juarez Autonomous University of Oaxaca	Universidad Autónoma Benito Juárez de Oaxaca (UABJO)
CAHE	Council for the Accreditation of Higher Education A.C.	Consejo Para la Acreditación de la Educación Superior A.C. (COPAES)
CAU	Chapingo Autonomous University	Universidad Autónoma Chapingo (UAChapingo)
CC	Clustering Coefficient	
COLMEX	The College of Mexico	El Colegio de México (COLMEX)
CRANSPI	Center for Research and Advanced Studies of the National Polytechnic Institute	Centro de Investigación y de Estudios Avanzados del IPN (CINVESTAV)
DC	Diameter Coefficient	
DE	Differential Evolution Algorithm	
DN	Density	
ICTS	Information And Communication Technologies	
IU	Intercontinental University	Universidad Intercontinental (UI)
JUAT	Juarez University Autonomous of Tabasco	Universidad Juárez Autónoma de Tabasco (UJAT)
JUSD	Juarez University of The State of Durango	Universidad Juárez del Estado de Durango (UJED)
MAUP	Meritorious Autonomous University of Puebla	Benemérita Universidad Autónoma de Puebla (BUAP)
MC	Modularity	
MCP	Maximum Clique Problem	
MHES	Mexican Higher Education System	
MITAHE	Monterrey Institute of Technology And Higher Education	Instituto Tecnológico y de Estudios Superiores de Monterrey (ITESM)
MNCST	Mexican National Council for Science And Technology	Consejo Nacional de Ciencia y Tecnología (CONACYT)
MWCP	Maximum Weighted Clique Problem	
NAUM	National Autonomous University of Mexico	Universidad Nacional Autónoma de México (UNAM)
NIC	National Institute of Copyright	Instituto Nacional del Derecho de Autor (INDAUTOR)
NPIT	National Program for the Improvement of Teachers	Programa de Mejoramiento del Profesorado (PROMEP)
NPI	National Polytechnic Institute	Instituto Politécnico Nacional (IPN)
OSIU	Online System Iberoamerican University	Sistema Universidad Iberoamericana (S.L. Ibero)
ODU	Open and Distance University	Universidad Abierta y a Distancia (UAYD)

(continued)

Table 10.19 (continued)

ACRONIM	Mean	Spanish (if it is indispensable)
OSLU	Online System la Salle University	Sistema Universidad Lasalle, Ac (S.L. Salle)
OSUVM	Online System of the University of the Valley of Mexico	Sistema Universidad del Valle de México (S.L. UVM)
PC	Postgraduate College	Colegio de Posgraduados (COLPOS)
PPDTST	Program For Professional Development Teacher, For The Superior Education	Programa Para el Desarrollo Profesional Docente, Para el Tipo Superior (PRODEP)
NRQP	National Register of Quality Postgraduates	Padrón Nacional de Posgrados de Calidad (PNPC)
PNU	Pedagogical National University	Universidad Pedagógica Nacional (UPN)
PU	Panamerican University	Universidad Panamericana (UP)
REGU	Regiomontana University	Universidad Regiomontana, Ac (UERRE)
SIT	Sonora Institute of Technology	Instituto Tecnológico de Sonora (ITSON)
SPE	Secretariat of Public Education	Secretaria de Educación Pública (SEP)
TUM	Technological University of Mexico	Universidad Tecnológica de México (UNITEC)
UAAF	University of the Army and Air Force	Universidad del Ejercito Y Fuerza Aérea (UDEFA)
UDCO	University of Colima	Universidad de Colima (UDC)
UMSNH	University of Michoacan of San Nicolas de Hidalgo	Universidad Michoacana de San Nicolás de Hidalgo (UMSNH)
UOG	University of Guadalajara	Universidad de Guadalajara (UDG)
UOGTO	University of Guanajuato	Universidad de Guanajuato (UDGTO)
UOM	University of Monterrey	Universidad de Monterrey (UDMON)
UOQROO	University of Quintana Roo	Universidad de Quintana Roo (UAQROO)
UOS	University of Sonora	Universidad de Sonora (UDSON)
UTAP	University of the Americas Puebla, Ac	Universidad de Las Américas Puebla, Ac (UDLAP)
VU	Veracruz University	Universidad Veracruzana (UV)
WDP	Winner Determination Problem	
WITHE	Western Institute of Technology And Higher Education	Instituto Tecnológico y de Estudios Superiores de Occidente (ITESO)

Appendix 10.B University Identifiers for Clustering Networks

This Appendix shows the numerical identifiers of the universities that are part of the networks modeled from the information of the EXECUM repository ar. Here, the first and third column shows the numerical identifier, while the second and fourth column the full names and respective acronyms in Spanish (to consult the English acronyms, please see the Table of Appendix 10.A).

Table 10.20   Numerical identifiers for universities in clustering networks

ID	University	ID	University
1	Benemérita Universidad Autónoma de Puebla (MUAP)	31	Universidad Autónoma de San Luis Potosí (AUSLP)
2	Centro de Investigación y de Estudios Avanzados del NPI (CRANSPI)	32	Universidad Autónoma de Sinaloa (AUS)
3	Colegio de Posgraduados (COLPOS)	33	Universidad Autónoma de Tamaulipas (AUT)
4	El Colegio de México (COLMEX)	34	Universidad Autónoma de Tlaxcala (AUTL)
5	Instituto Politécnico Nacional (NPI)	35	Universidad Autónoma de Yucatán (AUY)
6	Instituto Tecnológico Autónomo de México (ATIM)	36	Universidad Autónoma de Zacatecas (AUZ)
7	Instituto Tecnológico de Sonora (TIS)	37	Universidad Autónoma del Carmen (AUDC)
8	Instituto Tecnológico y de Estudios Superiores de Occidente (WITHE)	38	Universidad Autónoma del Estado de Hidalgo (AUSH)
9	Instituto Tecnológico y de Estudios Superiores de Monterrey (MITAHE)	39	Universidad Autónoma del Estado de México (AUMEX)
10	Sistema Universidad Anáhuac (O.S. Anáhuac)	40	Universidad Autónoma del Estado de Morelos (AUSM)
11	Sistema Universidad del Valle de México (O.S. UVM)	41	Universidad Autónoma Metropolitana (AMU)
12	Sistema Universidad Iberoamericana (O.S. Ibero)	42	Universidad de Colima (UOC)
13	Sistema Universidad la Salle, Ac (O.S. Salle)	43	Universidad de Guadalajara (UOG)
14	Universidad Autónoma Agraria Antonio Narro (AUAAN)	44	Universidad de Guanajuato (UOGTO)
15	Universidad Autónoma Benito Juárez de Oaxaca (AUBJO)	45	Universidad de Las Américas Puebla, Ac (UOAP)
16	Universidad Autónoma Chapingo (AUCHAPINGO)	46	Universidad de Monterrey (UOMON)
17	Universidad Autónoma de Aguascalientes (AUAC)	47	Universidad de Quintana Roo (AUQROO)
18	Universidad Autónoma de Baja California (AUBC)	48	Universidad de Sonora (UOS)
19	Universidad Autónoma de Baja California Sur (AUBCS)	49	Universidad del Ejército y Fuerza Aérea (UAAF)
20	Universidad Autónoma de Campeche (AUCAMP)	50	Universidad Intercontinental (IU)
21	Universidad Autónoma de Chiapas (AUCHI)	51	Universidad Juárez Autónoma de Tabasco (JUAT)
22	Universidad Autónoma de Chihuahua (AUCH)	52	Universidad Juárez del Estado de Durango (JUSD)

(continued)

Table 10.20 (continued)

ID	University	ID	University
23	Universidad Autónoma de Ciudad Juárez (AUCJ)	53	Universidad Michoacana de San Nicolás de Hidalgo (UMSNH)
24	Universidad Autónoma de Coahuila (AUCO)	54	Universidad Nacional Autónoma de México (NAUM)
25	Universidad Autónoma de Guadalajara (AUG)	55	Universidad Panaméricana (PU)
26	Universidad Autónoma de Guerrero (AUGUERRERO)	56	Universidad Pedagógica Nacional (PNU)
27	Universidad Autónoma de la Ciudad de México (AUCM)	57	Universidad Popular Autónoma del Estado de Puebla (PUASP)
28	Universidad Autónoma de Nayarit (AUN)	58	Universidad Regiomontana, Ac (REGU)
29	Universidad Autónoma de Nuevo León (AUNL)	59	Universidad Tecnológica de México (TUM)
30	Universidad Autónoma de Querétaro (AUQ)	60	Universidad Veracruzana (VU)

Appendix 10.C University Identifiers Networks Modeled with Information About Influence and Importance in Social Ntworks

Next, the numerical identifiers of the universities that are part of the networks modeled from the information obtained with the Social Analytics tool of XOVI are shown. Here, the first and third column shows the numerical identifier, while the second and fourth column the full names and respective acronyms in Spanish (to consult the English acronyms, please see the Table of Appendix 10.A).

Table 10.21 Numerical identifiers for universities in modeled networks with information about the influence on social networks

ID	University	ID	University
1	Universidad Autónoma del Estado de Hidalgo (AUSH)	27	Centro de Investigación y Estudios Avanzados (CRANSPI)
2	Universidad Jesuita de Guadalajara (JUG)	28	Universidad Pedagógica Nacional (PNU)
3	Universidad Autónoma de Yucatán (AUY)	29	Instituto Tecnológico Autónomo de México (ATIM)
4	Colegio de México (COLMEX)	30	Universidad Autónoma de Chiapas (AUCHI)
5	Universidad de Guadalajara (UOG)	31	Universidad Autónoma de Coahuila (AUDC)
6	Universidad Autónoma Metropolitana (AMU)	32	Universidad de Sonora (UOS)

(continued)

Table 10.21 (continued)

ID	University	ID	University
7	U. La Salle (O.S. Lasalle)	33	Universidad Iberoamericana (O.S. Ibero)
8	Universidad de Las Ame Ricas Puebla (UOAP)	34	Universidad Agraria Antonio Narro (AUAAN)
9	Universidad Autónoma de Chihuahua (AUCH)	35	Universidad Autónoma de Campeche (AUCAM)
10	Universidad Nacional Abierta y A Distancia de México (NUOD)	36	Universidad de Guanajuato (AUGTO)
11	Universidad Autónoma de Aguascalientes (AUAC)	37	Universidad Regiomontana (REGU)
12	Universidad Autónoma de Guerrero (AUGRO)	38	Universidad Autónoma del Estado de México (AUEMEX)
13	Universidad Anáhuac (O.S. Anáhuac)	39	Benemérita Universidad Autónoma de Puebla (MAUP)
14	Universidad Autónoma de Nuevo León (AUNL)	40	Universidad Autónoma de Guerrero (AUGRO)
15	Universidad de Quintana Roo (UQROO)	41	Universidad de Colima (UOCOL)
16	Universidad de Monterrey (UOM)	42	Universidad Tecnológica de México (TUM)
17	Universidad Autónoma de Baja California Sur (AUBCS)	43	Universidad Autónoma de Tlaxcala (AUTX)
18	Universidad Autónoma del Carmen (AUDC)	44	Universidad Autónoma Benito Juárez de Oaxaca (AUBJO)
19	Universidad Veracruzana (VU)	45	Universidad Autónoma de Ciudad Juárez (AUCJ)
20	Universidad Autónoma de Chapingo (AUCHAPINGO)	46	Universidad Nacional Autónoma de México (NAUM)
21	Instituto Politécnico Nacional (NPI)	47	Universidad Autónoma de Querétaro (AUQ)
22	Universidad Juárez del Estado de Durango (JUED)	48	Universidad Autónoma de San Luis Potosí (AUSLP)
23	Universidad Juárez Autónoma de Tabasco (JAUT)	49	Instituto Tecnológico de Sonora (TIS)
24	Instituto Tecnológico de Estudios Superiores de Monterrey (MITAHE)	50	Universidad Popular Autónoma del Estado de Puebla (PUASP)
25	Colegio de Posgraduados (COLPOS)	51	Universidad Autónoma de Morelos (AUSM)
26	Universidad de Michoacán (UOMICH)	52	Universidad Autónoma de Baja California (AUBC)

As mentioned in the body of this Chapter, there are some nodes that represent different information for a certain university (Specifically for ICTs use networks). For example, the NAUM, has five different types of node: OS, DSS, ICT, SK, and CD which represent different characteristics (for more information, see Sect. 10.3.2.2).

Table 10.22 Numerical identifiers for ICTs use networks

ID	University	ID	University
1	Universidad Nacional Autónoma de México (NAUM ICT)	11	Universidad Autónoma Metropolitana (AMU SK)
2	Instituto Tecnológico Autónomo de México (ATIM ICT)	12	Universidad Nacional Autónoma de México (NAUM DSS)
3	Universidad Autónoma Del Estado de México (AUMEX ICT)	13	Instituto Tecnológico de Estudios Superiores de Monterrey (MITAHE SK)
4	Universidad Nacional Autónoma de México (NAUM SK)	14	Centro de Investigación y Estudios Avanzados (CRANSPI ICT)
5	Universidad Autónoma Del Estado de México (AUMEX SK)	15	Centro de Investigación y Estudios Avanzados (CRANSPI SK)
6	Instituto Politécnico Nacional (NPI ICT)	16	Universidad Autónoma de Nuevo León (AUNL ICT)
7	Universidad Autónoma Metropolitana (AMU ICT)	17	Universidad Nacional Autónoma de México (NAUM CD)
8	Instituto Politécnico Nacional (NPI SK)	18	Instituto Politécnico Nacional (NPI CD)
9	Universidad Nacional Autónoma de México (NAUM OS)	19	Instituto Tecnológico Autónomo De México (ATIM SK)
10	Instituto Tecnológico de Estudios Superiores de Monterrey (MITAHE ICT)	20	Universidad Autónoma Metropolitana (AMU CD)

As in the previous case, there are several nodes to represent different characteristics for certain universities (Specifically in networks modeled with the information about the society perspective towards universities). For example, the AMU node has the following endings: GP, BP, V, NV, because the XOVI tool gets four types of data for this university (for more information see Sect. 10.3.2.2).

Table 10.23 Numerical University identifiers for networks modeled with information about society perspective towards universities

ID	University	ID	University
1	Universidad Autónoma Metropolitana (AMU GP)	12	Universidad de Guadalajara (UOG GP)
2	Universidad Nacional Autónoma de México (NAUM GP)	13	Instituto Tecnológico Autónomo de México (ATIM GP)
3	Universidad Nacional Autónoma de México (NAUM NV)	14	Universidad Autónoma del Estado de México (AUMEX GP)
4	Universidad Autónoma de Nuevo León (AUNL GP)	15	Instituto Politécnico Nacional (NPI 1)
5	Universidad Autónoma de Morelos (AUSM GP)	16	Instituto Tecnológico Autónomo de México (ATIM NV)
6	Universidad Nacional Autónoma de México (NAUM V)	17	Universidad Autónoma del Estado de México (AUMEX NV)
7	Universidad Autónoma Metropolitana (AMU NV)	18	Universidad Nacional Autónoma de México (NAUM BP)

(continued)

Table 10.23 (continued)

ID	University	ID	University
8	Universidad Autónoma Metropolitana (AMU V)	19	Instituto Tecnológico de Estudios Superiores de Monterrey (MITAHE GP)
9	Universidad Autónoma Metropolitana (AMU BP)	20	Instituto Politécnico Nacional (NPI NV)
10	Centro de Investigación y Estudios Avanzados (CRANSPI GP)	21	Instituto Tecnológico de Estudios Superiores de Monterrey (MITAHE NV)
11	Colegio de México (COLMEX GP)	22	Instituto Politécnico Nacional (NPI BP)

Appendix 10.D Identifiers for the Online and Distance Programs Offered by the NAUM

This appendix shows the numerical identifiers of the universities that are part of the networks modeled after the information obtained in the UNAM profile on the tableau platform. Here, the first and third columns show the numerical identifier, while the second and fourth columns show the full names of each program (online and/or distance programs).

Table 10.24 Identifiers for the online and distance programs offered by the NAUM

ID	Program	ID	Program
1	Nursing (online)	14	Geography (online)
2	Social Work (distance)	15	History (online)
3	Administration (online and distance)	16	Political Science and Public Administration (online and distance)
4	Accounting (distance)	17	Modern Language and Literature (English Letters)
5	Informatics (distance)	18	Sociology (online and distance)
6	Hispanic Language and Literature (online)	19	Psychology (online and distance)
7	Communication Sciences (online and distance)	20	German Teaching as LE (online and distance)
8	International Relations (online and distance)	21	Teaching English as LE (online and distance)
9	Pedagogy (online and distance)	22	Teaching French as LE (online and distance)
10	Law (online and distance)	23	Teaching Spanish as LE (online and distance)
11	Librarianship and Information Studies (distance)	24	Italian Teaching as LE (online and distance)
12	Economy (online and distance)	25	Sp. In Teaching Spanish as LE (online and distance)
13	Philosophy (distance)	26	Sp. In Animal Production (online)

References

AbuJarour, S., & Krasnova, H., (2018). *E-learning as a means of social inclusion: The case of Syrian refugees in Germany.* Proceedings of the Americas Conference on Information Systems (AMCIS 2018), New Orleans, LA, August 16–18, 2018.

Albert, R., & Barabási, A. L. (2000). Error and attack tolerance of complex networks. *Nature, 406*(6794), 378.

Albert, R., & Barabási, A. L. (2002). Statistical mechanics of complex networks. *Reviews of Modern Physics, 74*(1), 47.

Alonso, F., López, G., Manrique, D., & Viñes, J. M. (2005). An instructional model for web-based e-learning education with a blended learning process approach. *British Journal of Educational Technology, 36*(2), 217–235.

Al-Samarraie, H., Teng, B. K., Alzahrani, A. I., & Alalwan, N. (2018). E-learning continuance satisfaction in higher education: A unified perspective from instructors and students. *Studies in Higher Education, 43*(11), 2003–2019.

Altbach, P. (2015). The dilemmas of ranking. *International Higher Education, 42*, 1–2.

Antonio, A., & Tuffley, D. (2014). Creating educational networking opportunities with Scoop.it. *Journal of Creative Communications, 9*(2), 185–197.

Avello Martínez, R., & Duart, J. M. (2016). Nuevas tendencias de aprendizaje colaborativo en e-learning: Claves para su implementación efectiva. *Estudios pedagógicos (Valdivia), 42*(1), 271–282.

Badii, M. H., Guillen, A., Araiza, L. A., Cerna, E., Valenzuela, J., & Landeros, J. (2012). Métodos No-Paramétricos de Uso Común. *Revista Daena (International Journal of Good Conscience), 7*(1), 132–155.

Benlic, U., & Hao, J.-K. (2013). Breakout local search for maximum clique problems. *Computers and Operations Research, 40*(1), 192–206.

Builes, N. M. S., Castaño, P. A. J., & Zuluaga, N. M. V. (2018). Science as a "reality show": The mindset of academic research. *Revista de Orientación Educacional, 32*(61), 73–78.

Burov, O. Y. (2016). Educational networking: Human view to cyber defense. *Information Technologies and Learning Tools, 52*(2), 144–156.

Bustos, A., & Román, M. (2016). La importancia de evaluar la incorporación y el uso de las ICT en educación. *Revista Iberoamericana de evaluación educativa, 4*(2), 4–7.

Buxarrais-Estrada, M. R., & Ovide, E. (2011). El impacto de las nuevas tecnologías en la educación en valores del siglo XXI. *Sinéctica, 37*, 1–14.

Cabrera, L. G., Ortega-Tudela, J. M., Hita, M. A. P., Ruano, I. R., & Colón, A. M. O. (2010). La calidad en la docencia virtual: la importancia de la guía de estudio. *Pixel-Bit. Revista de Medios y Educación, 37*, 77–92.

Chang, F. P. C., & Ouyang, L. Y. (2018). Trend models on the academic ranking of world universities. *International Journal of Information and Management Sciences, 29*(1), 35–56.

Clauset, A., Arbesman, S., & Larremore, D. B. (2015). Systematic inequality and hierarchy in faculty hiring networks. *Science Advances, 1*(1), e1400005.

Conte, A., De Virgilio, R., Maccioni, A., Patrignani, M., & Torlone, R. (2016). Finding all maximal cliques in very large social networks. In *Edbt* (pp. 173–184), Bordeaux, France, 15–18 March, 2016.

Coordinación de Universidad Abierta y Educación a Distancia, M. (2016). *Tableau público NAUM.* https://public.tableau.com

Cortés, J., & Lozano, J. (2014). Social networks as learning environments for higher education. *IJIMAI, 2*(7), 63–69.

Daraio, C., Bonaccorsi, A., & Simar, L. (2015). Rankings and university performance: A conditional multidimensional approach. *European Journal of Operational Research, 244*(3), 918–930.

Davydova, N. N., Dorozhkin, E. M., Fedorov, V. A., & Konovalova, M. E. (2016). Research and educational network: Development management. *International Electronic Journal of Mathematics Education, 11*(7), 2651–2665.

De Bacco, C., Larremore, D. B., & Moore, C. (2018). A physical model for efficient ranking in networks. *Science Advances, 4*(7), eaar8260. https://doi.org/10.1126/sciadv.aar8260

De Witte, K., & Hudrlikova, L. (2013). What about excellence in teaching? A benevolent ranking of universities. *Scientometrics, 96*(1), 337–364.

De-los-Cobos, S. G. (2010). Búsqueda y exploración estocástica. Universidad Autónoma Metropolitana.

de-Marcos, L., Garcia-Lopez, E., & Garcia-Cabot, A. (2016). On the effectiveness of game-like and social approaches in learning: Comparing educational gaming, gamification & social networking. *Computers & Education, 95*, 99–113.

Díaz-García, I., Cebrián-Cifuentes, S., & Fuster-Palacios, I. (2016). Las competencias en ICT de estudiantes universitarios del ámbito de la educación y su relación con las estrategias de aprendizaje. *RELIEVE-Revista Electrónica de Investigación y Evaluación Educativa, 22*(1).

Dobrota, M., Bulajic, M., Bornmann, L., & Jeremic, V. (2016). A new approach to the QS university ranking using the composite I-distance indicator: Uncertainty and sensitivity analyses. *Journal of the Association for Information Science and Technology, 67*(1), 200–211.

Doğan, G., & Al, U. (2018). Is it possible to rank universities using fewer indicators? A study on five international university rankings. *Aslib Journal of Information Management, 71*(1), 18–37.

Dorigo, M., Maniezzo, V., & Colorni, A. (1996). Ant system: Optimization by a colony of cooperating agents. *IEEE Transactions on Systems, Man and Cybernetics, Part B (Cybernetics), 26*(1), 29–41.

Eid, M. I., & Al-Jabri, I. M. (2016). Social networking, knowledge sharing, and student learning: The case of university students. *Computers and Education, 99*, 14–27.

Erdös, P., & Rényi, A. (1959). On random graphs, I. *Publicationes Mathematicae (Debrecen), 6*, 290–297.

ExECUM-UNAM. (2017). *Dirección general de evaluación institucional.* http://www.execum.NAUM.mx.

Fadem, B. (2008). *High-yield behavioral science (high-yield series).* Hagerstwon, MD: Lippincott Williams & Wilkins.

Gockel, C., & Werth, L. (2011). Measuring and modeling shared leadership. *Journal of Personnel Psychology, 9*, 172–180.

Goeke, D., Moeini, M., & Poganiuch, D. (2017). A variable neighborhood search heuristic for the maximum ratio clique problem. *Computers and Operations Research, 87*, 283–291.

Gretzel, U. (2001). *Social network analysis: Introduction and resources.* Retrieved May, vol. 12, p. 2009.

Hernández, P. M., Leyva, S. L., Márquez, C. Z., & Cerda, A. B. N. (2014). Evaluación de la calidad de la educación superior en México: comparación de los indicadores de rankings universitarios nacionales e internacionales. *RIESED-Revista Internacional de Estudios sobre Sistemas Educativos, 2*(4), 35–51.

Hodges, J. L. (1990). Improved significance probabilities of the Wilcoxon test. *Journal of Educational Statistics, 15*, 249–265.

Hung, H. T., & Yuen, S. C. Y. (2010). Educational use of social networking technology in higher education. *Teaching in Higher Education, 15*(6), 703–714.

Karimi, F., Génois, M., Wagner, C., Singer, P., & Strohmaier, M. (2018). Homopily influences the ranking of minorities in social networks. *Scientific Reports, 8*, 11077.

Klašnja-Milićević, A., Vesin, B., & Ivanović, M. (2018). Social tagging strategy for enhancing e-learning experience. *Computers and Education, 118*, 166–181.

Kramer, A. D., Guillory, J. E., & Hancock, J. T. (2014). Experimental evidence of massive-scale emotional contagion through social networks. *Proceedings of the National Academy of Sciences, 11*(24), 8788–8790.

Laeeq, K., Memon, Z. A., & Memon, J. (2018). The SNS-based e-learning model to provide a smart solution for e-learning. *IJERI: International Journal of Educational Research and Innovation, 10*, 141–152.

Larner, W. (2015). Globalizing knowledge networks: Universities, diaspora strategies, and academic intermediaries. *Geoforum, 59*, 197–205.

Latora, V., Nicosia, V., & Russo, G. (2017). *Complex networks: Principles, methods, and applications.* Cambridge University Press. https://doi.org/10.1017/9781316216002

Liu, N. C. (2015). The story of the academic ranking of world universities. *International Higher Education, 54*, 2–3.

Márquez-Jiménez, A. (2010). Estudio comparativo de universidades mexicanas (ECUM): otra mirada a la realidad universitaria. *Revista iberoamericana de educación superior, 1*(1), 148–156.

Martínez Clares, P., Pérez Cusó, J., & Martínez Juárez, M. (2016). Las ICTs y el entorno virtual para la tutoría universitaria. *Educación XXI: revista de la Facultad de Educación, 19*(1), 287–310.

Moodle. (2018). https://moodle.org/. Consultado febrero de 2019.

Nee, C. (2014). The effect of educational networking on students' performance in biology. In *TCC worldwide online conference* (pp. 73–97).

Newman, M. (2010). *Networks: An introduction.* Oxford, UK: Oxford University Press.

Newman, M. E. (2006). Modularity and community structure in networks. *Proceedings of the National Academy of Sciences, 103*(23), 8577–8582.

Nogueira, B., & Pinheiro, R. G. (2018). A CPU-GPU local search heuristic for the maximum weight clique problem on large graphs. *Computers and Operations Research, 90*, 232–248.

Ordorika, I. (2015). Rankings universitarios. *Revista de la educación superior, 44*(173), 7–9.

Ordorika, I., & Rodríguez Gómez, R. (2010). El ranking Times en el mercado del prestigio universitario. *Perfiles educativos, 32*(129), 8–29.

Pempek, T. A., Yermolayeva, Y. A., & Calvert, S. L. (2009). College students' social networking experiences on Facebook. *Journal of Applied Developmental Psychology, 30*(3), 227–238. https://doi.org/10.1016/j.appdev.2008.12.010

Rust, V., & Kim, S. (2016). Globalization and new developments in global university rankings. In *Globalisation and Higher Education Reforms* (pp. 39–47). Cham, Switzerland: Springer.

Sadowski, C., Pediaditis, M., & Townsend, R. (2017). University students' perceptions of social networking sites (SNSs) in their educational experiences at a regional Australian university. *Australasian Journal of Educational Technology, 33*(5).

Sánchez Hervás, D., et al. (2017). *La reputación corporativa en la comunicación de las instituciones universitarias españolas a través de sus sitios webs.* Universidad Católica San Antonio de Murcia.

Sayama, H. (2015). *Introduction to the modeling and analysis of complex systems.* Open SUNY Textbooks.

Shore, K. A. (2018). Complex networks: Principles, methods, and applications. *Contemporary Physics, 59*, 223–224.

Silva, M., da Silva, S. M., & Araujo, H. C. (2017). Networking in education: From concept to action-an analytical view on the educational territories of priority intervention (teip) in northern Portugal. *Improving Schools, 20*(1), 48–61.

Social Analytics Tool, X. G. (n.d.). *Social analytics tool, XOVI gmbh.* https:// www.xovi.com/es/herramienta/social-analytics/

Storn, R., & Price, K. (1997). Differential evolution–a simple and efficient heuristic for global optimization over continuous spaces. *Journal of Global Optimization, 11*(4), 341–359.

Thomas, R. A., West, R. E., & Borup, J. (2017). An analysis of instructor social presence in online text and asynchronous video feedback comments. *The Internet and Higher Education, 33*, 61–73.

Ullman, J. D. (1975). NP-complete scheduling problems. *Journal of Computer and System Sciences, 10*(3), 384–393.

Vera Noriega, J. Á., Torres Moran, L. E., & Martínez García, E. E. (2014). Evaluación de competencias básicas en ICT en docentes de educación superior en México. *Píxel-Bit. Revista de Medios y Educación, 44*, 143–155.

Versteijlen, M., Salgado, F. P., Groesbeek, M. J., & Counotte, A. (2017). Pros and cons of online education as a measure to reduce carbon emissions in higher education in the Netherlands. *Current Opinion in Environmental Sustainability, 28*, 80–89.

Vohra, R. S., & Hallissey, M. T. (2015). Social networks, social media, and innovating surgical education. *JAMA Surgery, 150*(3), 192–193.

Watts, D. J., & Strogatz, S. H. (1998). Collective dynamics of 'small-world' networks. *Nature, 393*(6684), 440.

Wu, Q., & Hao, J.-K. (2015a). A review of algorithms for maximum clique problems. *European Journal of Operational Research, 242*(3), 693–709.

Wu, Q., & Hao, J.-K. (2015b). Solving the winner determination problem via a weighted maximum clique heuristic. *Expert Systems with Applications, 42*(1), 355–365.

Wur, Q. (2017). *Qs world university rankings by subject 2017, viewed 15 may 2017.*

Yang, J., & Leskovec, J. (2015). Defining and evaluating network communities based on ground-truth. *Knowledge and Information Systems, 42*(1), 181–213.

Zachary, W. W. (1977). An information ow model for conict and _ssion in small groups. *Journal of Anthropological Research, 33*(4), 452–473.

Zheng, J., & Liu, N. (2015). Mapping of important international academic awards. *Scientometrics, 104*(3), 763–791.

Zhu, J. (2014). Quantitative models for performance evaluation and benchmarking: data envelopment analysis with spreadsheets (Vol. 213). Springer.

Zornic, N., Maricic, M., Bornmann, L., Markovic, A., Martic, M., & Jeremic, V. (2015). Ranking institutions within a university based on their scientific performance: A percentile-based approach. *El profesional de la información, 24*(5), 551–566.

Index

© Springer Nature Switzerland AG 2020
A. Peña-Ayala (ed.), *Educational Networking*, Lecture Notes in Social
Networks, https://doi.org/10.1007/978-3-030-29973-6